AF643311

Cereals

NIPA® GENX ELECTRONIC RESOURCES & SOLUTIONS P. LTD.
New Delhi-110 034

www.nipaers.com
Browse, Search, Read & Buy...
Print Books
eChapters
eBooks
Publishers
Forthcoming
Subject Catalogues
New Titles
Online Resources on:
✓ Current Affairs
✓ Reasoning, Logic & Aptitude
✓ Competitive Examinations
✓ English Language Lab
✓ Writing & Pronunciation Tools
✓ Personality Development
✓ Online Programmes for Professionals
✓ Interview Preparation
Pay Using
PhonePe
Google Pay
paytm
PayPal
UPI
CCAvenue

Cereals

Processing and Nutritional Quality

Authors :

DR. SEWA RAM
Principal Scientist
Plant Biochemistry, Quality and Basic Sciences
Directorate of Wheat Research *(ICAR)*
Agrasain Marg, Karnal - 132 001, Haryana, India

DR. B. MISHRA
Vice-Chancellor
Sher-e-Kashmir University of Agriculture Science and Technology,
(Jammu & Kashmir) and
Formerly Project Director, Directorate of Rice Research, Hyderabad
Directorate of Wheat Research, Karnal - 132 001, Haryana, India

nipa®

NIPA® GENX ELECTRONIC RESOURCES & SOLUTIONS P. LTD.
New Delhi-110 034

NIPA® GENX ELECTRONIC
RESOURCES & SOLUTIONS P. LTD.

101,103, Vikas Surya Plaza, CU Block
L.S.C. Market, Pitam Pura, New Delhi-110 034
Ph : +91 11 27341616, 27341717, 27341718
E-mail: newindiapublishingagency@gmail.com
web: www.nipabooks.com

For customer assistance, please contact
Phone: + 91-11-27 34 17 17
Fax: + 91-11- 27 34 16 16
E-Mail: feedbacks@nipabooks.com

ISBN: 978-81-19072-37-8

Composed and Designed by NIPA®.

सत्यमेव जयते

डा. मंगला राय

सचिव एवं महानिदेशक

DR. MANGALA RAI

SECRETARY & DIRECTOR-GENERAL

भारत सरकार

कृषि अनुसंधान और शिक्षा विभाग एवं

भारतीय कृषि अनुसंधान परिषद्

कृषि मंत्रालय, कृषि भवन, नई दिल्ली–110001

GOVERNMENT OF INDIA

Dept. of Agricultural Research & Education

and Director General, Indian Council of

Agricultural Research

Ministry of Agriculture, Krishi Bhavan

New Delhi - 110001

FOREWORD

Cereals are the major source of energy and nutrition for larger part of the population and are consumed in different forms. Among cereals, wheat, rice, maize and barley are major sources of food and feed across the world. Improvement in the quality of different products of cereals has become a major issue internationally. This has gained impetus with the globalization of economy and competition in domestic and international market. Quality laboratories have been developed and good information generated for each cereal across the world. This will enhance the competitiveness in the cereal based industry and the profitability of farmers. Though cereals are excellent source of energy in the form of carbohydrates and vitamins, they are deficient in certain nutritional components such as micronutrients, essential amino acids in protein like lysine and tryptophan. This is further complicated by anti-nutritional factors present in grains which reduce the bioavailability of micronutrients and vitamins.

There are unique products of each cereal and the quality of which is governed by the specific chemical and physical constituents of the grain. The current understanding of the role of these constituents in determining the quality of each product has led to the tremendous improvement in the quality of each cereal along with higher yield potential. Excellent progress has been made in understanding the molecular basis of grain quality traits. This has further enhanced the efficiency of breeding in selecting desirable segregants using molecular marker approach. However, most of the information is not available easily to the readers at different levels. The publication **"Cereals : Processing and Nutritional Quality"** in its present form contains detailed aspects of the quality requirements of different products of cereals and also the chemical and molecular components governing these traits. It is an effort to provide information in the form of book containing all the aspects of processing and nutritional quality of cereals for the benefits of diverse groups of readers across academics and industry.

It is a pleasure to see the contents of the book which cover main aspects of cereal quality. The authors deserve appreciation for giving valuable information on cereal quality improvement. It is hoped that the publication will serve as reference material to plan for research activities in different cereals for the improvement of processing and nutritional quality and also for students to understand cereal quality.

Dated the 31st August, 2009
New Delhi

(MANGALA RAI)
Secretary, Department of Agricultural Research & Education and Director General, Indian Council of Agricultural Research, Ministry of Agriculture
Krishi Bhavan, New Delhi - 110 114

PREFACE

The food and nutritional security of about 6.7 billion population in the world depends upon the quantity and quality of cereals produced. Cereals are the cheapest source of energy and provide proteins, dietary fibre, minerals, vitamins and other nutritional factors to the human beings. Among cereals, rice, wheat and maize are the major sources of food and feed and contribute around 23%, 17% and 9% of the world calorie, respectively. These are grown across the world from temperate to tropical environments. Millions of people are involved in production, processing and marketing of these cereals. There are large numbers of wheat based products such as fermented pan bread, flat breads including chapati, biscuits, noodles and pasta products. Rice is generally cooked and eaten directly and also consumed such as breakfast cereals, baby food and in many other forms. Maize (known as corn) is the principal cereal crop in tropical and subtropical regions throughout the world. Virtually every part of maize has economic value as grains are used for human food, fermented to produce a vide range of food and beverages, fed to livestock and industrial uses in the production of starch, oil, sugar, cellulose and ethanol. Barley is used mainly for feed and malt purposes and little for human food. Recently cereals have been explored for their use in several industrial applications such as bio-fuels and biodegradable products. Therefore, improving quality of cereals for food, feed and industrial applications has paramount importance in world economy. This depends on understanding the chemical and genetic basis of industrial and nutritional quality traits. The chemical and molecular studies during the last 10 years have expanded the knowledge on cereal quality. This has demonstrated that there are some common genetic factors among cereals related to nutritional quality. However, the information is not accessible easily to the readers at different levels. Therefore, efforts have been made to provide information on quality of wheat, barley, maize and rice in this book for the benefit of researchers, students and industrial people.

Keeping above facts in mind the book has been divided into 13 chapters of which 10 chapters are related to wheat quality. The first chapter is related to wheat production, consumption and utilization. Separate chapters are provided on starch and proteins because they are main storage components in the grain. Wheat gluten proteins impart unique characteristics to dough to make it suitable for different products. The allelic variation in gluten proteins has strong relationship with gluten strength and extensibility and thus suitable in predicting wheat grain quality. Starch pasting properties have strong relationship with noodle quality in wheat and cooking quality in rice. Separate chapters have been provided on soft wheat products, fermented and flat bread, and durum wheat products. Since there are two international market classes of wheat termed as soft and hard; extensive studies have been conducted on grain texture and thus a separate chapter is provided. Wheat based industry demands specific type of flour for which chemical and rheological tests have been developed to explain the suitability of wheat for different products. Since large amount of genetic information has been generated on nutritional quality traits, a separate chapter has been given on improving nutritional quality. Similarly good information has been generated on molecular aspects of wheat quality improvement; a separate chapter has been added on molecular breeding and transgenics.

Individual chapter on rice, barley and maize has been provided to give a brief account of major traits covering both industrial and nutritional quality. Rice quality includes information on various types of rice available, cooking and eating quality traits and their molecular genetics as well as transgenic approach for improving nutritional quality. Recent studies demonstrated the utility of transgenic approach in improving bioavailability of micronutrients and also improving pro-vitamin A content in rice. The maize chapter includes information on milling, food and feed quality and maize as a source of biofuel. Detailed information is given on nutritional quality of maize including QPM. Chapter on barley includes information on malting and brewing quality and also food and feed quality. β-glucan content in barley has important role in food, feed and malting quality. Since β-glucans have health benefits, the relevance of barley in improving nutritional quality has been highlighted.

Efforts have been made so that the readers understand quality of major cereal crops at chemical, molecular, genetic and processing levels. Important references have been collected and reflected at the end of each chapter along with summary. This publication may serve as reference material to plan research activities in different cereals for the improvement of processing and nutritional quality. This will also help students to understand cereal quality which is going to be focused globally in coming years. Authors are thankful to those who have contributed directly or indirectly in bringing this book in this form. Readers are requested to give suggestions to improve this book further to make it more useful for wide range of stakeholders and readers.

Sewa Ram
B. Mishra

CONTENTS

❏❏❏

Chapter 1

World Wheat Production, Consumption & Utilization

Introduction

Wheat is an important cereal crop and is widely cultivated across the world. Wheat is consumed in different forms and major products are chapati, bread, biscuit, noodles and pasta products. Wheat has three groups of cultivated species namely bread, durum and dicoccum wheats. The common or bread wheats account for approximately 95% of world production. Nearly all of the remaining 5% cultivated varieties are durum wheats used for pasta products and couscous and dicoccums represent very little of the total wheat production. The bread wheat is divided into two broad classes based on grain texture; hard and soft. Hard wheats with high protein content and strong gluten are used for making bread and some noodles while soft wheats with lower protein content and weak gluten for biscuit, cakes and pretzels. Semi-hard wheats having some combinations of the above characteristics are utilized for making unleavened breads such as chapati, Asian steamed bread and some noodles. There can be large disparity within wheats of various classes; therefore to provide consistent quality, it is divided into grades. The grading is determined by the test weight, protein content, moisture and foreign material. The major wheat exporting countries have distinct grading patterns designed to provide wheat for specific purposes and consistent quality.

Total world wheat production has increased from 531 million tons in 1986/87 to 571 million tons in 2002/03. The European Union (103 million tons) was the largest producer of wheat in 2002, followed by China (90 million tons), India (70 million tons) and US (59 million tons). During the last several years world wheat production is hovering around 600 million tons. Other major wheat-producing countries are the Former Soviet Union, Canada, Australia, Turkey and Argentina. These nine countries produce about 74% of the wheat produced in the world. Because of the concentration of wheat production in a few countries, a large volume of wheat is traded in the world market. The total quantity of wheat traded in the world market was 91 million tons in 2002, which is about 26% of wheat produced in that year. World wheat trade is dominated by a few exporting countries: the United States, Canada, Australia, the EU, and Argentina. These countries have handled over 80% of wheat traded in the world market. While competition among exporting countries is strong, the world wheat market is not perfectly competitive. Australia and Canada use wheat boards to market their grain, while the EU relies on export subsidies to increase its market share. In addition, all major wheat exporters use credit guarantees and long-term preferential trade agreements to promote their exports.

World wheat production increased at a rate of 3.3 percent per year between 1949 and 1978. Increases at the start of this period were due to both an expansion of production area and increased per acre yields. However, starting in the 1960's, increased yields from improved varieties and a greatly expanded use of irrigation, pesticides, and fertilizers took on greater importance in sustaining the high rate of growth in world wheat production. In fact, the impact of new production technologies was so significant that their widespread adoption during this period became known as the "green revolution". In 1978, the U.S.S.R. harvested 27% of the total wheat in the world. The United States was the world's second largest producer followed by the European Economic Community and China. Canada, Australia, Argentina, India, Pakistan, and Turkey were the other major wheat producers. Wheat production trends shifted dramatically in the 1980's. Rate of increase in world wheat production slowed to 1.5% per year between 1982 and 1991. China maintained its rate of increase in production at 2.6% per year and became the world's largest wheat producer. Wheat production in India and Pakistan also increased at nearly 3% per year during this period. Smaller wheat producing countries in the "others" category increased production by 2.8% per year and accounted for a significant 16% of world wheat supply. Wheat production in Russia was very erratic during this period and its contribution to total world production fell to 16%.

Among the major wheat exporting countries, trends between 1982 and 1991 show a gradual reduction in total wheat production in the USA, Australia, and Argentina. Only the European Economic Community (EEC) maintained a high rate of growth in production (3% per year) and this was accomplished using intensive cropping practices. A report in 2003 by the Center for Agricultural Policy and Trade Studies Department of Agribusiness and Applied Economics North Dakota State University Fargo, North Dakota evaluated the U.S. and world wheat markets for the 2002-2012 period using the Global Wheat Policy Simulation Model (Fig. 1). This analysis is based on a series of assumptions about general economic conditions, agricultural policies, weather conditions, and technological change. Both the U.S. and world wheat economies

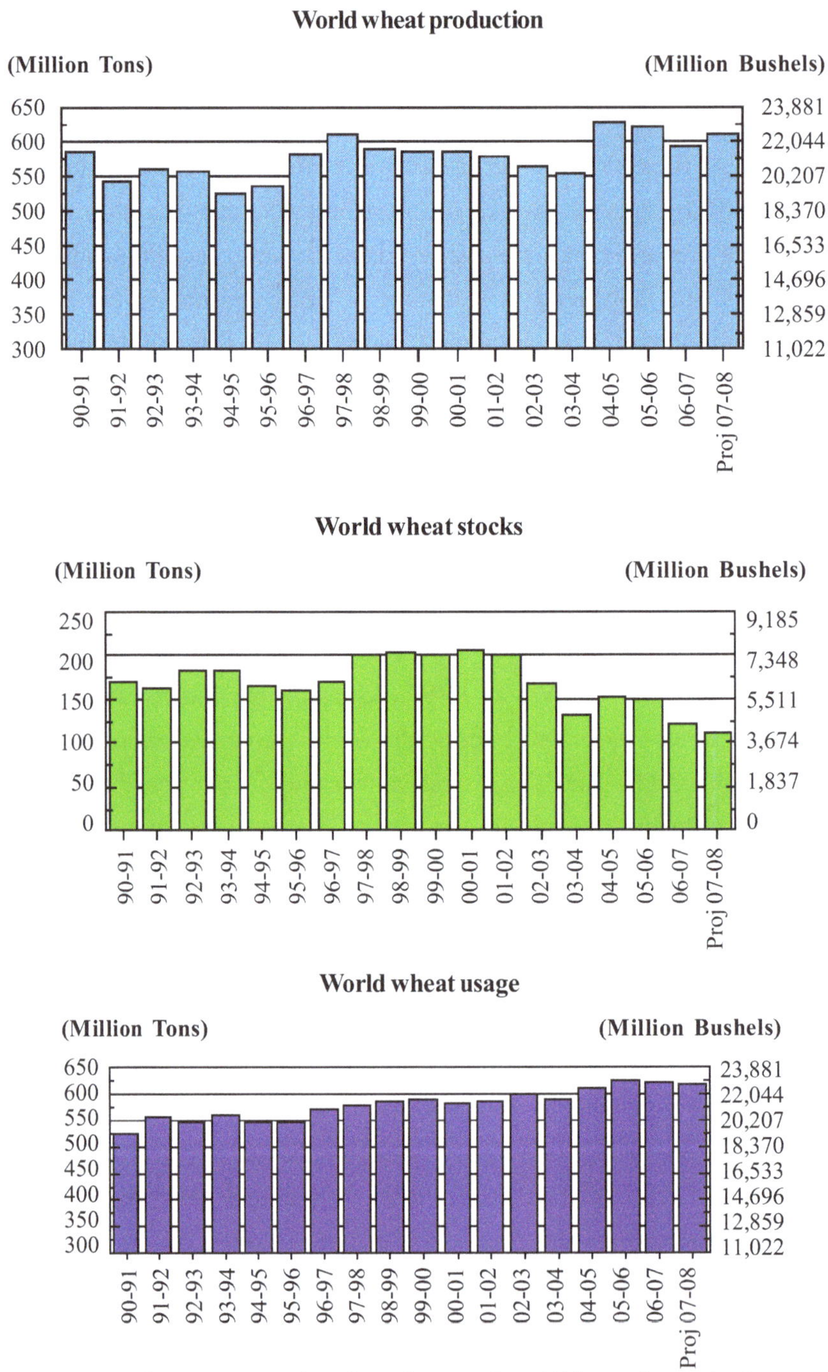

Fig. 1. Figures are taken from International Wheat Commission (North Dakota Wheat Commission 6/11/2007)

are predicted to improve for the next ten years. The report predicted total world wheat trade to increase by 16.8% from 79.2 million metric tons in 2002 to 93.4 million metric tons in 2012. Major wheat exporting countries as USA, Canada, European Union, Argentina and Australia will exhibit increase in production during the period 2003-2012. Most importing countries are predicted to increase their imports for both common and durum wheat. Among those countries, import demand for common wheat in China and Tunisia will grow faster than in other countries. Import demand for durum wheat in Algeria is predicted to grow faster than that for common wheat. Import demand for both common and durum wheat is largely based on optimistic income growth (2.5% to 6% annually) in developing and developed countries, which was provided by WEFA and Project Link. However, if the predicted income growth is not realized, import demand could grow slower than predicted and estimated prices could be lower.

Wheat Production and Consumption Trends in India

At the onset of green revolution in India during 1960's, wheat production increased and India became self sufficient in wheat production during 1980's and 90's with importing lesser quantity of wheat and started exporting during 2000-2005 (Fig.2ab). Wheat export started after large stocks of wheat accumulated by record production of around 76 million tons during 1999-2000 crop season. Since then wheat production in India has been fluctuating from 69 million tons to 78 million tons annually. The fluctuation in total production affects the buffer stocks which need to maintain at a level decided by the Government of India. It has become a great concern in India to enhance and sustain wheat production in the country to meet the rising demand of wheat and wheat based products (Fig 3). India has potential to enhance production and productivity of wheat by reducing yield gaps in farmer's fields and also enhancing yield potential. For this purpose Food Security Mission has been initiated to enhance wheat production by 8 million tones to 80 million tones by 2012 in India.

World Wheat Consumption

There are many factors that influence the pattern of wheat consumption in the world. The most important of these include: Price, Supply, Consumer income, Availability of substitutes and Politics. Market opportunities are influenced by a country's stage of

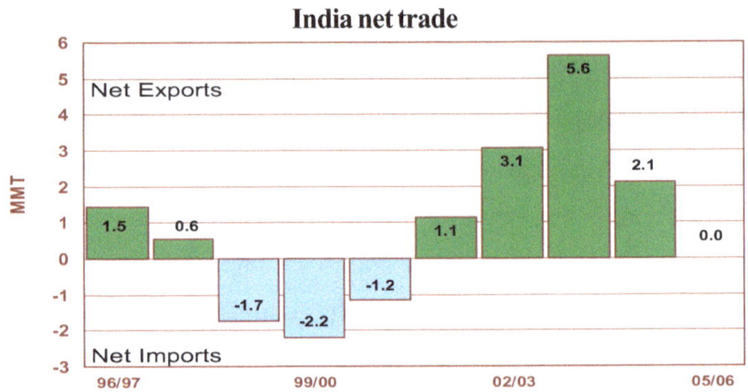

Fig. 2a.

Source : USDA, World Agricultural Supply and Demand Estimates; February 2006

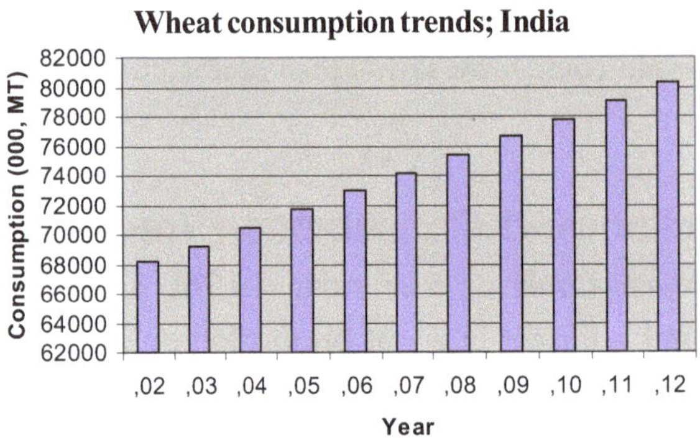

Fig. 2b.

Source : Agribusiness and applied economics report No. 517, August 2003

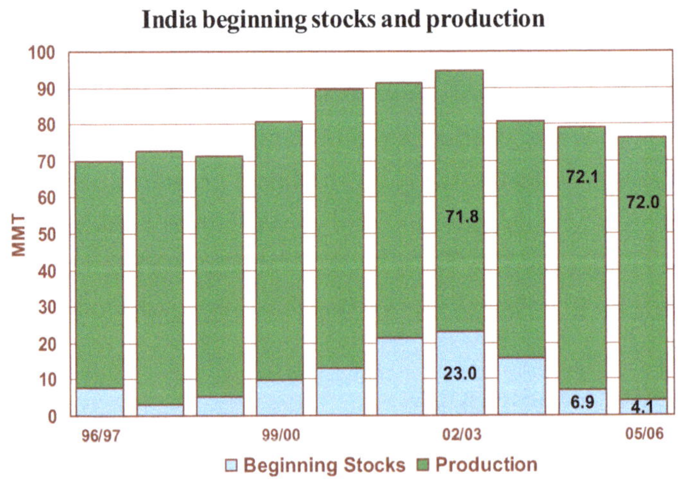

Fig. 3.

Source : USDA, World Agricultural Supply and Demand Estimates; February 2006

economic development. Therefore, wheat consumption should not be looked upon as a constant for any particular region or country. In a free market environment, increases in income of the poorest consumers generally results in a greater consumption of wheat for food. As consumer income increases wheat is gradually replaced by more expensive foods, particularly meat. As a result, wheat consumption in more affluent countries often depends upon its use for both human food and livestock feed.

World wheat production and consumption increased rapidly in the 1950's and 60's. The direct influence of price on wheat consumption was demonstrated in the early 1970's when a sharp price increase caused a significant shift away from the use of wheat for human and livestock feed. However, this shift was temporary and rate of wheat consumption quickly returned to the pre-1970 levels. Wheat consumption increased faster than production (1.9 versus 1.5%) from 1985 to 1991. Low production in 1988 caused world wheat reserves to fall to 65% of their 1986 level. The record production year of 1990 saw a small recovery, but there was a general downward trend in the level of wheat reserves from 1985 to 1991, especially in carryover stocks held by the USA. If world wheat reserves continued to be depleted at the 1985 to 1991 rate they would have disappeared entirely sometime between the years 2001 and 2005. The stage is set for a particularly volatile period in world wheat production and price adjustments, especially if consumption continues to exceed production, or if continued low prices encourage farmers to make significant further shifts in production away from wheat, or if adverse weather reduces production in any of the major wheat producing regions of the world.

World Wheat Imports

There have been major changes in the pattern of world wheat imports in the last fifty years. Western Europe shifted from a major wheat importer, which purchased over 50% of the world imports in 1949-50, to a major wheat exporter, which produced nearly 20% of the world wheat exports in 1988-89. In contrast, wheat imports by Japan increased during the 1950's and 60's making it one of the largest world markets by the mid-1970's. Developing countries also became a large market for wheat during this period. Total world wheat production slowly increased during the 1980's. However, an even larger increase in consumption by producing countries resulted in a slight decline in world wheat imports during this period. Russia and China, countries that started to import wheat on a large scale in the early

1960's, accounted for more than 30% of the world wheat imports in the 1980's. Both regions have large populations; however, they also have a good potential for wheat production. A large crop allowed China to significantly reduce its imports in 1992-93. Recent economic problems have had a significant effect on Russia's ability to import wheat.

Recent import patterns suggest that we should expect future wheat markets to be much more diverse than those we have traditionally seen. Developing countries are expected to continue to provide major opportunities for wheat exports. Consumption often outpaces production as standards of living and populations increase in these countries. This creates a large demand for food products, especially those produced from wheat. World wheat imports could expand to an even greater extent if the recent General Agreement on Trade and Tariffs (GATT) were to provide a less restricted environment for the movement of wheat into developed countries, especially the EEC. Changes of this nature would produce a more diverse, more quality conscious world wheat market where exporters with responsive marketing strategies and the ability to deliver a range of wheat quality types have an advantage.

World Wheat Exports

Five major exporters, Argentina, Australia, Canada, EEC, and USA, normally account for over 90% of the world trade in wheat. These countries also reluctantly maintain a significant proportion of the world wheat reserves as annual carryover. Among the "big five", Australian and Canadian farmers are the most dependent upon the international marketplace for the sale of their wheat. The global wheat market over the past 10-15 years has proven to be relatively static with low-level growth in consumption and little growth in the amount of trade. Along with the United States, Canada, the European Union and Argentina, Australia is one of the five major wheat exporters in the world. Although Australia represents only 3% of global wheat production, due to relatively small domestic consumption, Australia accounts for between 15-18% of total world exports.

Wheat Utilization

Wheat varieties are highly differentiated in terms of their agronomic and end-use attributes. Based on criteria such as kernel hardness, color, growth habitat, and protein content, wheat is divided into several classes. Color and hardness refer to physical properties of the wheat kernel. Based on the color of the outer layer of the kernel, common wheat varieties are described as white, amber, red, or dark, while the hardness of the kernel is used to characterize them as hard or soft. Growth habitat is an important agronomic feature of wheat varieties. Winter wheat is planted in late summer or fall and requires a period of cold winter temperatures for heading to occur. After using fall moisture for germination, the plants remain in a vegetative phase or dormancy during the winter and resume growth in early spring. In contrast to winter wheat, spring wheat changes from vegetative growth to reproductive growth without exposure to cold temperatures. In temperate climates, spring wheat is sown in spring. Since yields for winter wheat tend to be higher than for spring wheat, spring wheat is produced primarily in regions where winter wheat production is infeasible, where frozen soil kills the wheat plants, or where winters are too warm. Countries with mild winters, such as Argentina and Brazil, produce spring wheat but plant in the fall rather than in the spring.

Wheat is utilized in large number of wheat products. It is estimated that nearly two-thirds of the wheat produced in the world is used for food and the remaining one-third feed, seed and non-food uses. Bread, biscuit, noodles, chapati and pasta products are main products used in different parts of the world. With globalization of economy, increased urbanization in third world countries and economic development food habits are changing in developing countries. With that consumption of bread, biscuit and pasta products is increasing. There is some geographical distribution of different products and many of the products are local. For example, fermented leavened bread is mainly consumed in Europe, America, Canada, Australia and some other parts of the world. Noodles are mainly consumed in Eastern part of the world including China. Pasta products such as macaroni, spaghetti and fettuccine, though Italian in origin are now consumed world wide. Flat breads are consumed in North Africa, Middle East and Indian subcontinent. Major types include chapaties, roti, naan, paratha, poori, balady, pita and barabri. These breads typically have high crust to crumb ratio and limited crumb. They are generally baked at very high temperatures for very short time periods. Flat breads are typically produced from high extraction or whole wheat flours. Due to the higher bran content of the flours, white wheat is preferred. Generally weaker gluten is needed for making such breads as compared to pan or hearth breads. In India biscuit industry is the largest constituent of food processing industries and millions of people are involved directly or indirectly with the biscuit industry. The details of each class of end-product of wheat will be discussed in different chapters in the book.

References

Agribusiness and Applied Economics, Report No. 517, (2003).

Grain: World market and trade, USDA, January, (2007).

International Wheat Commission (North Dakota Wheat Commission) (2007).

Morris, C.F. and Rose, S.P. (1996). Wheat. In: Cereal Grain Quality. Edited by R.J. Henry and P.S. Kettlewell. Chapman & Hall. pp1-54.

USDA: World Agricultural Supply and Demand Estimates, February, (2006).

Wheat Production, Properties and Quality. Edited by W. Bushuk and V.F. Rasper (1994). Chapman & Hall.

❑❑❑

Chapter 2

BREAD QUALITY

Introduction

There are large numbers of wheat based products such as fermented pan bread, flat breads including chapati, biscuits, noodles and pasta products. Each product has its own specific requirements. For example fermented pan bread needs wheat with higher protein content, stronger and extensible gluten. Chapati requires hard wheat with high protein content and medium gluten strength. Waxy wheats have been found associated with improved shelf life of breads. White salted Japanese noodles need partial waxy wheats with soft grain characteristics and medium protein content and gluten strength. Partial waxy wheats with comparatively lower amylose content and higher starch paste viscosity improve texture of white salted noodles. Yellow alkaline noodles need comparatively harder wheat with medium strong gluten and without waxy trait. Pasta made from durum wheats requires stronger gluten, high protein content and higher content of yellow pigments. In this chapter detailed account of both fermented and flat breads will be given.

Fermented Bread

Fermented breads include traditional white pan, variety, hearth and sourdough breads and sweet goods including doughnuts. Leavened breads are characterized by a crumb texture that is light, airy and porous and chewy. This unique texture confers the wide popularity and appeal of leavened breads. These characteristics are determined mainly by the gluten quality. Gluten must be capable of holding gases released during fermentation. This is done by continuous gluten matrix that can stretch to accommodate the increase in volume during fermentation and baking. In addition to gluten proteins, other grain components such as starch, lipids and pentosans influence bread quality. The minimum basic formula for leavened bread includes flour, water, salt, yeast and some other sources of leavening. In addition large numbers of additives or improvers such as enzymes, emulsifiers and oxidants etc are used to increase dough processing quality, loaf volume, appearance, crumb softness and keeping quality. There are large numbers of processing variations used by bakers. The basic methods may vary from straight dough methods with varying periods of fermentation time, to sponge and dough, liquid ferments and short systems involving high speed mixing, such as Chorleywood bread process. Different steps such as mixing, fermentation, sheeting and molding, proofing and baking can vary in terms of specific conditions such as time and temperature.

Bread Baking Procedure

Detailed equipments and methods are beyond the scope of the book; however basic principals involved in the baking process will be discussed. Among the baking methods sponge–dough predominates followed by liquid ferment method. Straight dough procedure is generally used in retail operations and for the production of bread varieties, however recently its use is decreased. In the straight dough method of mixing all dough ingredients are mixed at one time, and prepared for a single fermentation process. Generally, the fermentation time for the straight doughs varies from 2-1/2 hours to 3 hours. These doughs are also punched after about 80% of the fermentation time has elapsed and given an additional 20% fermentation before make-up. Doughs produced by this method results in breads with course grain and texture and the crumb is not as soft as that produced by other methods. The baked loaf will not have as much volume as one made with the Sponge and Dough Method.

In the Sponge and Dough Method, there are two mixing periods and two fermentation periods. The sponge dough formula contains flour, water, yeast, and proteases. The sponge process adds tremendous flavor and texture to bread. It is the equivalent of an extra rise in the bread making process. In this process water, yeast, yeast food are blended and pumped into a mixer. Flour is pneumatically conveyed from storage bins into automatic overhead scales and dispersed into mixer. The sponge ingredients are then mixed and transferred into troughs and kept for fermentation at 30°C. The fermentation time varies from 3-5 hours depending upon the yeast level and percentage of flour used. The volume of the sponge increases four to five fold during fermentation and the mixer develops a characteristic spongy or web structure.

After the fermentation process is completed, the second part is called the Dough. Basically, the fermented sponge is thrown back into the mixer and the ingredients for the second part are added. The fermented sponge and all the second ingredients are mixed together to form the dough. Proper dough mixing produces a dough of optimal degree of development. The main factors affecting this reaction are the mixing time and optimal water absorption. Mixing time and water requirements are predicted by Farinograph and Mixograph instruments. After the gluten has been fully developed, the dough is dumped into a stainless steel dough trough and given a second fermentation time. As mixing progresses, complex interactions take place, net result being the formation of a thin gluten film that surrounds starch granules. Gluten is formed by the interaction of proteins through sulfhydryl-disulfide interchange, association of secondary bonds (hydrogen and hydrophobic), and the formation of lipid-protein complexes are involved. As compared to the sponge, the fermentation time of the dough is very short (15 to 20 minutes). The dough developed should have proper strength and should not be sticky. At the peak of development, dough has a silky, glossy appearance and elastic properties. Dough can be stretched into a thin membrane in which small bubbles and gluten strands can be seen.

The dough is then divided and scaled into pieces, rounded, molded and panning before baking. Dividing and scaling consists of cutting the dough (by hand or machinery) into loaf size pieces and weighing the pieces to insure uniformity. Because of the average loss of about 12 percent during make-up, proofing, baking and cooling, this loss must be taken into account. Rounding is the process of rounding the scaled pieces of dough into a round ball with smooth unbroken skin over its entire surface. The unbroken skin will retain the gas generated within the dough piece during the intermediate proofing period. Intermediate proofing is the process of giving the rounded dough pieces a short rest period (about 12 to 15 minutes) to recover from the effects of the dividing and rounding machine. The rest period allows the dough to relax after undergoing a great deal of punishment while being forced under a lot of pressure during extrusion from the divider pockets. Without the rest period, the dough would be tight and rubbery and would not go through the molder sheeting rollers without tearing. The intermediate proofing machine has small pockets or baskets covered with canvas traveling within an enclosed draft free area. Upon completion of the intermediate proofing period, the dough pieces are molded into the shape desired.

In the molder the dough passes through three distinct stages. Flattening is done in the head rollers of the molder. Second, the sheeting rollers sheet the dough into a flat piece of dough and the curling rollers and thread rollers twirl each piece of sheeted dough and give it a cylindrical shape. Next, the drum or pressure plate rolls and seals the loaf into it's final form. Conventional molders curl the dough in the same direction that it was sheeted. It is difficult to produce bread with a close uniform grain with this type of molder. For this reason some bakers twist two pieces of dough together or use cross grain molders. Cross grain molders curl the dough in the opposite direction from which it was sheeted. In this process, the dough is turned right or left after it has been sheeted and is then curled. Cross grain molding and twisting two pieces of dough together, prior to putting them into the baking pan, produce loaves with a close smooth grain and texture.

Pan proofing is the process of rolling the panned and racked molded dough pieces quickly into the dough proofing cabinet which is well insulated and maintained at a temperature of 95° F to 98° F and a relative humidity of 85%. Under these conditions the loaves are allowed to proof for about 45 to 60 min. At the end of the pan proofing period, the loaves are loaded quickly, but carefully into the oven for baking. In large commercial bakeries, the panned loaves are carried automatically on conveyor belts and loaded automatically into the oven. Oven temperatures vary from about 425° F to 450° F. During the first minutes of the baking process, carbon dioxide gas within the dough expands. This expansion causes a very rapid rise of the dough known as oven-spring. Fermentation is more vigorous and more rapid at this stage than at any previous stage. When the inside temperature of the loaf reaches 140° F. the yeast is killed and fermentation ceases. Alcohol produced during fermentation evaporates in the form of vapor. After the oven-spring the pliability of the dough gradually lessons and the dough becomes set and slowly changes to bread. Some of the moisture evaporates, the starch becomes gelatinized and more digestible, gluten and other proteins become coagulated. After the loaf sets, the intense heat dries out the part exposed to the air and causes a crust to form. The golden brown color of the crust is the result of chemical changes in the starch, sugar and milk known as a browning reaction (Mailard reaction) also known as carmelization.

Within the loaf, the crumb near the crust is subjected to a temperature as high as 300° F. The temperature gradually decreases towards the center of the loaf at which point it reaches about 212° F. At this temperature the gluten becomes stiff enough to give the loaf permanent form and retain it's structure. When the baking process is complete, the loaves are unloaded manually or automatically and either dumped on cooling racks or travel on enclosed air conditioned conveyor belts to the cooling area. The loaves are allowed to cool for about 1 hour at which point the internal temperature of the loaf drops to about 100° F and Is ready to be sliced and wrapped or bagged.

Basic Ingredients

Followings are the basic ingredients used in bread baking.

Flour

Wheat flour is the main constituent for bread making. It contains proteins, which, when combined with water form gluten. Gluten is the elastic material which holds the gas produced by the chemical reaction of the yeast enzymes on sugar. Gluten forms the framework of bread and also the cell structure of the interior of the loaf. Gluten is separated from starch by washing. The main gluten proteins are glutenins and gliadins. The gliadin is the sticky part and will form long tiny silky looking strands when touched with the finger. The glutenin on the other hand will look and feel like tough raw rubber. The flour should have a high content of quality protein to retain gas and contain enough natural sugars and diastatic enzymes to produce enough sugar from the starch for uniform gas production. Tolerance is the ability to produce a quality product for some time after the optimum fermentation time has elapsed and in the event that the dough was over mixed. The detailed structure and functionality of gluten proteins will be discussed in another chapter. Higher protein content and stronger gluten are preferred requirements for better loaf. Depending upon the requirements, bakers use vital gluten and other dough strengtheners to improve dough properties for bread. In US, hard spring wheat generally produces loaves with greater volume than winter wheats, but with slightly more open crumb texture. Millers combine the two types of wheat in their blend to improve interior loaf characteristics. Hard white wheat grown in some Western States is also used in bread production. It is slightly lower in protein than spring and winter wheats.

Different types of flours are produced during wheat milling. Basically, after the wheat has been cleaned and tempered, and the percentage of moisture within the grain has been adjusted, the wheat passes through several reduction rollers. The grounded middlings are separated into several streams of flour by sieving and bolting. The bran particles which have been removed are used for other purposes including animal feed. Whole wheat flour is produced from 100% of the wheat. Straight grade of flour is all of the flour produced after the outer portion (bran and germ) has been removed. Straight flour (~70% of the wheat) is similar to all purpose flour found in grocer stores. Patent flour is highly refined flour which remains after all the clear flours have been removed. Patent flours are produced from the innermost part of the wheat where the best quality proteins are located. Patent flour amounts to 40-60% of the whole meal flours, and the middlings that remain go into clear flours. Clear flour is the portion of the flour remaining after the patent flour streams have been separated. Clear flour generally contains higher percentage of protein than the other grades, but the quality of the protein is lower.

Enzymes in Flour

The two most important enzymes in flour are protease and diastase. Protease conditions the gluten, improving its elasticity and its ability to retain the gas produced during fermentation. Diastase changes some of the starch in flour to dextrins and maltose sugar. Some flours do not contain a sufficient amount of diastase enzyme due to poor climatic conditions under which the wheat was grown. Sprouted wheat can be added to the flour by the miller or the baker can add especially prepared diastatic malt in his formula. Details of enzyme action in improving bread quality are provided in enzyme section.

Flour Bleaching, Maturing and Enrichment

In order to produce quality breads from fresh milled flour it must be allowed to age or mature in storage for about a month. Bleaching and maturing agents are used to artificially age and whiten the flour. This results in flour that can be used immediately after milling to produce a quality product. Flour treatment includes bleaching, maturing, malting, and enriching. Bleaching removes yellow endosperm pigments using oxidizing agents like benzoyl peroxide. Chemical maturing improves flour strength and tolerance using oxidizing agents like potassium bromate, ascorbic acid, or azodicarbonamide. Standardizing amylase activity in flour is accomplished by adding malted wheat or barley flour or fungal alpha-amylase. Enrichment of flour is the process of replacing the vitamins and minerals removed during the milling process. Most of the vitamins and minerals are located in the outer portions of the wheat, and since most of the outer portions of the wheat are not used to produce the best grades of flour, these essential elements must be replaced. The exception is whole wheat flour.

Water

Water is a basic ingredient in bread baking. It would be impossible to produce a loaf of bread without water in some form. There are several types of water. Hard water produces better quality bread than any type of water. Soft water weekens the gluten during mixing and fermentation. This can be corrected to some degree by increasing the percentage of salt in the formula slightly and by using mineral yeast food in the formula. Alkaline water is harmful, because it not only weakens the gluten, but retards fermentation. Yeast likes a slightly acid medium to perform at its best. The weakening of the gluten and retarding effect on yeast can be corrected by using an acid ingredient such as vinegar (acetic acid) or lactic acid. Special types of mineral yeast food have been developed to correct this problem.

Yeast

Yeast is one of the essential ingredients in bread production. It is a one cell plant which multiplies by a process known as budding. Under the right conditions of water, sugars, warmth, and dissolved minerals, yeast causes fermentation. Yeast is available in compressed form which must be kept under refrigeration until it is ready to be used and active dry yeast which need not be refrigerated. Active dry yeast has an extended shelf life. Both will generally produce satisfactory results. However, a majority of large commercial bread bakeries use compressed yeast in their formulas. When using active dry yeast half as much active dry yeast is required as compared with compressed yeast.

Enzymes

Important enzymes in the yeast for bread production are invertase, maltase, zymase and protease. The protease enzyme in yeast is only active if the cell wall of the yeast is damaged in some way. That is the only time it can penetrate the cell wall of the yeast. There are always a few damaged yeast cells especially in dry yeast where some cells are damaged during the drying process, or in yeast that has become a little old. If the protease enzyme does penetrate the cell wall of the yeast, it will weaken the gluten. Invertase enzyme is an intercellular enzyme and converts sucrose, (cane or beet sugar) into dextrose and fructose, which are later attacked by the enzyme zymase. Zymase is the enzyme which changes the simple sugars, dextrose and fructose into carbon dioxide gas and alcohol, and several esters which result in producing the unique aroma flavor and taste of breads. Also, the gas produced causes the dough to rise. Maltase attacks malt sugar. The malt sugar is changed into two molecules of simple sugars such as dextrose. Fermentation will be discussed in more detail in a later paragraph on bread production procedures.

Salt

Salt is another essential ingredient in quality bread production. It has several functions. It regulates fermentation. Too little salt causes the dough to ferment too fast, while too much salt slows down the fermentation process. Salt is a very powerful ingredient and it must be scaled very carefully when weighing the ingredients for the dough. In the proper amount, salt produces good grain and texture, because it strengthens the gluten allowing it to fully mature during fermentation. It also prevents the growth of wild yeast and bacteria. A whiter crumb is produced and last but not least, it brings out the taste and flavor of bread.

Enriching Ingredients

Enriching ingredients such as sugar, milk, shortening, flavouring agents, mineral yeast foods are added to improve the texture and flavour of breads. Sugar is not considered an essential ingredient in bread baking. This is because flour contains a small amount of natural sugars and some starch is converted to sugar during the fermentation process. However, it does have many important functions. Some of the added sugar is converted to carbon dioxide and alcohol by the yeast enzymes. It provides necessary sweetness, helps to produce a golden brown color of the crust, improves the texture of the crumb, helps to retain moisture in the crumb and adds to the nutritional value of the bread. It has a stabilizing effect on fermentation, preventing wild fermentation. It improves crust color because of the lactose sugar it contains. The lactose sugar is not fermentable by baker's yeast. It also improves texture, crumb color, flavor and taste, and keeping quality of the baked loaf.

Shortening acts as a lubricant in the dough, making the dough more pliable, prevents stickiness, and reduces the amount of dusting flour necessary during the make-up process. When shortening is used, the dough expands more easily and smoothly. In the baked product it makes the crust tender, improves the keeping quality and produces a crumb that is soft and chewy. Mineral yeast food was developed to stabilize water by adding mineral salts, which are essential in dough fermentation and conditioning. Mineral salts condition soft water, and acid salts neutralize the alkalinity of alkaline water which is harmful to yeast and gluten. Doughs made with mineral yeast foods are not sticky and are very soft. They go through machines with less trouble and will mold up free of air pockets. Use of mineral yeast foods results into greater loaf volume, improved texture and crumb color, better crust color, and greater uniformity. In addition, poppy seeds, sesame seeds and rye powder are used as flavouring agents.

Staling of Bakery Products and Mold Infection

There are three ways in which bakery products stale. They are starch retrogradation (firming of the crumb) and by getting infected by molds and rope. In simple terms, staling of crumb is the process in which starch molecules shrink upon cooling and become firm. Anti-staling ingredients such as emulsifiers and enzymes are used in baking industry to improve bread quality. Some of the more common emulsifiers used are monoglycrides, calcium steroyl lactylate, and sodium steroyl lactylate.

Enzymes in Bread Making

The use of enzymes in biotechnological processes is part of a long established tradition. Today, enzymes play a central role in baking technology and the applications of enzymes in baking are expected to grow at a much faster rate. The addition of enzymes to bread dough was described for the first time more than 100 years ago when malted barley was used as a source of amylase activity which converts starch into sugars to serve as a substrate for fermentation. Fifty years later lipoxigenase was described as a tool for bleaching the bread crumb and to improve machinability. Since then the use of different enzymes has increased in bread making. Several commercial enzyme preparations are used in baking industry as amylases, proteases, xylanases and oxidases and lipases etc. In today's modern society, consumer's food needs as well as preferences have changed substantially. In order to have more wholesome and fibre-rich foods, the baking industry is shifting towards producing whole wheat bread with limited use of chemicals. Since enzymes are produced from natural ingredients, they will find greater acceptance by the consumers than when chemicals are used. However, production of such type of breads poses certain technological difficulties. In order to improve the consumers' acceptance to whole wheat bread with improved body, texture, flavour and other desirable properties, appropriate technology is required to be developed. Microbial enzymes can be of great help in overcoming these problems and they can be added as processing aids/additives. The added enzymes can be easily inactivated by the baking time–temperature treatment. Enzymes are already present in bread wheat flour as discussed in previous section.

In general, enzymes act on biopolymers present in wheat (polysaccharides and proteins) or sometimes the combinations thereof. Amylases are used for attacking damaged starch or gelatinized starch generating sugars for yeast fermentation and for slowing down the firming of the bread crumb i.e. the aging of the bread. Amylases are used to improve loaf volume, flavour and shelf life. Proteases are used for shortening mixing time and for obtaining more flexible doughs. Proteases have effects on dough softening, anti-staling, and gluten strengthening, improved loaf volume and dough machinability properties. Oxidases are though to act predominantly on the disulphide bridges which are essential for building and stabilizing the protein network. Lipases act on lipids and have effects on improved dough machinability, improved loaf volume, improved crumb softness, white crumb colour and anti-staling. Pentosanases (hemicellulase, xylanases) acting on non-starch polysaccharides, are able to change water absorption and improve loaf volume, flavour, texture and freshness.

As the consequence of the hydrolytic action of xylanases, free sugars such as pentoses could be released. Free sugars can be used by microorganisms for fermentation. Moreover, xylo-oligosaccharides released by endoxylanases could form healthier food. The healthy effects of xylo-oligosaccharides are mainly related to their effects on the gastrointestinal flora. The non-starch polysaccharide in the endosperm cell walls of wheat constitute up to 75% of the cell wall dry matter weight. There are 3–4% (w/w) pentosans in normal wheat flour, partially soluble and partially insoluble. Arabinoxylan occur as minor components of wheat flour (2–3% dry basis) and can be divided as water extractable (WE-AX) and water unextractable (WU-AX). Water unextractable arabinoxylans, which make up 70% of wheat endosperm cell walls, can be hydrolyzed by endoxylanases and cause them to loose their strong water holding capacity.

Protein Quality and Quantity

Wheat storage proteins are the major determinants of bread quality. Both protein content and quality influence bread loaf and hence wheat protein content can demand a premium price in the marketplace. Protein measurement is routinely done at elevators and throughout the grain marketing system on all received grain. NIR instruments have provided inexpensive, rapid determinations of wheat moisture and protein contents. Protein content is important to the milling and baking industry because it can predict baking performance. There are genotypic differences in protein content and quality. Studies demonstrated the linear relationship between protein content and bread loaf volume. But relationship was stronger if protein content was increased in the flour of a genotypes. This means if protein content is kept constant in different flours from different varieties with differences in protein quality, the loaf volume does not increase linearly (Fig 1). Therefore, both protein content and quality are important in bread making. Drought conditions can stress grain and produce high protein wheats. Yet flours milled from this wheat bake poor quality breads. Protein quality is the difference.

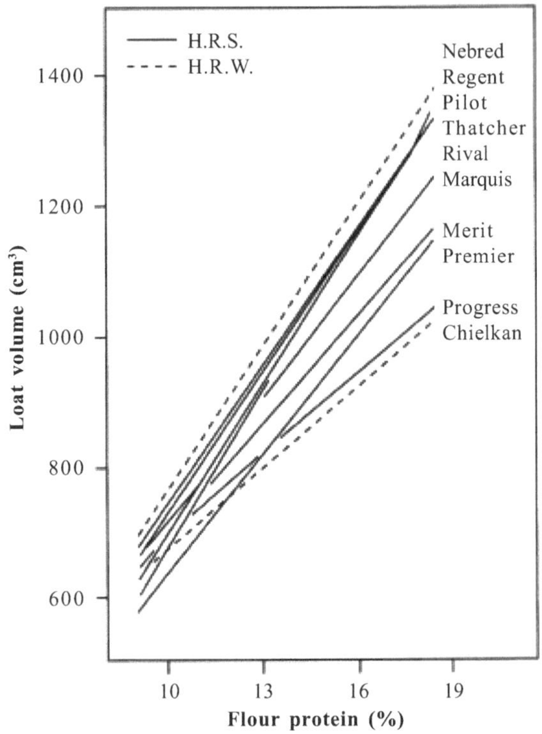

Fig. 1. Relationship between flour protein content and loaf volume demonstrating the effect of quality and quantity of proteins on loaf volume. (From Wheat Quality Elucidation: The Bushuk Legacy, 2002, Edited by Perry K.W. Ng and Colin W. Wrigley).

High protein content alone is not enough. The proteins must have functional properties for water uptake (absorption) and mixing strength to insure proper gluten protein development. Gluten is a complex network formed by the glutenins and gliadins in which starch molecules are entangled. Gluten proteins are not formed unless water is present and mixing occurs. The effect of gluten protein subunits of dough strength has been studied extensively. Understanding the molecular basis of action of gluten proteins including high molecular weight glutenins, low molecular weight glutenins and gliadins in determining gluten strength and bread making quality is in progress. Some of the genes have been characterized and used in the development of transgenics. Molecular markers have been identified and are being used breeding. These studies successfully assisted in developing varieties with improved bread making characteristics. The details of these aspects will be discussed in different chapters. In brief, during the bread making process gluten proteins are responsible for the following:

- Gas retention during fermentation which results in good loaf volume and loaf softness.
- Bread crumb strength so that peanut butter can be successfully spread without tearing the bread slice.
- Bread crumb texture that is soft to the touch, yet does not stick to the roof of the mouth.

Therefore, the strength and the extensibility of gluten proteins are important determinants of bread quality. Hence the measurement of gluten strength in early generation breeding lines as a prediction test for gluten quality is imperative for a successful wheat breeding program. During last 100 years large numbers of tests have been developed for measuring gluten strength using flours of different amounts. All these rheological tests and chemical evaluations have been used in predicting bread quality. Equations have been developed and had a great assistance to baking industry in deciding the use of water and the mixing time. The details of these tests will be provided in other chapters.

Flat Bread

Based on the density, breads are divided into three main groups; 1) those with high specific volume such as pan bread, 2) those with medium specific volume such as French and rye bread and 3) those with low specific volume such as pita and lavosh breads and flat breads of Northern Europe, Middle East and Indian subcontinent. Thus flat breads are dense in texture, mostly crust and little crumb. Crust is thin and light with brown and dark spots. Crumb is small in quantity and coarse and dense. There are large numbers of different types of flat breads consumed through the world. Major types include chapati, tortilla, roti, naan, paratha, poori, balady, pita and barabri. In recent years, popularity of flat breads has increased in all parts of the world because of nutritional benefits attached to these breads.

Based on the cross sections of these products, they can be divided into two groups, single layered and two layered. The distinctive groups are created by the processing method. In the single layered flat bread, dough pieces are baked immediately after sheeting or at relatively lower temperature for a longer time. For production of double layered flat bread, sheeted dough pieces are allowed to proof prior to baking. Proofing allows the dough to relax, aerate and develop a thin skin. During baking thin skin changes to a pale coloured extensible crust, and the stream from free water in the dough, along with the pressure from CO_2 and other volatile substances developed during fermentation, force the top and bottom crusts apart and create two layers.

Single layered flat breads are the most diverse products. They can be either leavened or unleavened. Unleavened bread includes chapati, arepa, tortillas, paratha, puri and Scandinavian flat breads. The leavened single layered flat breads are divided into dough and batter-based products. Barbari, naan, pizza crust and tanoor are few examples of diverse and widely consumed dough-based products. The batter-based flat bread includes blintzes, crepes, crumpets, dosai, kisra, rogag and mechanically leavened pancakes. Typical double-layered flat breads are baladi and Arabic breads. Other names for Arabic flat breads include pita, shami, Labanese and maafrood. These are leavened products containing sour-dough or yeast as leavening agents. Generally flat breads have common quality requirements and details of some the breads are given below. Readers are suggested to go through selected references provided in the end of the chapter for details of other types of flat breads.

Chapati

Chapati the unleavened flat bread is most common flat bread of the world. This form of flat bread is mainly consumed in India, Pakistan, Afghanistan, Bangladesh, Mongolia and Tibet. In India 85 % of the wheat produced is consumed in the form of chapati. Chapati is prepared from a mixture of whole meal (atta) and water to make stiff dough. Some small salt or oil can be included in the formula. Cohesive dough is formed by kneading the ingredients with hands. Water is added during kneading the dough at regular intervals. The amount of water is determined by the quality of the flour. Water requirements depend on the chemical constituents in the flour as well as the amount of starch damage. Higher starch damage lead to higher water absorption. After a short resting time, dough pieces are cut and rolled into dough balls. Dough balls are then flattened and sheeted to a desirable thickness and baked on both the sides on a hot plate. For experimental purpose in the laboratory, chapati thickness is kept constant as 2 mm.

Chapati is baked in griddle heated to around 230°C on both the sides till desired texture is developed. Therefore, quality of chapati depends on the experience of the person. For example different people produce chapati with different texture characteristics because of the amount of water used and the time taken in kneading the dough differ among themselves. Slow and steady addition of water and kneading of the dough brings the required consistency for chapati making. Depending on the quality of flour whether weak or strong the quantity of water is adjusted. Convenience in spreading the patty for making chapati is used as a criterion for deciding the consistency of the dough. Little amount of flour is spread on the patty to reduce sticking during spreading. Though for laboratory purposes, the amount of water for chapati making can be estimated by Farinograph consistency line of 500BU. During baking puffed ball is developed which on cooling collapses in flat chapati. Puffing is done due to accumulation of steam generated during baking.

Chapati quality is assessed by subjective parameter such as taste, colour, appearance, aroma, pliability and keeping quality. Objective assessment of chapati pliability and shear strength can be measured by texture analyzer. Generally quality of freshly baked chapati is good for all types of wheats used. However, differences are pronounced during storage. Chapati should be soft and pliable and maintain pliability during storage. Varietal differences are there in keeping quality of chapati. Chapati with very strong gluten becomes leathery within few hours of storage. Chapati made from soft wheat with weak flour becomes brittle during storage in cloth. Therefore, gluten is an important component in wheat which has a major role in chapati pliability. Medium strong wheats with higher protein content and hard grain texture with high starch damage and high water absorption capacity are preferred for chapati making. Scanning electron microscopic studies have demonstrated that large gluten strands cover starch granules in good chapati while gluten is comparatively shorter and starch granules are exposed in poor chapati. Greater film forming ability of gluten in hard wheat flour dough is manifested in long bulky starch strands interwoven with protein matrix in its chapati crumb. Higher moisture retention and starch gelatinization as a consequence of greater film forming ability of gluten in hard wheat flour results in pliable and soft textured chapati (Srivastava *et al.* 2002).

Most of the varieties developed in India are hard to very hard as measured by SKCS. Molecular analysis also supports the genetically hard texture of Indian wheats. Most of the varieties have null mutation in puroindoline A. This indicates that selection pressure was exerted during breeding wheat varieties in India towards hard grain texture. In addition to hard texture, selection pressure has been towards lustrous, amber and bold grain characteristics. Amber colour is desirable because whole meal is used in chapati which contains bran particles. Bran particles impart their own colour to the flour, dough and chapati. That is why red coloured grains are not accepted by chapati consumers. Red grained impart undesirable dark colour to the chapati. Phenol test has been developed to screen early generation breeding lines for selecting genotypes. In this test grains are dipped in 1% phenol for 2-4 hours and the change in colour is observed. Based on the intensity of the colour, grains are classified into 10 groups from 1-10 scale with increasing intensity. Grains with low intensity or with out colour are selected because white or light colour of chapati is desired by consumers. Over all hard, medium strong wheats with low phenol reactions are selected for chapati making.

Tortilla

Tortillas are ranked as fastest growing segment in U.S. baking industry. Almost one tortilla per person each day is consumed in U.S. In households, tortilla is prepared by mixing flour with salt and water and cohesive dough is developed. After a short resting time, it is divided into small pieces and then rounded and sheeted followed by baking on hot plate. In the United States, the expanding market for a wide range of ethnic foods, including corn and wheat flour tortillas, led to the development of these industry from small, family oriented manual operations to highly automated large production lines. Based on the dough sheeting operations, three commercial processing methods are employed. These are hot press, hand stretching and die-cut methods. Hot press method is most popular. In this method, dough pieces are sheeted into thin layers by a hydraulic pressing device on a heated conveyer plate at temperatures ranging from 175°C to 240°C. Tortilla dough pieces are then baked in a three tier oven. The baking temperature ranges from 191°C to 260°C and the time from 17 to 40 seconds. Good quality tortilla should be soft, opaque, flexible without cracking when folded light colour and well puffed. When tortilla becomes stale, the texture turns firmer, less extensible and less rollable. Cracking and breaking of tortilla can be delayed by using flour with high protein quality or by adding vital wheat gluten. Good quality tortillas are large in diameter, opaque (99%-100%) and should have long shelf life (more than 2 weeks). Commercially available tortilla flours are mostly milled from hard red winter wheats developed for bread quality with moderate to strong protein quality and dough strength.

Balady Bread

Balady bread produced in Egypt and other Arab countries is special type of product of wheat. Bread accounts for 70% of the average Egyptian's daily intake of calories and protein. The most common type is two-layer, flat, sourdough bread, called balady bread. Twenty-five percent of balady bread produced is discarded due to poor quality. To reduce the waste, the government of Egypt has put high priority on research into the quality of balady bread. The bread is prepared by mixing flour (80-90% extraction) with water, salt and yeast. The dough is mixed in the usual manner and kept for several hours for fermentation. Dough pieces are then made and deposited on a tray covered with thin layer of bran. Dough ball is then pressed into a pancake shape and kept for fermentation for 30 min. The bread is then hearth baked at 500° C for 1 min. When the loaf is cut into two pieces, the crumb and crust are separated forming characteristic pocket. The cavity formed can be filled with beans, vegetables and other fruits. On an average, the crust and the crumb each constitute 50% of the loaf.

Arabic Bread

Another type of pocket bread in Arab countries is Arabic bread. It includes shamey, Syrian bread, kohbz and msfrood etc. Traditionally the bulk dough is fermented by air borne wild bacteria and yeasts. Baker's yeast and sodium bicarbonate is also sometimes included in the formulation. The bread is made by following steps: mixing, primary fermentation scaling (dividing), intermediate proofing, two dimensional sheeting, final proofing and baking. The formula includes flour, water (40-55%), salt (15%), yeast (1%) and sugar (0.5%). Unlike balady bread which is baked at very high water absorption (75 to 85%), the Arabic bread is made with lower water absorption, and therefore steam expansion in the very hot oven may not be the primary factor in ballooning effect that causes the separation into two layers.

Barbari Bread

Barbari bread is an oval leavened bread popular in Iran. Fedeh is a similar product that is popular in Turkey. It is probably one of the thickest and heaviest flat breads. Dough mass is prepared by mixing flour (72-77% extraction), water (60%), salt ((2.0%), yeast (1%) and soda (0.35%), and allowed to ferment for 2 to 3 hours. Ingredients are mixed to the desired consistency and fermented for 2 hr. It is made into balls of approximately 900 g and flattened into oval shape and rested for 20 min. A gelatinized paste of the flour and water (10% w/v) is prepared by boiling flour water mixer and then cooling down to room temperature. One or two tablespoons of this paste are spread evenly on the sheeted dough with fingers by forming a number of grooves. These grooves prevent the separation of the top and bottom crusts of barbari bread. After application of the paste, dough is allowed to ferment for an extra 20 min. The sheeted dough is then baked at 220° C for 8 to 12 min on the hearth of a brick oven.

Flour Quality Requirements of Flat Breads

A number of attempts have been made to identify quality requirements of different flat breads. Generally hard white wheat flours with higher starch damage and intermediate gluten strength and flour protein content are used for making flat breads. Flours with extraction rate of 85% or above are used in most of the flat breads. Protein content and gluten strength are important factors in making different types of breads. Traditionally, flat bread production in North Africa, the Middle East and Southern Asia are adapted to the weak strength of the local wheats. Therefore, the use of high protein strong wheat may not be suitable for flat bread baking. Balady, for example, must be over mixed to break down the gluten network for optimum bread production. Water absorption is another parameter important for flat breads. Unlike pan bread, in which the optimum water absorption is from 60-65%, water absorption in flat breads ranges from 35% in matzo to 85% in Iranian sangak.

For Arabic breads optimum product quality is obtained from hard wheat flour of intermediate protein content (10-12 %) and dough strength and starch damage above 6%. Hard wheat results in significant starch damage during milling. This allows higher water absorption and satisfactory pocket formation during dough mixing and baking. Both protein levels and starch damage have strong correlation with crust colour of Arabic breads. Williams 1998 reported that weak dough caused problems in sheeting, and the resultant product became stale and hardened faster than made with strong flour. Dough made with very strong flour on the other hand contract after sheeting and result in product of poor texture and bitability.

For chapati, hard wheats with medium gluten are preferred. Soft wheat generally produce chapati of low pliability and becomes brittle on storage Very strong flours produce chapati with leathery characteristics and hence unsuitable for chapati making. In general, flour for good chapati making should have protein content in the range of 10-13%, water absorption of not less than 68% and elastic and extensible gluten. In addition dough should not be sticky and browning should not occur during storage. Medium strong flours are easy to handle during kneading by hands.

Similar to other flat breads, hard wheat flour with moderate protein content produce superior quality hard pressed tortillas. A combined effect of starch properties, protein quality and flour ash content are related significantly to tortilla quality. Flour water absorption measured by Farinograph gives good indication of tortilla quality.

Summary

A major portion of hard wheat produced in the world is consumed in the form of fermented and flat breads. Various types of fermented breads such pan, variety, hearth and sourdough are produced. The quality of these breads is mainly governed by the quality of gluten. Generally strong and extensible gluten and high grain protein content are preferred for better bread quality. In addition, other components such as oxidizing enzymes and starch quality influence bread making quality. All of these components including gluten have been studied in greater details and the information has been used in breeding to enhance bread quality. All of these individual components will be discussed in other chapters. The consumption of flat bread is increasing in the world because of their higher nutrition value as nutritional quality components are retained in whole wheat flour. The quality requirements of flat breads depend on the type of flat bread whether they are single layered or two layered and also the preference of consumers in the region. Though, general requirements of flat breads are medium strong gluten and higher protein content and higher water absorption.

References

Faridi, A. (1988). Flat Breads. In.Wheat Chemistry and Technology, edited by Y. Pomeranz. AACC. St Paul Minnesota, USA.

http/www. Bakingandbakingscience.com

Perry K.W. Ng and Colin W. Wrigley (2002). Wheat Quality Elucidation: The Bushuk Legacy. American Association of Cereal Chemists, 3340 Pilot Knob Road, St. Paul, Minnesota, USA.

Qarooni, J. (1996). Flat Bread Technology, Chapman and Hall.

S.P. Cauvain and L.S. Young (1998). T echnology in bread making. Blackie Academic & Professional, an Imprint of Chapman & Hall, London, SE1 8HN, UK.

Srivastava, A.K., Meyer, D., Haridas Rao, P. and Seibel, W. (2002). Scanning electron microscopic study of dough and chapati from gluten-reconstituted good and poor quality flour. *J. Cereal Sci.* 35: 119-128.

W. Bushuk and V.F. Rasper. (1994). Wheat Production, Properties and Quality. Blackie Academic & Professional, an imprint of Chapman & Hall, Wester Cleddens Road, Bishopbriggs, Glasgow, UK.

Y. Pomeranz. (1988). Wheat Chemistry and Technology. AACC 3340 Pilot Knob Road, St. Paul, Minnesota, USA.

❑❑❑

Chapter 3

Soft Wheat Product Quality

Introduction

For many years it has been an elusive task to provide a precise definition as to what constitutes quality in wheat. Each country has its own traditions in bread and other bakery products. Quality will mean different things to a miller, a baker, a grain handler or a plant breeder. As the baking industry becomes increasingly specialized, specific qualities are sought. Even among those wheat varieties generally recognized as having desirable quality traits, the influence of environment can be substantial, and this necessitates a high level of skill on the part of grain buyers to survey the current year's crop and blend the wheat for a specific end-use. This genotype X environment effect on quality also limits the capacity of end-users to source identity-preserved wheat from a narrow geographic area. Those of us engaged in the field of plant breeding are concerned primarily with those traits that can be manipulated genetically, and we study their stability under a range of climatic conditions. Constraints in time, cost, and sample availability usually necessitate the use of predictive tests during early generations, followed by milling and baking tests on advanced lines that have met a number of agronomic, disease and predictive quality criteria.

There are quite large differences in grain composition and processing quality among wheat cultivars within a species. Hence, one cultivar may be suitable to prepare one food type but unsuitable to prepare a different one. Quality differences among wheat cultivars have gained even more importance in grain trading due to important global economic and social trends. Recently, many countries have adopted, or are in the process of adopting, free-market economies, impacting positively on the income of the population, particularly of that concentrated in urban areas. Concomitantly, several of these countries are experiencing a trend towards increased urbanization and increased demand for traditional and new convenient, processed wheat-based foods. In India for example, the demand for biscuit, cakes, pretzels, noodles, bread and pasta products is increasing @5% per annum. Millions of people are involved in milling, baking and pasta products industry with huge turnover per annum.

As a consequence, wheat processing industries now require various distinct wheat supplies possessing specific grain quality attributes. For example hard wheats with strong gluten and high protein content are used for bread making and soft wheat with low protein content and weak gluten for cookie making. Often chemical improvers are used to improve the dough properties for these products. Most of these chemicals have been banned because of their toxic effects. Therefore, only alternative is to develop varieties suited to meet the requirements of these end-use products. This needs better understanding of genetics of quality traits and their molecular basis to accelerate breeding for producing wheat varieties for each end-use product separately. Micro level tests requiring lesser amount of materials (< 1 g), lower technical expertise, high throughput, low cost and explaining larger variability in the quality of end-products are needed to expedite the breeding programme for developing varieties. In this chapter, details are provided on quality requirements of soft wheat products and relationship with rheological, chemical and genetic traits.

Soft Wheat Products

Grain texture is the major determinants of end product quality in wheat. Based on grain texture wheat is classified into two distinct classes as hard and soft. Hard wheats are used for bread and chapati making and soft wheat for biscuit, cakes, pretzels and noodles. Hard wheats need more force to crack and have higher starch damage during milling which affects the flour quality MacRitchie, (1980). Higher starch damage is detrimental to the biscuit making quality and some starch damage is required for bread making quality. Soft wheats produce flour of low water absorption, fine particle size and low protein content suitable for biscuit making as compared to bread flour from hard wheat Hoseney *et al.* (1988). Hardness of a given cultivar is genetically controlled and is not directly correlated with grain protein content. Therefore grain hardness is amenable to genetic modification. The details of genetic and molecular basis of grain hardness are given in the chapter on grain hardness.

There are many different kinds of products made from soft wheats. The most important among them are biscuits, cookies, crackers, wafers, pretzels, pastries, waffles, oriental noodles and cakes. Soft wheat flour is also used as a thickener for soups and soup mixes, in the manufacture of crumb for coating fish and meat products. All of these products have better appearance and better eating quality when made from soft rather than hard wheat flour. Though, each product may have some different requirements than other products made from soft wheats. For example pastry and crackers flours have distinct quality requirements. Quality ranges are presented for flours that may be intended for a contemporary pastry flour and for a flour with somewhat stronger gluten protein characteristics for the production of crackers or for export where utilization will be in various international breads. The stronger protein cracker wheats are not semi-hard wheats. They only have more protein content and should have stronger gluten. Some categories are more important to individual millers and bakers, depending on their objectives. However, a high quality soft wheat cultivar will satisfy the ranges of all categories. The highest quality cultivars will be within range of as many tests within each category as possible.

Products made using pastry flour include, cookies, cakes, pie crust, donuts, layered pastries, wafers, cones, batters, pancakes, waffles, soup thickeners, muffins, and biscuits. In general, the desirable data ranges for pastry flour represent soft kernel texture, no field sprouting, high milling qualities, medium protein content, low alkaline water retention, high test weight and grain condition, medium protein strength, and baking qualities that are typical of soft wheats having those characteristics. Products made using cracker/export flour include chemically leavened crackers, fermented dough crackers, pretzels, and various international pan and flat breads. In general, the desirable data ranges for cracker/export flour represent medium kernel texture, no field sprouting, high milling qualities, high protein content, medium alkaline water retention, high test weight and grain condition, strong protein strength, and baking qualities that are typical of soft wheats having those characteristics. These characteristics should be target goals for soft wheat breeders developing cracker/export quality soft wheats.

Biscuit/Cookie Quality

Biscuit quality is evaluated by measuring the softness, diameter and spread factor of cookies baked as per the AACC sugar snap cookie test. Larger cookie diameter and higher spread factor along with soft texture are used as the main quality attributes Nemeth *et al.* 1994; Yamamoto *et al.* (1996) for evaluating biscuit making quality. During baking process the dough expands and flows because of leavening gases and gravitational force. The final size of the cookie depends upon the rate of the flow and the set time when the expansion stops. Doughs of good cookie-making quality flour flow much faster than those made from poor quality flour Hoseney *et al.* (1988). The set time is determined by the amount of the water in the dough that is free to act as solvent and the strength of the dough. Therefore, ultimate quality of cookie depends upon the chemical constituents in the flour responsible for water holding and the quality and quantity of the gluten proteins determining dough strength. Generally soft wheats with weak gluten and low protein content are preferred for biscuit making (Gaines 1991, Souza *et al.* 1994). Protein content although reported to be influenced by few genes, is much more influenced by the environment. Protein quality in turn is determined mainly by the genetic components as glutenins and gliadins. Details are provided on other chapter on gluten proteins.

Most of the study of what happens during cookie baking and of the functions of the ingredients has been conducted with a sugar-snap cookie formula. The formula for sugar-snap cookie is high in sugar and shortening and low in water. As a result of those ratios and the creaming step, only part of the sugar is dissolved during mixing. Thus, only about half of the sucrose is dissolved during mixing, and the remainder is dissolved during baking. Ammonium and sodium bicarbonates are use as leavening agents. Carbon dioxide produced from bicarbonate is used as leavening gas. The advantages of bicarbonate are: it is tasteless, nontoxic, and readily available with low cost. It is relatively stable and stores well. Because the salt is alkaline, it produces no gas at room temperature unless the pH is lowered by an acid. At higher temperature, the bicarbonate disproportionates to form Na_2CO_3, H_2O and CO_2. When the dough is baked, it flows and expands as a result of leavening. The final size depends upon the duration and rate of flow of the dough. The higher spread is considered better for biscuit making and the lower spread for poor quality. Because all cookies are cut to the same size, the difference between cookies produced from good quality flours and those from poor quality flours does not express until the dough is heated.

Soft wheat products are often less dense than typical products containing hard wheat; however, soft wheat have a more uniform internal structure, softer bite, greater tenderness and more desirable height or spread characteristics. Some characteristics, especially protein content, influence other quality characteristics. Protein content is highly influenced by the environment and thus standard samples are used for comparison. Test samples that do not meet that standard will usually appear atypical and inferior in many test results. They also are very difficult to evaluate for quality without comparison with a known standard wheat grown at the same location under optimum conditions. Cultivars should be subjected to similar and typical agronomic production practices. They also should be planted and harvested at optimum dates for their location.

Flour Constituents

The principal components of flour are starch, proteins, pentosans and lipids. Though much information is not available, there are reports concerning the role of these components in cookies or sweet biscuits. Other factors such as 1B/1R translocation lines have also been reported to have higher water absorption and lower cookie diameter. The role of major flour constituents is discussed below.

Starch

Starch is present as intact or damaged granules in the flour. Higher starch damage is detrimental to biscuit making quality as damaged starch absorbs about three times more water than undamaged granules. Damaged starch in soft wheats usually forms a small proportion of the total starch as measured by AACC procedure. Starch does not gelatinize during the cookie baking process. Gelatinization of starch granules occurs when starch is heated in the presence of water; high sugar and low water (20% f.b.) content in cookie dough does not allow this to occur, consequently the starch present in the cookie as such flour has little effect on cookie spread. Some cookie manufacturers use chlorinated flour and/or modified starch in their formulations. In this case, starch granules are damaged and will gelatinize even in the dough stage. In addition to starch damage, starch composition is also important in determining water absorption. Starch is composed of two components as amylose and amylopectin. Waxy wheats with reduced amylose content have higher water absorption. Waxy wheats have been reported suitable for improving noodle (Japanese) quality as well as keeping quality of pasta products and bread. However, waxy wheats may not be suitable for cookies/biscuits because of higher water absorption. Waxy wheats have been characterized in greater detail and used in breeding for the improvement of noodle and bread quality. The information will be provided in other related chapters.

Protein Content and Quality

Gluten development is little in the process of manufacturing biscuits. Higher protein content makes the product hard in bite and also creates problem in shrinkage during cookie setting. Therefore, low and week gluten is preferred for biscuit making. Observations also suggest that protein has a role to control the breakdown of wheat endosperm during milling. Protein content of 8-10% is often specified for cookie flours. Hardness of the biscuits is increased with the increase in protein content. However, other observations showed that protein content did not have major effect on cookie baking quality of soft wheats. Protein quality influences cookie quality and therefore, by manipulating gluten strength cookie quality can be improved. Generally chemical improvers (such as sodium meta-bisulphite) are used to reduce strength of the dough for better biscuit making quality. Thus by reducing gluten strength in wheat the biscuit making can be improved. There is large variation in gluten strength in wheat and exceptionally low gluten strength has been reported in one of the Indian land race of wheat, Nap Hal. Nap Hal contains Glu-D1 double null trait where 5+10 subunits coded by the Glu-D1 locus are absent. Further studies reported the presence of soft grain characteristics in Nap Hal controlled by puroindolines coded at ha locus 5D short arm. Therefore, Nap Hal can be used as a source of genes for both low gluten strength and soft grain characteristics.

Pentosans

Pentosans, although present as only 2-3% of the flour, are important for their ability to absorb upto 10 times of their weight of water. In association with other hemicellulosic substances, they form the basic structure of the endosperm cell walls in wheat grain and are found in the starch tailings fraction of the flour. Pentosans have the potential to interact with amyloplast membranes and have been studied over the years as contributors to both end-use functionality in the form of water binding constituents of flour and in relation to endosperm texture. Total pentosans are usually divided into water soluble or water insoluble fractions. Water insoluble pentosans are typically associated with cell walls whereas water-soluble pentosans are found interior to endosperm cell walls probably in association with membranes. This association of pentosans with amyloplast membranes may serve to modify starch protein matrix adhesion and therefore have an effect on texture. Therefore pentosans may have role in water absorption of flours indirectly by affecting grain hardness and directly by its water binding capacity.

Bettge and Morris 2000 reported that among soft wheats samples, pentosans showed significant hardness modifying effect that carried over into end-use quality. Among hard wheat samples, pentosans had a minimal role in modifying grain hardness. They reported that 53-76 % of the variation in grain texture of soft wheats was due to pantosans along with protein content. Total pentosans together with flour protein accounted for 87% of the variation in cookie diameter for soft wheat samples. Membrane associated pentosans were found more accountable for variation in alkaline water retention capacity. Ram *et al.* 2004 reported strong negative correlation between sucrose retention capacity (indirectly related to pentosans content) and cookie diameter in soft wheat samples. The information can be used in breeding for improving biscuit making quality by utilizing interclass hybridizations of both hard and soft genotypes.

Solvent Retention Capacity Tests

Soft wheat products characteristics are related to the relatively low water absorption, fine granulation and low protein content. For example Yamazaki 1954 observed a strong correlation between the hydration characteristics of flour and cookie baking quality and developed the AWRC test. Alkaline Water Retention capacity is the amount of water held by the flour under alkaline conditions. Biscuit quality of wheat flour depends primarily upon the chemical constituents in the flour responsible for water holding and the quality and quantity of the gluten proteins determining dough strength. Generally soft wheats with weak gluten and low protein content are preferred for biscuit making. Based on the above quality characteristics required for cookie-making solvent retention capacity tests were developed and adopted by AACC as method 56-11 (2000).

Solvent retention capacity is the weight of solvent held by flour after centrifugation and expressed as the percentage of flour weight (14% mb). In this test four different solutions as lactic acid, sucrose, water and sodium carbonate are utilized for providing information on chemical and physical aspects of wheat samples. A combined pattern is produced useful for predicting commercial baking performance. Generally lactic acid SRC is associated with glutenin characteristics, sodium carbonate SRC with starch damage, sucrose SRC with pentosan content and gliadin characteristics and water SRC with all of those four constituents. In one of the studies Rams *et al.* (2004) SRC profile could explain 70 % variability in Biscuit spread factor.

Thus SRC profiling provides additional, complementary information to the AWRC and assist in identifying the proper baking and processing characteristics of flours. Earlier the information on AWRC along with SDS-sedimentation volume, 2 gram mixogram tests, flour swelling volume for starch characteristics and single kernel characterization system for testing grain texture was used for improving cookie baking quality. Since baking tests are time consuming, labour intensive, expensive and require large quantity of sample, small-scale tests developed (using 1 g flour) can be used to predict the biscuit making quality of early generation breeding material.

Farinograph Water Absorption

Functional properties of flour including water absorption are measured using Mixograph and Farinograph and are used to differentiate wheat flours of good and poor baking quality. In both the instruments during mixing first water is brought into contact with flour particles that facilitate water absorption to develop dough and a mixing curve formed. The determination of optimum water needed to obtain a certain dough consistency is essential to give dough with optimal handling characteristics suitable for rheological testing as well as product quality. However, Farinograph and mixograph tests are time consuming, labour intensive and expensive; small-scale tests are required to predict the end-use quality of grain from early generation breeding material. Further, there is a major problem in finding optimum water absorption of flour for use in mixograph and Farinograph.

Recently prediction equation for water absorption was developed using SRC profiles of flours from 192 wheat genotypes Rams *et al.* (2005). The data exhibited that 83 % variability in Farinograph water absorption can be explained by SRC profile using 1 gram flour. Since large numbers of diverse genotypes were used in the estimation of various parameters in this investigation, high correlations observed between SRCs and functional properties including water absorption have obvious implication in breeding programme for the improvement of wheat cultivars.

Cake

Cake is a baked batter made from sugar, eggs, shortening, milk and leavening mixed together in such a way to produce a fluffy, fine grained baked product. There are two basic types: Batter type which contains high percentage of shortening and the foam type contains little, if any shortening at al. From these two basic mixes, many types of cakes can be made. The quality of the cake depends on the type of ingredients used, mixing method, batter temperature, baking etc. Each ingredient used in cake baking is employed for the specific characteristics it has and/or the result it has on the finished product. If these effects are understood, the ingredients may be selected with the assurance that the products produced will be good. Cakes like cookies are quite high in sugar and shortening. The major difference between the two is that cake formula tends to have much more water. When baked, cakes set to give a light, tender product while cookies collapse during baking. The difference appears to be controlled by the concentration of sugar in available water. If the concentration of sugar is too high, the starch does not gelatinize during baking. If the starch is not gelatinized, the cookie or cake collapses. This happens in sugar cookies to give a relatively flat product. The effect of flour and other constituents on cake quality is discussed below.

Flour

Flour is the primary structure builder in most cakes. The gluten formed during mixing coagulates and assists in supporting the heavy weight of sugar and shortening while baking. Cake flour used in cake baking is made from soft wheats. Cake flour should have low protein content (7 to 9%). Cake flour should be properly bleached, because bleaching helps to carry more sugar and shortening as well as water during mixing. Cake flour should also have pH around 5.2 which is slightly acid. Acidity helps to mellow or soften the gluten.

Sugar

Sugar is used in cake as a sweetener. Sugar is considered a tenderizer because of its tenderizing effect resulting from the softening action on flour proteins. Sugar lowers the carmelization point of the batter, allowing the cake crust to color at a lower temperature. Sugar helps to retain moisture left in the baked cake after baking thereby keeping the cake moist and edible for several days. A portion (about 50 percent) but not all of the sugar may be replaced with syrup. When this is done, the liquid content of the syrup must be deducted from the liquid going into the mix. The greater the percentage of sugar contained in the formula, the longer the cake will stay fresh. The amount to use will be covered later under formula balance. For optimum results, sugar must be completely dissolved in the batter. This is the reason why in balancing a cake formula, especially in high ratio cakes (cakes containing greater quantities of sugar and water than flour in the formula) water must exceed the sugar in the formula.

Shortening

Shortening incorporates air in the cake batter during mixing. This air helps to obtain volume in the baked cake. It also tenderizes the cake. In making high ratio cakes, an emulsified shortening must be used to form an emulsion. Regular shortening is not capable of forming an emulsion when large amounts of sugar and liquid are used. If not enough liquid is used to dissolve the sugar, the cake will collapse in the center.

Eggs

Eggs furnish structure, moisture, flavor, color, and food value to the cake. Their most valuable contribution is structure. The proteins in the egg coagulate during baking and assist the flour as a structure builder. Egg is the only ingredient that can be used to regulate the toughening action in a cake. If weak flour is used, the eggs can be increased. If the percentage of shortening (a tenderizer) called for in the formula is increased, the eggs must be increased also. It is important to know the percentages of fat, moisture, and protein content of eggs when balancing cake formulas.

Milk Solids

Milk solids have a binding effect on the protein of the flour, thereby increasing the toughness in a cake. A portion of the total solids in milk contain lactose sugar, which carmelizes at a low temperature (270 to 275° F). It is used to control crust color. It, along with the proteins in milk, adds food value and flavor to the cake, and helps to retain moisture in the cake. If liquid milk is used in place of dry milk powder, it will be necessary to know the liquid content of the milk in order to make adjustments to the formula so the formula can be properly balanced.

Leavening

Cakes are leavened mainly in three ways. Incorporation of air during mixing, chemically leavened and vapor pressure created in the oven. The manner of leavening depends upon the type of cake being made in regard to the richness of formula, consistency of batter and baking temperature. Cakes low in water and high in enriching ingredients get a larger amount of leavening during mixing and require less chemical leaveners than cakes made from lean formulas high in liquids. In addition to leavening the cake, chemical leaveners control the eating qualities of the cake. Excess soda, for example results in an undesirable "soapy taste". Baking Soda is necessary to produce the rich red color in Devils Food Cake. However, one must be careful not to use too much soda in the formula. To produce a chocolate cake having a brown crumb color rather than a red crumb color, baking soda must be left out. Some types of cakes require no chemical leavening. Examples are the true Pound Cake and the Basic Sponge Cake. These two cakes are leavened entirely by physical means such as air incorporated in the batter during mixing and vapor pressure created during baking. This is the reason why these types of cakes require a longer mixing period and precise control of temperature of the batter during mixing. Salt is used primarily for taste and to tone down the sweetness.

Liquids

Liquids in some form are required in every cake formula. The liquid may be in the form of water, liquid milk, eggs or any other ingredient which contain water. Water has several functions in cake production. It develops the gluten, dissolves the sugar, makes the function of baking powder possible, regulates the batter consistency, and controls the temperature of the batter. It is possible to carefully regulate the water portion of the formula by figuring the liquid content of any liquid ingredient used in the batter. The amount of water going into the cake formula is partially controlled by the type of shortening used. An emulsified type shortening will carry considerably more water in the mix, thus allowing the use of more sugar such as in High Ratio Cakes. The total liquids (Liquid in the form of water and the liquid contained in the eggs) should always equal or exceed the weight of the sugar in the formula, because all of the sugar in the formula must be dissolved to produce a quality cake.

Formula Balance

In order to create a cake batter that will produce high quality cakes, certain amounts of the different ingredients have to be put together in a definite sequence at controlled mixing speed, time and temperature. The general relationship of ingredients that have to be brought into balance, differ according to the type of cake to be produced. In other words, the formula balance for batter cakes differs considerably from that of the foam type cake (Sponge cake and Angel food cake).

Batters

Principal products made from simple batters containing soft wheat flours include wafers, waffles and pancakes. Wafers are baked from a fluid batter containing 35 to 40% dry solids. The bulk of these solids is supplied by soft wheat flour. Other ingredients used for making wafers are oil, salt and sodium bicarbonate. The batter is baked between pairs of heated metal plates. Each plate is engraved with crisscross lines or some other suitable pattern, which is transferred to the wafer during baking and facilitates the adhesion of fillings. The baked wafer consists of porous starch gel.

Batters used in the manufacture of waffles and pancakes are much richer than that used for wafers. Appreciable amount of fat (25-35% of flour weight) and sugar (10-25%) are added, together with eggs, milk solids and baking powder. Waffle batter is baked between deeply indented plates, to produce a thick product with a crisp outer surface and a soft chewy interior. It is normally eaten immediately while hot or deep frozen for subsequent reheating. Pancakes are baked on flat surface and are turned over partway through the baking process.

Crackers

Crackers are divided into three main classes as soda crackers, sprayed and savory crackers. Leavening can be accompanied either by yeast fermentation or chemical leavening. A forth group, with higher levels of added sugar, e.g., graham crackers, is invariably chemically leavened. Most yeast leavened crackers are based on sponge-and-dough fermentation process. A wide variety of crackers can be produced with minor variations in ingredients or processes. The fermentation process changes the rheology of the cracker dough as well as influencing the flavour, colour and texture of the final product. To produce a light cracker with a flaky but not brittle texture, moderately strong wheat is required. The flour characteristics required for cookies and cracker production are provided in the (Table 1). This indicates that flour with bright and creamy colour having low ash content, low protein content, low water absorption, low damaged starch, fine granulation, medium mixing requirements and satisfactory dough handling properties.

Table 1: Characteristics of flour used for cookies/crackers production (Faridi *et al.* 1994)

Product type	Typical ranges			
	Protein (%)	Ash (%)	Viscosity (M)	Spread factor (w/t)
Cookies				
Wire cut	7.0-8.0	0.38-0.42	20-40	8.5-10.0
Rotary	8.0-9.0	0.38-0.42	30-50	7.5-9.0
Deposit	7.5-8.5	0.38-0.42	25-45	8.0-9.5
Crackers				
Soda-sponge	9.0-10.0	0.41-0.45	55-70	6.5-8.0
Soda-dough	8.5-9.5	0.39-0.43	40-55	7.5-9.0
Snack	8.5-9.5	0.40-0.44	40-55	7.5-9.0

Noodles

Wheat flour noodles are an important part in the diet of many Asians. It is believed that noodles originated in China as early as 5000 BC, then spread to other Asian countries. Today, the amount of flour used for noodle making in Asia accounts for about 40% of the total flour consumed. In recent years, Asian noodles have also become popular in many countries outside of Asia because of globalization of economy and changing food habits. Additionally noodles are easy to prepare and processing operations include mixing raw materials, dough sheeting, compounding, sheeting / rolling and slitting. This series of processes remains constant among countries for all noodle types. Noodle strands are further processed to produce different kinds of noodles. About three parts of flour are usually mixed with one part of salt or alkaline salt solution to form crumbly dough. The dough is compressed between a series of rolls to form a dough sheet. The sheeted dough is then slit to produce noodles. The noodles are now ready for sale, or are further processed to prolong shelf life and to modify eating characteristics.

In the preparation of instant fried noodles, the steaming process causes the starch to swell and gelatinize. The addition of alkaline salts (kan sui, a mixture of sodium and potassium carbonates) in some Chinese type noodles gives them a yellow color and a firmer, more elastic texture. Asian noodles are different from pasta products in ingredients used, the processes involved and their consumption patterns. Pasta is made from semolina (coarse flour usually milled from durum wheat) and water, and extruded through a metal die under pressure. It is a dried product. After cooking, pasta is often eaten with sauces. Asian noodles are characterized by thin strips slit from a sheeted dough that has been made from flour, water and salt (common salt or alkaline salt). Noodles are often consumed in soup. Eggs can be added to each product to give a firmer texture.

Classification of Asian Noodles

Though, wide differences exist between countries, noodles are classified into two main types as Chinese and Japanese type. Chinese type noodles are generally made from hard wheat flours, characterized by bright creamy white or bright yellow color and firm texture. Japanese noodles are typically made from soft wheat flour of medium protein. It is desirable to have a creamy white color and a soft and elastic texture in Japanese noodles. The most apparent difference between the two types is the addition of alkali which gives alkaline noodles their characteristics yellowness. Inclusion of alkali also gives a characteristic aroma and flavour and firm and elastic texture to the noodles. It also modifies the dough, starch pasting and cooking characteristics. The other differences between white salted and yellow alkaline noodles are given in the table.

According to the width of the noodle strands, Japanese noodles are further classified into four types as So-men (very thin, 0.7-1.2 mm wide), Hiya-mughi (thin, 1.3-1.7 mm wide), Udon-standard (1.9-3.8 mm wide) and Hira-men (flat, 5.0-6.0 mm wide). Since the smaller size noodles usually soften faster in hot water than the larger size, So-men and Hiya-mughi noodles are usually served cool in the summer, and Udon and Hira-men are often eaten hot in the cool seasons. Other noodle types also have their own typical size. Based on processing, noodles are classified into fresh, dried, boiled and steamed. None of the approaches discussed above are sufficient to define each noodle type. For instance, boiled noodles contain fully cooked and parboiled types. Parboiled types include both hokkien and Chinese wet noodles. In addition, wet noodles are parboiled in most of Asia, but are fresh, uncooked noodles in Japan. Therefore, a possible nomenclature should incorporate key aspects such as formulation and basic processing to fully describe the nature of each noodle type (Table 2).

Table 2: Major classes of Asian noodles[a]

	General class name	
	Salted	Alkaline
Colour	White or creamy white	Yellow
Major raw material	Flour, water sodium chloride	Flour, water alkaline salts, e.g., sodium carbonate
Countries consumed	Japan, Korea, Northern China	Malaysia, Singapore, Indonesia, Thailand, Taiwan, Hong Kong, Southern China, Japan
pH of noodles	6.5-7	9-11
Organoleptic qualities	Soft, elastic texture	Firm, elastic texture

[a]Kruger *et al.* 1996.

Noodle Processing Technology

The basic processing steps for machine-made noodles are outlined in Fig. These steps involve mixing raw materials, resting the crumbly dough, sheeting the dough into two dough sheets, compounding the two sheets into one, gradually sheeting the dough sheet into a specified thickness and slitting into noodle strands. Noodle strands are further processed according to noodle types.

Mixing Ingredients

Mixing is often carried out in a horizontal or vertical mixer for 10-15 min. Since the horizontal mixer seems to have better mixing results, it is more commonly used than the vertical one in commercial noodle production. Mixing results in the formation of a crumbly dough with small and uniform particle sizes. Since water addition level is relatively low (vs. bread doughs), gluten development in noodle dough during mixing is minimized. This improves the dough sheetability, sheeted dough smoothness and uniformity. Limited water absorption also slows down noodle discoloration and reduces the amount of water to be taken out during the final drying or frying processes. Flour proteins, pentosans and starch (especially damaged starch) determine the flour water absorption level. Even so, the water absorption level in noodle dough is not so sensitive to processing as is that in bread dough. Variations in noodle dough water absorption among different flours is generally within 2-3%, and this is usually determined by dough handling properties. Flour particle sizes and their distribution affect the time water penetrates into the flour. Large particle flours require a longer time for water to incorporate and tend to form larger dough lumps. It is desirable to have relatively fine and evenly distributed particle size flours to achieve optimum dough mixing.

Dough Resting

After mixing, the dough pieces are rested for 15-30 min. before compounding. This step is to relax the dough for easy reduction in the subsequent sheeting operation. Dough resting helps water to penetrate into dough particles evenly; protein mellows and becomes more extensible resulting in smoother and less streaky dough after sheeting. Resting produces cooked noodles with a firmer eating quality. In commercial production, the dough is rested in a receiving container while being stirred slowly. The resting time takes about 30-40 min.

Sheeting and Compounding

The rested, crumbly dough pieces are divided into two portions, each passing through a pair of sheeting rolls to form a noodle dough sheet. The two sheets are then combined (compounded) and passed through a second set of sheeting rolls to form a single sheet. The roll gap is adjusted so that the dough thickness reduction is between 20-40%. The combined dough sheet is often carried on a multi-layer conveyor belt located in a temperature and relative humidity controlled cabinet.

Sheeting, Slitting and Waving

Further dough sheeting is done on a series of 4-6 pairs of rolls with decreasing roll gaps. At this stage, roll diameter, sheeting speed and reduction ratio should be considered to obtain an optimum dough reduction. It is during the sheeting stage that gluten is fully developed and a uniform protein matrix must be formed for better quality. Noodle slitting is done by a cutting machine, which is equipped with a pair of calibration rolls, a slitter, and a cutter or a waver. The final dough sheet thickness is set on the calibration rolls according to noodle type and measured using a thickness dial gauge. The sheet is cut into noodle strands of desired width with a slitter. Noodles can be either square or round in shape by using various slitters. Noodle strands are cut into a desirable length by a cutter. At this stage, Chinese raw noodle, Japanese udon noodle, Chuka-men and Thailand bamee noodle making is complete (Fig. 1). For making instant noodles, noodle strands are waved before steaming and cutting.

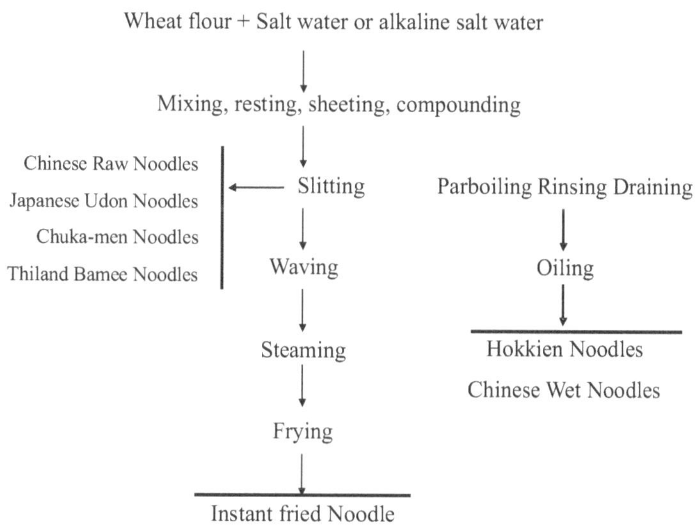

Fig. 1. Basic processing steps involved in noodle making.

Cooking Noodles

Cooking processes include parboiling, boiling, and steaming. Hokkien noodles and Chinese wet noodles are usually parboiled for 45-90 seconds to achieve 80-90% gelatinization in starch. The noodles are then coated with 1-2% edible vegetable oil to prevent the strands from sticking together. Parboiled noodles have an extended shelf-life (2-3 days) and high weight gain (60-70%). They are quickly re-cooked by boiling or stir-frying prior to consumption. Japanese udon noodles are boiled for 10-15 min., rinsed and cooled in running water, steeped in dilute acidic water before packing, and further steamed for more than 30 seconds in a pressurized steamer. This type of noodle usually has a shelf-life of 6 months to one year. It is also called longevity noodle. Several steps can be taken to assure optimal cooking: (a) the weight of cooking water is at least 10 times that of the uncooked noodles, (b) the size of the boiling pot is properly chosen, (c) the pH of the boiling water is 5.5-6.0, (d) the cooking time is precisely controlled to give optimal results to the product, and (e) the cooking water temperature is carefully maintained at 98-100° C throughout the boiling process.

In making instant noodles, the wavy noodle-strands are conveyed to a steamer to cook the noodles. As mentioned earlier, the purpose of steaming is to gelatinize the starch and fix the noodle waves. The steaming time varies according to noodle size, but can be determined by squeezing a noodle strand between two clear glass plates. If the white noodle core disappears, the noodles are well cooked. Steam temperature, steam pressure, and steaming time are key process factors affecting the product quality.

Drying Noodles

Noodle drying can be achieved by air drying, deep-frying or vacuum drying. The air drying process has been applied to many noodle types, such as Chinese raw noodles, Japanese udon noodles, steamed and air-dried instant noodles, and others. Air-drying usually takes 5-8 hours to dry regular noodles (long and straight) and 30-40 min. to dry steamed and air-dried instant noodles. Drying by frying takes only a few minutes. Vacuum drying of frozen noodles is a newer technology making it possible to produce premium quality products. For the manufacture of regular dry noodles, raw noodle strands of a certain length are hung on rods in a drying chamber with controlled temperature and relative humidity. Air drying usually involves multistage processes since too rapid drying causes noodle checking. In the first stage, low temperature (15-20° C) and dry air are applied to reduce the noodle moisture content from 40-45% to 25- 27%. In the second stage, air of 40° C and 70-75% relative humidity is used to ensure moisture migration from the interior of the noodle strands to outside surfaces. In the final stage, the product is further dried using cool air.

For the manufacture of air-dried instant noodles, wavy noodle-strands are first steamed for 18-20 min. at 100 °C, then dried for 30-40 min. using hot blast air at 80° C. The dried noodles are cooled prior to packaging. Air-dried instant noodles have a low fat content so some people prefer them. They also have a longer shelf-life because little fat rancidity is involved. Steaming appears to be very critical to this type of noodle since it affects the water rehydration rate of the product. However, slow output of the process and lack of pleasant shortening taste and mouthfeel make the product less popular in Asia compared with instant fried noodles. Drying by frying is a very fast process. Water vaporizes quickly from the surface of the noodles upon dipping into the hot oil. Dehydration of the exterior surface drives water to migrate from the interior to the exterior of the noodle strands. Eventually, some of the water in the noodles is replaced by oil. Many tiny holes are created during the frying process due to the mass transfer, and they serve as channels for water to get in upon rehydration in hot water. It usually takes 3-4 min. to cook or soak instant fried noodles in hot water before consumption.

Noodles Evaluation

The evaluation (scoring) of noodles focuses mainly on three characteristics—process performance (machining), noodle color and noodle texture. The process effect is generally weighted higher for instant noodles due to more steps involved and high speed production. Noodle color is particularly important for Chinese raw, Japanese udon, Chuka-men and Thailand bamee because of the lack of heat treatment, which allows more rapid darkening. As for Malaysian hokkien noodles, color is also very important because it is evaluated on both parboiled and uncooked noodles. Hokkien noodles have a typical shelf-life of 2-3 days. Although good noodle color is required, desirable texture is essential in all the markets. Other quality characteristics are weighted lower, but they can be very critical to overall noodle performance. For example, both Chinese wet noodles and Malaysian hokkien noodles are sold in a parboiled form hence the cooking weight gain (%) is very important quality attribute to noodle makers. If a noodle can take up more water within a fixed cooking time and maintain its texture characteristics, it will be a more desirable and profitable product. Each noodle type has its own evaluation sheet due to a different focus on the noodle quality preferences. Within the categories of processing, noodle color or noodle texture, there are a number of evaluation items.

Noodle Color

All noodle types require good brightness. Color can be either white or yellow depending on the absence or presence of alkali salts. The natural yellow colour is attributed to the presence of naturally occurring flavones in flour. Flavones are colourless at acidic pH and are present in large amount in germ and bran. The degree of yellowness in noodles is dependent on the alkali used in the formulation and the wheat variety. Flours of low extraction rate contain less flavone resulting into less yellow noodle. Yellowness and brightness of noodles is affected not only by alkali, but also by bran specks, specific flour proteins, proteases and polyphenol oxidases, level of damaged starch, flour particle size and added flour bleaching agents. Raw noodles become darker and more yellow with time. Darkening is because of the presence of bran specks and enzymatic browning by polyphenol oxidases. Polyphenol oxidases exist in multiple forms and concentrated in bran. These enzymes oxidize phenols to quinines which are subsequently converted to dark coloured pigments by self polymerization or by interaction with proteins. Noodle darkening is arrested by boiling or by steaming due to enzyme inactivation. Minimal noodle darkening within 48 hours is desirable. This may not be a problem for the instant noodles because they are dried and the color is very stable.

Noodle Texture

Contrary to color, noodle texture characteristics are more complicated and less understood. There is a distinction in noodle bite between the Japanese type and other noodle types in that the Japanese type is softer, while others are harder or firmer. Chinese raw, wet and instant fried, chukamen, Malaysian hokkien, Philippine instant fried and Thailand bamee noodles are hard in bite, while Korean instant fried noodles are firm in bite. The hard bite noodles require high protein flour, while the firm bite noodles require medium protein flour with strong starch. Korean instant fried noodles are somewhat similar to Japanese udon in that both require flours of high peak viscosity and large breakdown measured by an amylograph. However, the flour protein content of Korean instant noodles (bag type) is (9.0-10.5%), higher than that of Japanese udon noodle flour (8.0-9.5%). Thus, the Korean instant fried noodle is also harder.

Assessment of noodle eating quality is usually based on sensory evaluations. Taste panels are made for evaluating fresh alkaline noodles. The members of the taste panel are chosen based on their ability to discriminate textural characteristics. Alkaline noodles should be firm and elastic and with smooth and non sticky surface. Samples are scored using a numeric scale (usually 1-10) with reference to the standard sample which is located at a predefined point on the scale. Firmness is defined as the force required to bite a noodle strand with teeth and chewiness or elasticity as the time to chew 10g noodle to a consistency small enough to swallow. Surface smoothness is evaluated during mastication or by the appearance of the noodle surface. However subjective assessment of noodle quality has some disadvantages. Taste panels need extensive training and can not test large numbers of samples at a time. This has led to the development of objective assessment methods for measuring noodle firmness such as texture analyzer, TA-TX2 (Stable Microsystems, UK). Noodle strands are compressed using flat blade and the compression force and recovery are utilized in assessing the noodle quality. The method has shown strong correlation with sensory evaluation of firmness.

Noodle Quality Requirements

In many cases, physical quality measurements of wheat and wheat test methods are similar and independent of end products made. For example, wheat should be clean and sound, high in test weight, and uniform in kernel size and hardness. These characteristics result in efficient milling and high flour extraction. The Falling Number test is done to determine wheat sprout damage level. Wheat kernel hardness, diameter, weight and their distribution can be measured using a Single Kernel Characterization System. Wheat kernel hardness deserves particular attention since it affects the tempering conditions, flour starch damage level, flour particle distribution and milling yield. Damaged starch not only absorbs more water but may also reduce noodle cooking and eating quality. Accordingly, noodle wheat should not be too hard, and milling processes should be controlled to avoid excess starch damage. The uniformity of wheat kernel hardness appeared to improve milling performance.

Low ash content in flour is always an advantage for noodles since flour ash is traditionally viewed as causing noodle discoloration. One of the important noodle flour specifications is ash content, although there is no guarantee that low ash flour can always make desirable color noodles. The presence of the enzyme polyphenol oxidase (PPO) in the flour is believed to be partially responsible for noodle darkening. Thus, it may be useful to measure the activity of this enzyme in the wheat. Different noodle types require different protein contents and dough strength (discussed later). Generally speaking, Chinese type noodles need hard wheat of high protein content and strong gluten, and Japanese noodles require soft wheat of medium protein content and higher protein quality. Wheat protein content is often determined by nondestructive method of NIR and gluten strength by a sedimentation test.

Flour Quality Characteristics

The above discussion of wheat sources and quality requirements provides a valuable yardstick in aiming for desired flour quality. However, each noodle type requires its own specific flour quality criteria. Flour protein, ash content and flour-pasting characteristics are major specifications. Protein content varies according to the noodle type to achieve the desired eating quality. Generally, flour protein content has a positive correlation with noodle hardness and a negative correlation with noodle brightness. Thus, there is an optimum flour protein content required for each noodle type. Japanese udon noodles require soft wheat flour of 8.0-9.5% protein. Other noodles require hard wheat flours of high protein content (10.5-13.0%), giving a firmer bite and springy texture. Flour ash content has been rated as one of the important specifications because it affects noodle color negatively. Flour ash content is largely determined by the ash content in the grain. Wheat with an ash content of 1.4% or less is always an advantage. Most noodle flours require ash content below 0.5%, but premium quality noodles are often made from flours of 0.4% or less ash.

Starch pasting characteristics (as measured on the amylograph or Rapid Visco Analyzer) also play an important role. The ratio of amylose to amylopectin content determines a starch's pasting characteristics. Flour amylose content between 22-24% is often required for Japanese type noodle making. Measurement of the pasting viscosity of flour or wholemeal also relates to noodle quality, and eliminates a starch isolation step. However, the presence of excessive alpha-amylase activity (breaks down starch) in the flour or wholemeal will undermine the prediction results because even a small quantity of the enzyme is likely to reduce the paste viscosity. The addition of certain alpha-amylase inhibitors into the test solution has been shown to improve the correlation between the viscosity of flour or wholemeal and the eating quality of Japanese type noodles.

Dough properties measured by sedimentation test, and Farinograph and Extensigraph measurements are often also included in noodle flour specifications because they affect noodle processing behavior and noodle eating quality. High sedimentation volume is correlated positively with Chinese style noodles that require a firm bite and springy texture. Extensigraph parameters measure the balance of dough extensibility versus elasticity. Too much extensibility results in a droopy dough, while too much elasticity causes difficulty in controlling final noodle thickness. Farinograph stability time has shown a positive relationship with Chinese raw noodle texture and tolerance in hot soup. It should be cautioned that a noodle dough is much lower in water absorption than bread dough (28-36% versus 58-64%). Rheological tests, initially developed to evaluate bread dough performance, may not be applicable to noodle dough evaluation.

There is a need to develop new tests specifically for relating a noodle dough's rheological properties to eating quality.

Starch Structure and Content

As discussed in previous paragraph, starch properties are important in determining the quality of Japanese white salted noodles (Udon). Higher starch and flour paste peak viscosity, lower gelatinization and peak temperatures, higher break down and higher swelling power and volume result in better Asian salted noodles. Flour swelling volume is significantly correlated with noodle quality and has been used as a means of screening Breeder's lines in a programme aimed at producing good Japanese noodle wheats. Lower amylose contents are related to higher starch swelling properties and better noodle qualities. In addition higher proportion of A-type starch granules, lower protein content and higher protein quality and soft grain texture result into better salted noodle quality. The optimum amylose content seemed to be about 22% for good quality noodles, with starch of higher or lower amylose content coming from flour of lower noodle-making quality. Recently the effect of chemical properties on starch physical properties has been studied. Granule-bound starch synthase present in starch granules has been identified as being related to the noodle-making potential of wheat flour. In wheat, there are three forms of the enzyme arising from separate genes on chromosomes 4A, 7A and 7D. The form corresponding to the gene on chromosome 4A is absent in many of the wheats, with better noodle making quality. Details are given in other chapter on starch quality. It should also be noted, however, that it has also been found to be absent in many varieties which are not normally considered suitable for manufacture of high quality Japanese noodles.

Factors other than amylose content also contribute to noodle quality, as some flours of relatively poor noodle quality also contained close to the optimum amount of amylose. This was demonstrated by Batey *et al.* 1997, wherein the pattern of oligosaccharides obtained from digestion of starch with amylase correlated with noodle quality. Their results confirmed those of earlier workers that starch properties are related to Japanese noodle quality, but they also showed the complex nature of that relationship. For example, both the amylose content and the amylopectin structure appear to be important for noodle quality, but they do not appear to be related to each other. Wheat varieties of good Japanese noodle-making quality (based on sensory testing) give relatively small amounts of oligosaccharides with a degree of polymerisation (DP) of 5 or greater, while varieties of poor noodle-making quality give much greater amounts of these larger oligosaccharides. There is a significant negative correlation between the amount of DP5 oligosaccharide and noodle eating quality. The correlation between the amount of this oligosaccharide and starch paste viscosity was much smaller. The size of the HPLC peak corresponding to this oligosaccharide may be used as a rapid method of screening for noodle-making quality. Because oligosaccharides of greater than DP4 should have at least one branch point, the results suggest that the structure of the amylopectin in starch of good noodle-making wheats has relatively few branch points close together. The best noodles are made from flour whose starch tends to have the amylose content on the lower end of the range and the amylopectin containing a low proportion of branch points.

Rapid Visco-Analyzer (RVA)

Cooked noodles rely on the gelatinization properties of starch for their physical structure and eating quality. High pasting viscosity of starch from wheats has been reported associated with good Japanese noodle-making potential. It is measured by either the amylograph or Rapid-Visco Analyser (RVA). Some correlations between aspects of Yellow Alkaline Noodle (YAN) quality (particularly smoothness and firmness) and starch paste viscosity have also been reported. However viscosity of flour and whole meal has poorer correlations with noodle quality. The RVA was initially developed as a tool for the rapid evaluation of sprout damage. Extension of this technology by incorporating controlled heating and cooling, enabled development of an instrument capable of measuring the viscosity of flour or starch pastes during gelatinization in a manner similar to the amylograph. The advantage of using the RVA for this purpose was its relatively small sample size (3–4 g) and the speed at which the test could be conducted (10–20 min). The capability of easily programming the heating-cooling cycle gave a flexibility that permitted optimization of conditions to suit a particular purpose. Although the pasting properties of starch are better correlated with noodle quality than corresponding measurements on flour and wholemeal, it would be impracticable to isolate starch from each sample before testing. Thus, any viscosity test must be applied to flour or, preferably, wholemeal samples.

One ubiquitous component in wheat is α-amylase, which is found in small quantities even in sound grain and is likely to have an effect on the paste viscosity of flour and wholemeal, particularly the latter. Inhibition of α-amylase activity before and during viscosity measurements has been reported for acid, chelating agents such as ethylenediamine tetraacetic acid (EDTA) and heavy metal salts such as mercuric chloride and silver nitrate. Silver nitrate has also been used to inhibit amylolytic activity during flour swelling volume tests of breeding lines to overcome the effects of sprout damage (Crosbie and Lambe 1993), and in RVA measurements on landraces of wheat and improved correlation with Japanese noodle quality. Large variations are exhibited among wheat varieties in RVA paste viscosity. This indicated that genotypes are available having desired flour paste viscosity to improve noodle making quality of wheat cultivars. Details of these tests are provided in another chapter.

Summary

There are two distinct classes of wheat produced in the world. The important soft wheat products are biscuits, cookies, crackers, wafers, pretzels, pastries, waffles, oriental noodles and cakes. The quality requirements of these products have been identified. Generally low to medium strong gluten and lower protein content are preferred for most the soft wheat products. Other components of the grain such as starch, pentosans and lipids also affect quality of these products. Based on the chemical characteristics of the grain various microlevel tests have been developed which are useful in predicting quality of early segregating generations. Among the soft what products, Asian noodles are becoming popular in the world. Research on Japanese udon noodles is ahead of other noodle types. Process properties, noodle color and noodle texture are the three key quality attributes in the evaluation of a wheat flour for any noodle making. In terms of noodle color, brightness is required, and whiteness or yellowness is essential depending on the noodle type. Noodle texture, however, is more complicated in the characterization of each noodle type, and progress can only be made to understand this property by involving Asian flour and noodle industrial representatives. Instrumental measurements of noodle color and texture are important in establishing their relationship to sensory characteristics of noodles.

References

Batey, I.L., Curtin, B.M. and Moore, S.A. (1997). Optimization of Rapid Visco Analyzer test conditions for predicting Asian noodle quality. *Cereal Chem.* 74: 497.

Bettge, A.D. and Morris, C.F. (2000). Relationships among grain hardness, pentosan fractions and end-use quality of wheat. *Cereal Chem.* 77(2): 241-247.

Crosbie, G.B. and Lambe, W.J. (1993). The application of the flour swelling volume test for potential noodle quality to wheat breeding lines affected by sprouting. *J. Cereal Sci.* 18: 267-276.

Faridi, A., Gaines, C. and Finney, P. (1994). Soft wheat quality in production of cookies and crackers. pp 154-168. In : Wheat: Production, Properties and Quality. W. Bushuk and V. F. Rasper, eds. Chapman & Hall: Glasgow Scotland.

Gaines, C.S. (2000). Collaborative study of methods for solvent retention capacity profiles (AACC Method 56-11). *Cereal Foods World* 45(7): 303-306.

Guttieri, M.J., McLean, R., Lanning, S.P., Talbert, L.E. and Souza, E.J. (2002). Assessing environmental influences on solvent retention capacities of two soft white spring wheat cultivars. *Cereal Chem.* 79(2): 880-884.

Hoseney, R.C., Wade, P. and Finley, J.W. (1988). Soft Wheat Products. pp 407-456. In: Wheat Chemistry and Technology. 3rd edn Vol. II. Y. Pomeranz ed. American Association of Cereal Chemists, St. Paul, MN.

Hou, G. and Kruk, M. (1998). Asian Noodle Technology, Technical Bulletin, Wheat Marketing Center, Portland.

Kruger, J.E., Matsuo, R.B. and Dick J.W. (1996). Pasta and Noodle Technology. American Association of Cereal Chemists, St. Paul, MN.

MacRitchie, F., (1980). In: *Advances in Cereal Science and Technology,* pp 271-326. Vol III Y Pomeranz, Editor, American Association of Cereal Chemists, St. Paul, MN.

Nagao, S. (1996).Processing Technology of Noodle Products in Japan. In: Pasta and Noodle Technology (Kruger, Matuso and Dick). American Association of Cereal Chemists, St. Paul, MN.

Nemeth, L.J., Williams, P.C. and Bushuk, W. (1994). A comparative study of the quality of soft wheats from Canada, Australia and the United States. *Cereal Foods World.* 39: 691-700.

Ohm, J.B., Chung, O.K. and Deyoe, C.W. (1998). Single-kernel characteristics of hard winter wheats in relation to milling and baking quality. *Cereal Chem.* 75:156.

Ram, S. and Singh, R.P. (2004). Solvent retention capacities of Indian wheats and their relationship with biscuit making quality. *Cereal Chem.* 81(1): 128-133.

Ram, S., Dawar, V., Singh, R.P. and Jag Shoran ('2005). Application of solvent retention capacity tests for the prediction of mixing properties of wheat flour. *J.Cereal Sci.* 42 (2): 261-166.

Souza, E., Kruk, M. and Sunderman, D.W. (1994). Association of sugar-snap cookie quality with high molecular weight glutenin alleles in soft white spring wheats. *Cereal Chem.* 71: 601-605.

Yamamoto, H., Worthington, S.T., Hou, G. and NG, P.K.W. (1996). Rheological properties and baking qualities of selected soft wheats in the United States. *Cereal Chem.* 73: 215-221.

□□□

Chapter 4

Durum Quality

Introduction

Durum or macaroni wheat (*T. turgidum var. durum*) is an important crop used chiefly for the production of pasta in Western Europe and North America, and two layered bread, single layered bread, burghul, couscous, pasta, frek and feed in other parts of the world (Table 1). In North Africa and the Middle East, durum is also used for producing flat breads. On a global basis only a small amount of durum wheat is used for baking of pan-style and hearth-style bread. Also, durum wheat for baking is often used in blends with common wheat. In India, durum occupies approximately 2.0 mha area and production is estimated to be around 2.5-3.0 million tones annually amounting to 3-4% of the total wheat production. The consumption of pasta is increasing at the rate of around 5% annually and currently it is estimated to be around 10 Lakh tones. The demand for pasta products may increase further due to increasing population and improving standard of living.

Table 1: Durum use and quality requirements in different parts of the world (Abdalla *et al*, 1992)

Country/Region	Use	Requirements
America, Europe, Australia, North Africa	Pasta	Medium to strong gluten, high yellow pigment
North Africa	Couscous	Medium to strong gluten, high yellow pigment
Middle East	Unleavened Bread	Medium to strong gluten
Near and Middle East	Bulgur	Hard grain, Yellow pigment
Andrean Region	Mote	Hard grain

Pasta is increasingly becoming popularized because it is simple to prepare, easy to transport and has excellent storage properties. It is nutritionally healthy and can be prepared in a variety of ways. The transformation of durum wheat into pasta is a straightforward operation. The semolina is hydrated, mixed, pressed and finally extruded and cut into desired shape before being dried. Breeding durum wheat for pasta and couscous (which has similar quality criteria to pasta) must remain the prime consideration in durum wheat variety development, because those two products account for over 80% of durum wheat consumption worldwide.

Mineral Content

Durum wheat is similar in composition to common wheat. The percentage of starch, protein, minerals, lipids and amino acids are roughly equivalent. One of the notable differences is the mineral distribution in the kernel. This difference may contribute to the superior cooking quality of pasta from durum semolina. The straight grade flour milled to 75% extraction from hard red spring wheat normally contains about 0.50% ash. Durum wheat flour, on the other hand, milled to the same extraction contains about 0.75 to 0.80% ash. A more striking difference is in the ash content of gluten, about 0.25% in the flour of hard red spring wheat and about 1.0% in that of durum wheat R.R. Matsuo, (1994). Durum semolina contains 1.7 times more potassium, 2.0 times more magnesium and 1.5 times more phosphorus than red spring flour. What functional role potassium, the mineral constituent in the highest concentration, plays has not been elucidated. More studies have been carried out on phosphorus in phytic acid and in phospholipids. Phytic acid is the main phosphorus containing compound in wheat, accounting for 72 to 78% of the total phosphorus in hard wheats and about 66% in durum wheats. In durum semolina phytic acid comprises only 2.7 to 6.0% of the total phytate in wheat, while in hard wheat flour, the percentage ranges from 15.1 to 45.5%.

Pasta Processing

The process of conversion of durum into pasta product is a simple process in which durum kernels first milled to produce semolina. Semolina is then hydrated into mixing chamber in a predetermined ratio. Vitamins and minerals are commonly dry blended at the durum mill before delivery to the pasta plant. For mixers open to atmosphere, the hydrated semolina particles pass through rotary seal into an evacuated extrusion chamber. In some presses, the mixing chamber operates under vacuum. The process of mixing and extrusion under vacuum minimizes pigment oxidation and removes trapped air bubbles. In the extrusion chamber, the hydrated semolina pieces are formed into a cohesive plastic mass. The extrusion pressure created by the screw can be as high as 150 atm. During extrusion, a considerable amount of heat is generated due to friction. The extrusion chamber must be cooled with circulation water to maintain the dough temperature around 50°C. The dough at the end of extrusion worm is split into a number of streams to equalize pressures along the long spaghetti die face. If the pressure is uneven, the extrusion pattern becomes irregular, resulting in excessive trimming. In the processing of short goods, the circular disc must be designed to ensure uniform extrusion; otherwise, the length of the pieces will be non-uniform. The curtains of spaghetti are then subjected to blast of air for surface drying, called "case hardening" and conveyed into the first of several drying chambers.

Drying is the most critical phase in processing. Moisture must be removed at a uniform rate to prevent moisture gradients within the strands that can cause checking or cracking. First stage of drying is the predryer, where pasta moisture is lowered from 30% to 20%. In conventional systems, the main dryer is controlled at about 55°C, with relative humidity initially at 95% and decreasing to 70%. The final stage is the conditioning, where the final moisture content of around 12% is established and the product is "cured". The total time varies from 15 to 28 hours, depending on the spaghetti diameter and on the drying system. The dramatic change in pasta drying technology during the later part of the 20th century has had a profound effect on raw material standards. First came the introduction of high temperature (HT) drying (60°C to 85°C). HT drying was rapidly accepted by the pasta-making industry because of improved hygiene and improved cooking quality, compared to low temperature (LT) drying. Now ultra-high temperature (UHT) drying (85°C to 110°C) has become common, with drying times as short as ~ 4-5 hr for long goods and ~ 2-3 hr for short goods Pollini (1996). The influence of drying technology on pasta properties brings traditional criteria for durum wheat selection into question. For example, there is evidence that under HT and UHT drying conditions protein strength has less influence on pasta cooking quality than under LT drying conditions D'Egidio *et al.* (1996). Enzymatic browning due to peroxidase has been shown to affect pasta color regardless of whether pasta is dried at LT or HT, but the interrelationships between varieties, growing environment, drying procedure and semolina extraction rate are not well understood.

Milling Quality

The milling quality is mainly represented by high extraction rate (semolina yield). There are many factors that the miller must consider when evaluating the processing potential of durum wheat. Physical condition (grade) is one of the factors determining the semolina milling quality. It includes degree of damage, presence of impurities as foreign material and the test weight. Presence of foreign materials decreases the semolina yield and imparts undesirable speckiness in the semolina. The test weight (weight per volume) is widely used as in primary durum wheat grading specification and milling quality indicator. Several studies have demonstrated that test weight is related positively to semolina yield. Thin, shrunken and damaged kernels have lower test weight and reduced semolina yield and increased semolina speckiness. Test weight is affected by the variety (shape and size of the grain) and the environment. Considerable genetic variation exists in the semolina yield of durum wheat. Millers try to maximize the yield of semolina and minimize flour yield because semolina commands higher price.

The endosperm/bran ratio of durum wheat is largely dependent on cultivar, whereas endosperm friability is greatly influenced by agro-climatic conditions. Use of ash content as a criterion for determining semolina purity can influence the cultivar ranking because of the differences in grain ash content and of the distribution of minerals in the endosperm. It appears that the ease with which endosperm is freed in the form of semolina during milling could he a criterion for predicting semolina quality. Hard Amber Durum (HAD) wheat with 75% or more of hard and vitreous kernels of amber color is in greatest demand by millers. Lower ash content is preferred by the industry. High semolina ash content is normally associate with a longer extraction time, which can produce a duller semolina colour because of the presence of high ash outer endosperm particles. Further pigment degradation during pasta processing is indirectly affected by semolina ash content. Bright yellow colour is important to make good quality pasta products. This colour is the result of natural carotenoid pigments present in the seed Troccoli A *et al.* (2000). The details are given in the other section of yellow pigments.

Grain moisture content represents another aspect of quality. The percentage of other components such as protein and starch are inversely related to moisture content. In addition grain with high moisture content (>14%) may get heated during storage with sprouting, fungal damage and deteriorating conditions. Vitreousness is another aspect of the kernel which affects semolina yield. Vitreousness implies that the endosperm has a glassy, translucent appearance rather than a mealy or floury nature. Generally vitreous areas of the endosperm have higher protein content than the mealy areas.

Pasta Cooking Quality

Cooking quality has greater importance to consumers and hence to producers, breeders and processors. Cooking quality includes the time of cooking, swelling of pasta strands after cooking, firmness and resilience of cooked pasta, surface condition or the extent of disintegration of cooked product, smoothness of the cooked product and the aroma and taste. Therefore, the appearance and texture of pasta should be acceptable after cooking, and the pasta should have a pleasing flavour and mouth feel when eaten. Pasta strands should not broken apart or released excess solids into the cooking water, should not be sticky but exhibit some firmness. Although sensory evaluation is used for measuring pasta quality, objective procedures have also been developed. Basically the procedures monitor the response to a force on the cooked pasta of either pushing or pulling or both to measure such factors as firmness, tenderness, elasticity, compressibility or stickiness. The combination of these factors gives a good estimation of the relative textural characteristics of cooked product. The relative turbidity of cooked water is an indication of the amount of breakdown of pasta during cooking and is referred to as cooking loss. Major progress has been made in understanding the physicochemical basis of the pasta cooking quality.

Quality Requirements

Durum wheats are preferred raw material for pasta processing, although common wheats can also be used to make a less satisfactory product. The characteristics that determine the industrial or technological value of durum wheat can be grouped into two broad categories as milling performance and pasta quality. Milling performance measures the proportion of semolina of desired degree of refinements that can be extracted from wheat. Pasta quality depends upon the ability of the extracted semolina to give pasta with a required appearance, resistance to breakage and tolerance to cooking.

During the last two decades, some progress has been made in our understanding of the chemical and molecular basis of functional quality of durum pasta products. Cooking quality is determined by two independent parameters: visco-elastic behavior and the surface condition of the cooked pasta. The surface of freshly extruded pasta is a continuous protein film, while the inner portion is a compact structure of starch granules embedded in an amorphous protein matrix aligned in layers parallel to the protein film. During the cooking of good-quality pasta, the protein absorbs water and swells more rapidly than starch. In poor quality pasta, the proteins aggregate in discrete masses rather than in a continuous matrix. Hydration of the protein fraction of pasta before the beginning of starch gelatinization appears to be important to produce a firm, good quality cooked pasta. It is generally accepted that strong gluten with high elastic recovery gives greater cooking stability and higher cooked firmness score, whereas pasta made from weak gluten with low elastic recovery is prone to deteriorate rapidly and becomes soft with over cooking. Consequently, gluten strength is recognized to be one of the most important quality criteria in durum breeding.

Gluten Strength: Genetic Basis

Gluten strength is governed by the quality and quantity of glutenins and gliadins, the main storage proteins. Glutenins are polymeric and their monomeric units are divided into high molecular weight and low molecular weight subunits based on their migration on SDS-PAGE. Gliadins are divided into α, β, γ and ω groups based on Acid-PAGE separation. The role of HMW glutenin subunits has been studies extensively in bread wheat for bread making quality and discussed in other chapter, however fewer reports are available in durum wheat. The reason may be the predominant HMW-GS alleles present in durums have low glu-1 ratings based on Payne scores (mostly null at the *Glu-A1* locus and the frequent presence of bands such as 20, 6+8 etc., at *glu-B1* with few contrasting genotypes available for investigation.

Early studies during 1970's demonstrated the usefulness of gliadins γ-45 and γ-42, encoded at the *Gli-B1* locus, as markers of good and poor pasta quality, respectively. However, subsequent studies reported that LMW-GS encoded at the *Glu-3* loci tightly linked to the *Gli* loci are responsible for the differences in the quality. The LMW-GS are more likely to be the causal factors for technological quality, therefore, more efforts should be made towards characterizing them and corresponding genes. Despite their important effects on the qualitative properties of durum wheat, only a limited numbers of low molecular weight (LMW) glutenin genes have so far been characterized. LMW glutenins are divided into B, C and D subunits according to their mobility in SDS-PAGE. B and C subunits are coded by *Glu-3* loci located on the short arm of group 1 chromosomes. Subunits coded by *Glu-B3* locus have been reported associated with gluten strength in durum wheats. Genes encoding B-LMW glutenins are tightly linked to Gli-1 loci encoding γ gliadins γ (gamma) gliadins 42 and 45 are considered genetic markers for differences between poor and good pasta making quality. The large amount of variation in LMW types led to the conclusion that the majority of B subunits are inherited as two groups controlled at *Glu-A3* and *Glu-B3* loci which are tightly linked to *Gli-A1* and *Gli-B1* loci respectively. *Glu-B2* and *Glu-B4* are separate loci coding for some intermediate mobility b subunits occurring on the short arm of chromosome 1B.

Good pasta making quality is associated with specific allelic form of LMW glutenins named LMW-2. LMW-2 comprises a group of polypeptides encoded by *Glu-B3* locus tightly linked to *Gli-B1* locus coding γ and ω gliadins. Most commonly grown durum wheat cultivars have either the LMW-2/45 (plus γ-gliadin 35) or LMW-1/42 (plus gliadins 33, 35 and 38) allelic forms, later being associated with poor pasta making characteristics. There are indications that the better quality associated with the presence of LMW-2 in durum wheat is mainly due to the fact that subunits are more abundant than the LMW-1 subunits and that the structural differences may play minor role. There are minor structural differences between LMW-1 and LMW-2 types. The main difference between the LMW -1 and LMW-2 protein groups is the presence of slow moving *Glu-B3* coded LMW-GS in the latter. This slow moving LMW-GS corresponds to the 42K LMW-GS in most genotypes and is consistently the most abundant LMW-GS polypeptide. Allelic variants of LMW-2 groups have also been described and in all cases these are associated with good technological properties (Brites and Carrillo 2001).

Molecular analysis of the genes encoding gluten proteins is of particular interest both for basic and applied research because they belong to highly polymorphic multigene families and their encoded proteins are those mainly responsible for the qualitative properties of durum wheat dough. The corresponding genes should be characterized and sequence specific PCR primers designed. This will help in the breeding programme to develop durum wheat cultivars suited to meet the requirement of pasta quality. Finally in future this will lead to the development of transgenics for particular trait as more recently the transformation protocols have been standardized in durum wheat (Barcelo and Lazzeri 1995, He *et al.*1998).

Gluten Strength Measurements

Miograph, farinograph and sedimentation tests are used to screen durum wheats for gluten strength. Glutenin has the important property of swelling in various non-reducing solvents (dilute acetic acid, lactic acid and SDS) and the swelling volume appears to be directly related to quantity and quality of glutenin. Therefore, small-scale tests used to predict bread-making quality are based on glutenin swelling capacity or directly on insoluble glutenin content. The Zeleny and SDS sedimentation tests measure the sedimentation value of a suspension of flour in dilute lactic acid and in SDS-lactic acid solvents, respectively. The sedimentation value depends mainly on the amount and the swelling characteristics of the glutenin, since other proteins such as gliadin are soluble in the Zeleny and SDS test solutions. High SDS sedimentation values have been associated with stronger gluten and superior pasta making quality in durum wheat. Though, SDS sedimentation volumes are influenced by environment, crop year, and their interaction with cultivars, SDS sedimentation volumes are highly heritable and can be used for selecting among early generation progeny (Matuz, 1998).

The SDS sedimentation test is a modification of the Zeleny sedimentation test that requires only a small sample of whole meal flour, is simple to perform and is highly reproducible. Since its development, wheat breeders have explored the feasibility of using the SDS sedimentation test for durum wheat cultivar improvement and during this process several slight modifications were made to the procedure (Dick and Quick, 1983). While there are limits to the amount of sample that can be used with the SDS sedimentation test, it would seem that the sample weight can be adjusted for individual convenience. It has been casually suggested that either whole meal or flour can be used in the SDS sedimentation test. Whole meal is more desirable than flour due to ease of preparation. Cultivars with different protein quality, as expressed by their gluten characteristics, should be differentiated by the SDS sedimentation test. Though SDS sedimentation test assesses protein quality, however, results also are influenced by protein concentration. The effect of protein concentration on SDS sedimentation volume may be eliminated by multiplying SDS sedimentation volume by the grain protein concentration of the sample divided by 10.

Mixograph Characteristics

The mixograph is a type of recording mixer, albeit a smaller, simple piece of equipment than the farinograph. In farinograph, dough is developed by pack-squeeze type of gentle kneading and shearing action and by folding and stretching action of mixing pins in mixograph. During mixing first water is brought into contact with flour particles that facilitate water absorption to develop dough and a mixing curve formed. The mixing curve contains a peak at which dough is fully developed and after that breakdown starts. Although the curve obtained from the mixograph results from the developing dough exerting force on the mixing pins, the curve and interpretation of it are not the same as for a farinogram. Mixograph records the increase in stress as dough is mixed to its maximum resistance and the subsequent decrease in stress on over-mixing. A number of measurements are taken that relate to dough strength and stability of which mixing time (MT), peak dough resistance (PDR) and resistance breakdown (RBD) are used most widely. The small amount of flour and ingredients required, have made possible to use this device to test the effect of addition of protein fractions on dough properties. Mixograph procedures have also been improved by computerization, which facilitates data handling and allows objective interpretation of the mixogram. Mixing curves are obtained by mixing a fixed amount of semolina and water in the mixograph.

These mixograms are then compared to a series of standards obtained from the same instrument. Mixograms are rated on a scale of 1 to 8, with the higher values indicating strong mixing characteristics (Fig. 1). Weak-gluten semolina produces a curve similar to curves 1-3 and strong gluten semolina curves are more like curves 5-8.

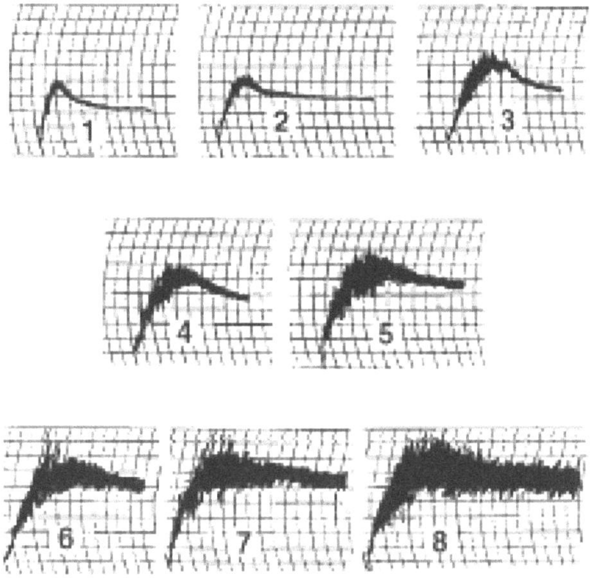

Fig. 1. Reference mixograms for Durum Wheat

Yellow Pigment (β-Carotene Content)

The genetic improvement of durum wheat involves not only the desirable agronomic traits but also its capability to yield good quality products (Semolina or pasta) with higher nutritional value. Nutritional value of durum wheat depends upon the quantity of beta-carotenes, precursor of vitamin A, responsible for yellow colour in semolina and pasta products. Carotenoids are natural compounds that have many health benefits to human beings. Carotenoids reduce the oxidative damage to biological membrane by scavenging peroxi-radicals such as those involved in certain human diseases and in the ageing process. Furthermore, carotenoids protect cells and organisms against the harmful effects of light and air. β-carotenes have also been implicated in reducing the risk of cancer in human beings. Vitamin A deficiency also increases susceptibility to malaria and diarrhoeal disease, and reduces the bioavailability of micronutrients, including iron, important to the health of all family members (WHO, 2002).

Genetically increasing the levels of vitamin A and its precursors in dietary staples is especially relevant for poor families living in isolated rural areas for whom vegetable sources are often in short supply. Therefore, an improvement in the endogenous quantity of beta-carotenes should enhance the nutritional value of pasta products. There are reports in other crops as rice, maize and sorghum etc where efforts have been made in manipulating β-carotene content by molecular techniques; however, little information is available in wheats. Evaluation of thousands of durum wheat germplasm lines from national and international sources as well as advanced entries in coordinated programme in India indicated the availability of lines with higher yellow pigment content and gluten strength. The wide variability in yellow pigment content was observed ranging from 3 ppm to >10 ppm.

Genetic studies showed that wheat endosperm yellow colour is highly heritable (67–90%) and under polygenic control Parker *et al.* 1998, Elouafi *et al.* (2002). Quantitative trait loci (QTL) for endosperm yellow colour accounting for a large proportion of the genetic variation (50%–60%) have been mapped on chromosome arms 7AL and 7BL. Smaller QTLs were detected on chromosomes 2A, 3A, 5A, and 5B (Parker *et al.* 1998, Elouafi *et al.* 2002). Later BAC clones were identified having genes responsible for carotenoid biosynthesis (Cenci *et al.* 2004). They reported the isolation of BAC clones containing genes coding for three different enzymes of the carotenoid biosynthesis pathway: phytoene synthase (PSY), phytoene desaturase (PDS), and ζ-carotene desaturase (ZDS). Primers were designed on the basis of wheat ESTs similar to the sequences of these three genes in other species, and used to screen a BAC library from *Triticum turgidum* var. *durum* ($2n = 28$, genomes AABB). Eight, six, and nine 384-well plates containing at least one positive clone were found for PSY, PDS, and ZDS, respectively. BACs selected for each of these genes were then divided in two groups corresponding to the A and B genomes of tetraploid wheat, based on differences in the length of the PCR amplification products, conformation-sensitive gel electrophoresis (CSGE), or cleavage amplification polymorphisms. Positive clones were then assigned to chromosomes using a set of D genome substitution lines in *T. turgidum* var. *durum* 'Langdon'. PSY clones were localized on chromosomes 5A and 5B, PDS on chromosomes 4A and 4B, and ZDS on chromosomes 2A and 2B.

Biosynthesis of Yellow Pigments

Briefly, the process involves the synthesis of a 40 carbon atom chain (C40) backbone, its desaturation, cyclization, and subsequent modifications (e.g., hydroxylation and epoxidation). The two first steps are common for all carotenoid pathways, whereas cyclization can produce β or ζ rings having different properties. Phytoene synthase (PSY) catalyses the reaction making phytoene (C40) from two molecules of geranyl-geranyl pyrophosphate (C20); phytoene desaturase (PDS) catalyses the double desaturation at the 11–12 and 11'–12' symmetrical positions of phytoene producing ζ-carotene; and ζ-carotene desaturase (ZDS) catalyses the reaction producing lycopene by desaturation of 7–8 and 7'–8' symmetrical positions of ζ-carotene. Double lycopene cyclization can produce β, ε or α-carotene, having β – β, ε–ε, and β –ε rings, respectively. Subsequent modifications transform α -, β -, and ε-carotene into lutein or other xanthophylles (Fig. 2).

Lipoxigenase and Pasta Colour

Yellow pigment content in the pasta products depends upon (i) the natural carotenoid pigments present in the seeds; (ii) their residual content after milling and after the storage of either grain or semolina; (iii) their oxidative degradation by lipoxygenase (LOX) during pasta processing; (iv) the oxidative balance among different compounds implied in this reaction; and (v) the processing conditions. Among them LOX activity has shown significant correlation with final yellow pigments in pasta products. The LOXs are non-heme iron containing dioxygenase that catalyse the oxidation of polyunsaturated fatty acids containing a *cis,cis*-1,4-pentadiene system, producing conjugate *cis,trans*-diene hydroperoxides. Radicals produced during the intermediate steps of this reaction are responsible for oxidative degradation of carotenoid pigments. LOX purification studies (Barone *et al.* 1999) have revealed a monomeric structure of about 95 kDa molecular mass, with a high affinity for linoleic acid as a substrate and an optimal pH for hydroperoxidation and bleaching activities near to dough pH (Barone *et al.* 1999). Thus, good conditions for pigment degradation during pasta processing are produced. Further studies demonstrated the existence of two to three (Pastore *et al.* 1999) semolina LOX isoforms. The proteins have a N-terminal b-barrel domain and a larger catalytic domain containing a single atom of non-heme iron (Brash 1999). The metal is liganded to conserved histidines and to the carboxyl group of a conserved isoleucine at the C terminus of the protein. As the iron is non-heme, lipoxygenases appear virtually colourless to the eye. The enzymes are usually in the ferrous (inactive) form when isolated. Oxidation to the active ferric enzyme is required for catalysis.

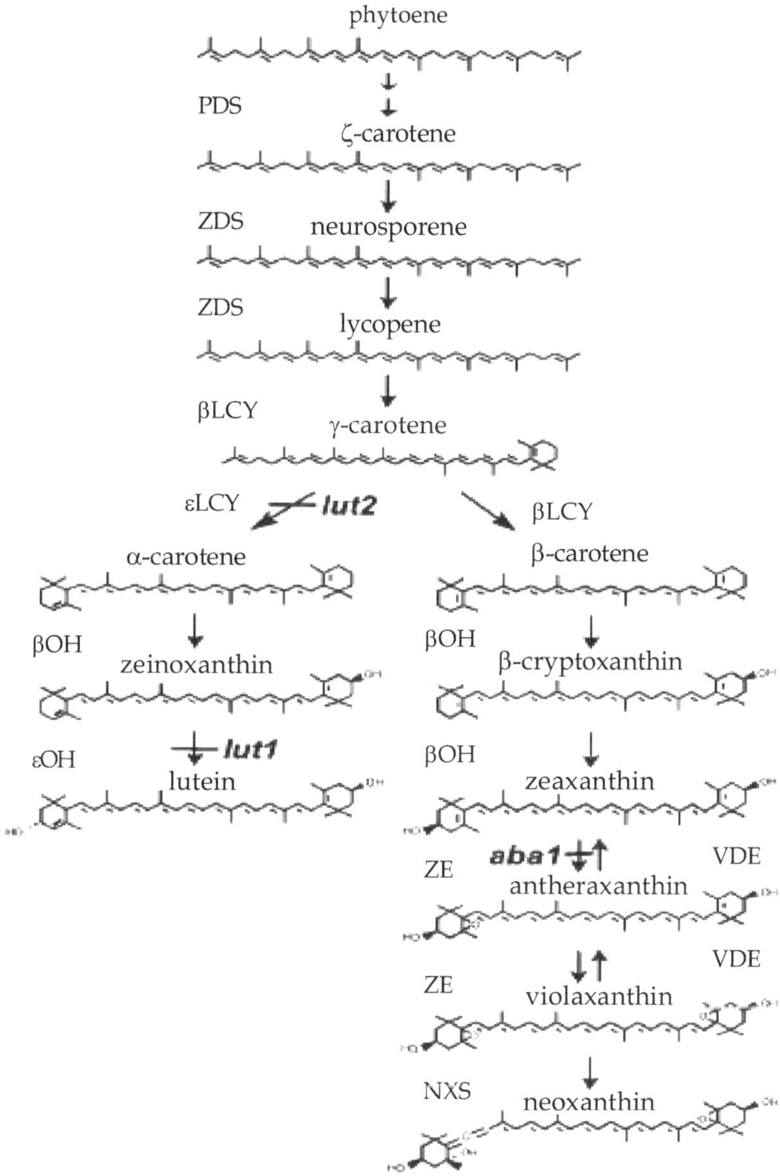

Fig. 2. Carotenoid Biosynthetic Pathway. (Park *et al.* 2002) The commonly held view of the carotenoid biosynthetic pathway in plants is a series of four desaturations to form all-*trans*-lycopene from phytoene. Lycopene is subject to two cyclization reactions to form β or γ carotene, which are modified further to form the various xanthophylls. PDS, phytoene desaturase; ZDS, ζ-carotene desaturase; β–LCY, β-cyclase; ε–LCY, ε-cyclase; βOH, β-hydroxylase; ε–OH, β-hydroxylase; ZE, zeaxanthin epoxidase, NXS, neoxanthin synthase; VDE, violaxanthin deepoxidase.

The β-carotene bleaching reaction occurrs only in presence of linoleic acid, confirming the cooxidative nature of enzymatic pigment degradation. Therefore, the prevention of their degradation during pasta making is a necessary goal to attain. It is known that many compounds which act as physiological antioxidants, such as the β-carotene, in addition to the α-tocopherol and the L-ascorbate, are able to inhibit the LOX from various systems. Similarly, it has been demonstrated that β-carotene Trono *et al.* (1999), α- tocopherol Pastore *et al.* (1999) and L-ascorbate (Pastore *et al.* 1999) inhibit both the linoleate hydroperoxidation and the β-carotene bleaching in durum wheat semolina. Particularly, it is shown that externally added β-carotene can act as inhibitor of the LOX-catalysed linoleate hydroperoxidation and an inverse relation between the % carotenoid loss and the initial carotenoid content in semolina from durum varieties showing similar LOX activity Trono *et al.*, (1999) was found. So, an increase of the amount of β-carotene, α-tocopherol and L-ascorbate compounds by external addition or breeding could be helpful to reduce pigment loss during pasta processing and to improve pasta quality.

The concurrent polyunsaturated fatty acids and pigment oxidation by LOX during durum wheat storage or processsing is a complex phenomenon where the availability of the substrates and the balance between antioxidative and perooxidative factors play an important role in determining the quality of final products. A significant proportion of the carotenoid pigments are lost during milling (approximately 8%) or during pasta processing (approximately 16%) Borrelli *et al.* (1999). This large reduction in yellow pigments during pasta processing is mainly due to LOX activity. Borrelli *et al.* (1999). The reaction rate of LOX is particularly high at the beginning of dough mixing when its primary substrates, oxygen and lipid, are present in the highest amounts, and the mixing itself helps their contact and enhances the incorporation of oxygen in the dough. The involvement of LOX in colour loss is demonstrated by positive correlation between the decrease of β-carotene content after pasta formation and LOX activities in semolina. In addition to this, the hydroperoxidation and bleaching activities of LOX are highly correlated demonstrating that the bleaching might be ascribable to a co-oxidative action by LOX. However, there are reports indicating that the loss of yellow index, another measure of the colour, is highly correlated with the β-carotene loss but not with the LOX activities. This can be explained by the fact that the physical expression of colour is also affected by factors different from LOX activity, such as peroxidase and polyphenoloxidase activities, as well as ash content, that are involved in brownness of products and increase the colour loss after pastification.

Genetics

Cereal LOX genes have been studied in more detail in barley than in wheat. Two LOX isoenzymes have been purified from germinating barley grains and mapped on barley chromosomes. Wheat LOX isoenzymes were first assigned to chromosomes 4A (*Lpx-A1*), 4B (*Lpx-B1*), 4D (*Lpx-D1*), 5A (*Lpx- A2*), 5B (*Lpx-B2*), and 5D (*Lpx-D2*) using nulli-tetrasomic lines. Mapping studies in wheat showed colinearity of LOX genes between wheat and barley suggesting that they are orthologous. These mapping results indicate that the wheat *Lpx-2* locus corresponds to the barley *LoxC* gene (homoeologous group 5), whereas the wheat *Lpx-1* locus corresponds to barley *LoxA* or *LoxB* locus (homoeologous group 4). Hessler *et al.* (2002) sequenced several wheat fragments and based on their higher similarity to the barley *LoxA* gene assigned them to the *Lpx-1* locus. Recently, Carrera *et al.* (2007) identified wheat sequences on homoeologous group 4 corresponding to the barley *LoxB* gene, which they proposed to designate *Lpx-3*. Two different genes related to the barley *LoxA* on chromosome 4B were also identified and designated as *Lpx-B1.1* and *Lpx-B1.2*. The identification of lines lacking one or more LOX isoenzymes, which can be employed in breeding programmes, could be useful in improving pasta quality. Another approach can be the development of transgenic wheat lines in which the LOX activity is down-regulated by antisense or co-suppression.

Summary

Pasta is increasingly becoming popularized because it is simple to prepare, easy to transport and has excellent storage properties. It is nutritionally healthy and can be prepared in a variety of ways. More than 80% of durum is consumed in the form of pasta and couscous. The characteristics that determine the industrial or technological value of durum wheat can be grouped into two broad categories as milling performance and pasta quality. Milling performance measures the proportion of semolina of desired degree of refinements that can be extracted from wheat. Pasta quality depends upon the ability of the extracted semolina to give pasta with a required appearance, resistance to breakage and tolerance to cooking. Good quality pasta products need stronger gluten and higher grain protein content. The role of individual protein subunits in imparting gluten strength has been identified. In addition yellow colour is another desirable trait of good quality pasta products. Yellow colour is imparted by the presence of carotenoid pigments. In addition to imparting yellow colour, carotenoids pigments have many health benefits to human beings. Recently molecular markers have been identified associated with higher carotenoids in durum. Lipoxigenase (LOX) activity has shown significant correlation with final yellow pigments in pasta products. LOX genes have been characterized. In conclusion genetic components for determining pasta quality have been characterize at molecular level that is very useful in breeding.

References

Abdalla, O., Dieseth, J.A. and Singh, R.P. (1992). Breeding Durum Wheat at CIMMYT. In: *Durum Wheats: Challenges and Opportunities,* Wheat Special Report No. 9, Rajaram, S., Saari E.E. and Hettel, G.P. (eds) CIMMYT, Mexico DF, pp. 1-13.

Barone, R., Briante, R., D'Auria, S., Febbraio, F., Vaccaro, C., Del Giudice, L., Borrelli, G.M., Di Fonzo, N. and Nucci, R. (1999). Purification and characterization of a lipoxygenase enzyme from durum wheat semolina. *J. Agr. Food Chem.* 47: 1924-1931.

Borrelli, G.M., Troccoli, A., Di Fonzo, N. and Fares, C. (1999). Durum wheat lipoxygenase activity and other quality parameters that affect pasta colour. *Cereal Chem.,* 76: 335-340.

Brites, C. and Carrillo, J.M. (2001). Influence of High Molecular Weight (HMW) and low molecular weight glutenin subunits controlled by Glu-1 and Glu-3 loci on durum wheat quality. *Cereal Chem.* 78: 59-63.

Carrera, A., Echenique, V., Zhang, W., Helguera, M., Manthey, F., Schrager, A. Picca, A., Cervigni, G. and Dubcovsky, J., (2007). A deletion at the Lpx-B1 locus is associated with low lipoxygenase activity and improved pasta color in durum wheat (*Triticum turgidum* ssp. durum). *J. Cereal Sci.* 45: 67–77.

Cenci, S. Somma, N., Chantret, J., Dubcovsky, and Blanco, A. (2004). PCR identification of durum wheat BAC clones containing genes coding for carotenoid biosynthesis enzymes and their chromosome Localization. *Genome* 47: 911–917.

Elouafi, I., Nachit, M.M. and Martin, L.M. (2002). Identification of a microsatellite on chromosome 7B showing a strong linkage with yellow pigment in durum wheat (*Triticum turgidum* L. var. *durum*). *Hereditas* 135: 255–261.

Hessler, T.G., Thomson, M.J., Benscher, D., Nachit, M.M. and Sorrells, M.E. (2002). Association of a lipoxygenase locus, Lpx-B1, with variation in lipoxygenase activity in durum wheat seeds. *Crop Sci.* 42: 1695–1700.

Matuz, J. 1998. Inheritance of SDS sedimentation volume of flour in crosses of winter wheat (*Triticum aestivum* L.) *Cereal Res Commun.* 26 :203-210.

Park, H., Kreunen, S.S., Cuttriss, A.J., DellaPenna, D. and Pogson, B.J. (2002). Identification of the Carotenoid Isomerase Provides Insight into Carotenoid Biosynthesis, Prolamellar Body Formation, and Photomorphogenesis. *The Plant Cell.* 14: 321–332.

Parker, G.D., Chalmers, K.J., Rathjen, A.J. and Langridge, P. (1998). Mapping loci associated with colour in wheat (*Triticum aestivum* L.). *Theor. Appl. Genet.* 97: 238–245.

Pastore, D., Trono, D., Padalino, L., Simone, S., Valenti, D., Di Fonzo N. and Passarella, S. (2000). Inhibition by α-tocopherol and L-ascorbate of linoleate hydroperoxidation and β-carotene bleaching activities in durum wheat semolina. *J. Cereal Sci.* 31(1): 41-54.

Pollini, C.M. (1996). THT Technology in the Modern Industrial Pasta Drying Process. In : Pasta and Noodle Technology pp. 59-74. Kruger, J.E., Matsuo, R.R., and Dick, J.W. American Association of Cereal Chemists, St. Paul, Minnesota.

Troccoli, A., Borrelli, G.M., De Vita, P., Fares, C. and Di Fonzo, N. (2000). Durum Wheat Quality : A multidisciplinary concept. *J. Cereal Sci.* 32: 99-113.

Trono, D., Pastore, D. and Di Fonzo, N. (1999). Carotenoid dependent inhibition of durum wheat lipoxygenase. *J. Cereal Sci.* 29: 99-102.

❑❑❑

Chapter 5

DOUGH RHEOLOGY AND CHEMICAL TESTS

Introduction

Dough is a complex mixture of flour constituents, water, salt and other ingredients depending upon the requirements of end-products. It is the principal intermediate stage in the transformation of wheat into different end-use products. Rheological properties of dough determine the behaviour of dough during mixing, rounding and molding used in making different products and also effect on the quality of finished product. Therefore, understanding the various features of dough rheology and their relations with product quality is important. The importance is enhanced further because of mechanization and automation of baking industry. Dough is composed of continuous phase in which gas cells are dispersed. The continuous phase is called "dough phase". It consists of starch granules embedded in protein matrix along with some lipids, cell remnants and yeast cells in fermented dough. Gas production and retention is the most comprehensive quality requirements of the dough for bread making. The ability of the wheat dough to retain gas during fermentation and oven rise is unique. This depends on the ability of the dough to be stretched into thin membranes because of protein phase.

Rheological Test

Rheology is the study of the flow and deformation of materials. When the flour is mixed with water and other constituents depending upon the end product, mechanical stress is exerted to develop dough. The rheological properties of a dough are derived from the relationship between the stress on a material and the corresponding strain (deformation) exerted on it and the time. In rheological experiments deformation force is either held constant or varied with time. The deformation may be caused by shear or extension force. Stretching experiments are more useful in predicting dough quality. Generally to measure rheological behavior, a well controlled, well defined deformation or strain is applied to a material over a given time and the resulting force response is measured (or vice versa), to give an indication of material parameters such as stiffness, modulus, viscosity, hardness, strength or toughness of the material (Dobraszczyk and Morgenstern, 2003).

Rheological tests measure the mechanical properties (viscosity, elasticity, consistency, extensibility etc.) of dough under various deformation conditions. Most of these tests are used to predict baking performance and behavior of dough during processing before baking. Dough is a non-linear viscoelastic fluid that is shear thinning, work hardening and possibly exhibits small yield stress. Physical dough testing devices are used to evaluate strength and performance characteristics of flour under mechanized conditions. The conviction is that rheological properties of dough affect its behavior during processing, has led to the design of methods and instruments for measuring rheological or similar properties. Instruments and test methods designed for this purpose have been reviewed by Dobraszczyk (2001).

Rheological techniques are generally characterized according to the type of strain imposed e.g. compression, extension, shear, torsion and relative magnitude of the imposed deformation. Descriptive rheological measurements are done by Mixograph, Farinograph, Alveograph and extensograph. Among them Mixograph and Farinograph are the two most widely used types of recording dough mixers. Both instruments record the power that is needed to mix the dough at a constant speed. The record consists of an initial rising part, which shows an increase in resistance with mixing and is interpreted as dough development time. The point of maximum resistance is generally identified with optimum dough development and is followed by a second part of more or less rapid decrease in consistency and resistance to mixing. The rheological test, which is used to measure biaxial extension, is Alveograph. The principals of these measurements and the functional relations with dough quality are described below.

Mixograph

The Mixograph is a rapid tool used to measure the mixing characteristics of flour. Since Swanson and Working (1933) introduced the basic Mixograph system, recently, sample size has been reduced to improve its utilization in breeding programmes. Mixograph data have been used to differentiate wheat flours of good and poor baking quality. The 10g Mixograph has been specifically used to evaluate the quality of early generation lines in wheat breeding programmes because it utilizes a small sample amount and requires a short analysis time. Mixograph parameters have been considered very useful in selecting good quality wheat lines in hard wheat breeding programmes because they are significantly affected by the genetic effects. Conventional Mixograph parameters that have been used in quality evaluation include mixing time, water absorption, and mixing tolerance.

A mixogram is conducted with 10 g of flour and the appropriate amount of water to give optimum absorption. The Mixograph is a recording mixer reflecting the resistance of the dough during mixing. Mixograph curve suggests mixing time requirements, tolerance and optimum water absorption. Depending on the dough strength of the flour, the curve reaches to a maximum either well defined peak or a plateau. After further mixing, a decrease in mixing curve is recorded and the breakdown starts. Dough breakdown behavior is reflected in the tail of the curve when mixing continues beyond the mixing peak and is commonly referred to as mixing tolerance. The rate of breakdown shows the stability of the dough and its sensitivity to mechanical treatment. For desirable bread-type flour, from Hard Red and White, Winter and Spring selections with adequate flour protein, a mixogram should have the following properties: high water absorption, moderately long mixing requirements (3-6 min.), strong gluten strength, good dough mixing tolerance, minimum oxidation requirements. Good dough mixing tolerance, as seen on the mixogram would indicate that a bread dough should be elastic after mixing. Extremely long mixing time is considered undesirable because the power and time requirements would be excessive in a commercial bakery. In addition, extremely long mixing time is often associated with tough, bucky doughs which do not perform properly in a mechanized bakery.

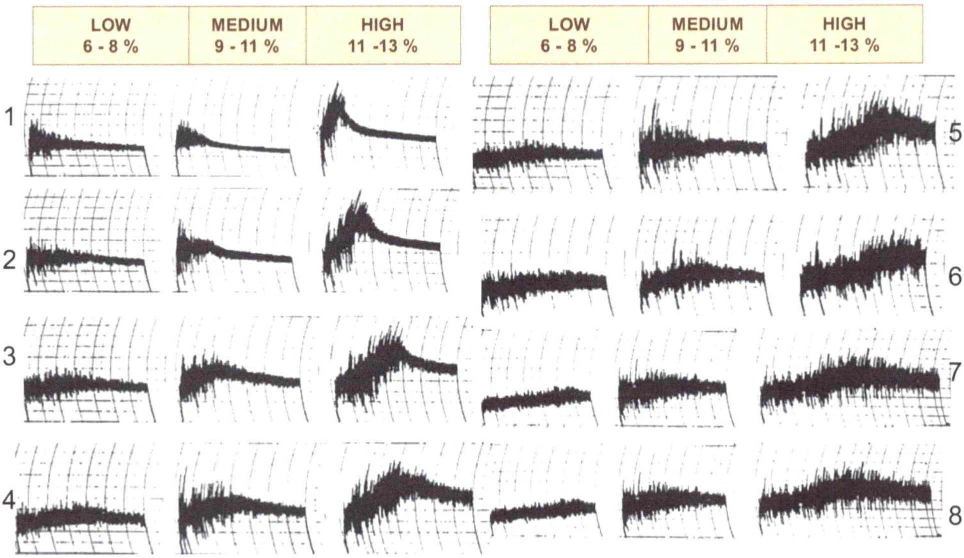

Fig. 1. Eight different types of mixograms based on protein content and quality

Reference mixograms have been developed to compare mixograms generated from any wheat sample (Fig. 1) (The Mixograph Handbook). (The Mixograph Handbook). Depending upon the shape of the curve and the protein content, 1-8 mixogram types have been developed. Protein content is divided into three categories: low (<9%), medium (9-11%) and higher (11-13%). The mixogram of a 11% protein flour may be classified as high if the height of the curve reflects that of a flour with a 13% or better protein; meaning the protein in the flour is of higher quality. Tolerance is scored from zero to 9 (weak to strong). Tolerance is measured between the 3rd and 4th minutes after the mixogram reached the peak (Fig. 2). This is determined by intersecting lines crossing the middle of the mixogram envelop front to back. Generally higher the protein content and the quality, higher the peak climbs and exhibits higher tolerance. The broadness of the peak and thickness of the tail is an indication of dough tolerance (sensitivity to mechanical treatment). Larger the value of band width, greater is the tolerance and therefore quality.

Type mixograph between 3-4 minutes window after peak

L-	=	less than three squares below the midpoint of tail
L	=	3 squares below the midpoint of the tail
LM	=	3 1/2 squares below the midpoint of the tail
M	=	4 squares below the midpoint of the tail
MH	=	4 1/2 squares below the midpoint of the tail
H	=	5 squares below the midpoint of the tail
H+	=	5 1/2 squares below the midpoint of the tail

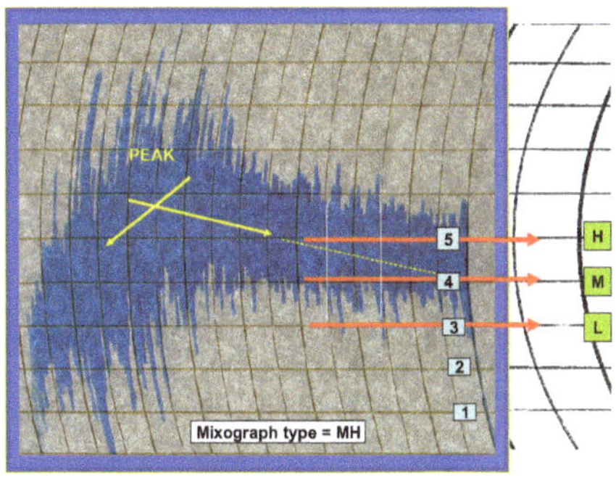

Fig. 2. The measurement of mixing tolerance using Mixograph curve.

Mixograph Parameters and their Functional Relations with Quality

Application of computer technology in Mixograph procedure has improved its control and data processing The computer analyzed Mixograph parameters had significant correlations with the conventional analysis, suggesting that they could be used to evaluate flour quality instead of conventional analysis. The small amount of flour and ingredients required, have made possible to use this device to test the effect of addition of protein fractions, gluten or specific high molecular weight glutenin subunits on dough properties. For example, while studying the effect of gliadins on mixing properties of flour, Fido *et al.* (1997) found that all groups of gliadins resulted in decreased mixing time and peak resistance, and an increase in resistance breakdown. Khatkar *et al.* (2002) also reported the similar effect of addition of gliadin subgroups on Mixograph parameters. Flour protein content had highly significant and high correlations with Mixograph peak height. Gluten quality, expressed as baking volume, can be predicted by Mixograph parameters alone. Soft wheat patent flours exhibiting longer Mixograph peak times and shorter peak heights have been reported to produce bigger diameter sugar snap cookies (Hou *et al.* 1996). On the other hand Yamamoto *et al.* (1996) reported the positive correlation among Japanese sponge cake volume and Mixograph peak time and stability when influence of flour protein content was eliminated. But it was found to be negatively correlated with Mixograph peak height.

The relationship between Mixograph parameters and indices of wheat grain quality have been reported by several workers. They found that the grain hardness, as expressed by the starch damage percentage, was highly correlated with Mixograph peak height. The width and height of the mixogram was related to the proportion of unaggregated proteins before the mixing peak and to polymeric proteins after the dough consistency reached a maximum. The studies concentrating on composition-derived alterations of Mixograph parameter were conducted by Bekes *et al.* (2001). Only one compositional parameter was altered at a time and all other were kept constant. Authors reported that mixing time and Mixograph peak resistance increased, and resistance breakdown decreased, with an increase in the protein content. At constant protein content, increases in the glutenin to gliadin ratio increased mixing time and Mixograph peak resistance, and decreased resistance breakdown. If both protein content and the ratio of polymeric to monomeric protein were kept constant, higher proportions of HMW-GS in the polymeric protein resulted in increased mixing time and peak resistance, and decreased resistance breakdown.

Farinograph

Hankoczy, a Hungarian investigator, originally developed the Brabender Farinograph, next important recording dough mixer. Farinograph measures plasticity and mobility of dough subjected to a prolonged, relatively gentle, mixing action at constant temperature. It determines the dough properties by measuring the resistance of dough against the mixing action of paddles (blades). It is generally used to determine water absorption required to make a dough, to evaluate the effects of mixing ingredients on mixing properties and to check blending requirement (Fig.3). Results are also used to predict processing effect, including mixing requirements for dough development, tolerance to overmixing, and dough consistency during production. In addition Farinograph is used to predict finished product texture characteristics. For example strong dough properties are related to firm product texture. Weak flour has lower water absorption and short stability time than strong gluten flour.

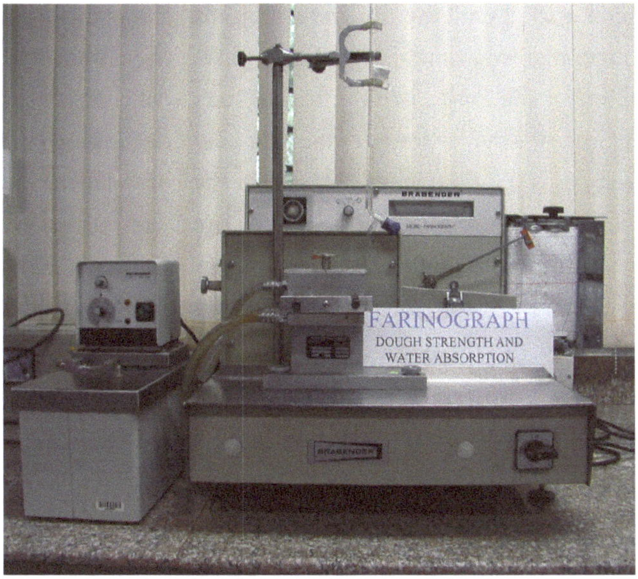

Fig. 3. The Farinograph is used for measuring water absorption and dough strength.

The amount of water added to flour with 14% moisture to produce a maximum consistency of 500 FU is called Farinograph water absorption. In addition to water absorption, the shape of the Farinogram with a maximum consistency of 500 FU is used to characterize flour properties. Variation in Farinograph water absorption is generally recognized to be effected by two major factors, protein content and starch damage. Absorption increases as protein content increases and gluten quality improves. Starch damage content has a positive correlation with Farinograph water absorption but has negative correlation with the dough development time. Flours having the higher flour protein content and starch damage, exhibit higher Farinograph absorption value. Pentosan content has also been reported to have relatively large effect on flours water absorption. Farinograph water absorption has been reported to be positively associated with protein content and negatively to protein quality. Recently equations have been developed to predict Farinograph water absorption based on the chemical constituents of the grain. Details are given in the section on chemical constituents below.

Farinograph curve

Farinograph curve contains peak time, arrival time, departure time, stability time and mixing tolerance index. Peak time indicates dough development time; beginning with the moment water is added until the dough reaches maximum consistency. This gives an indication of optimum mixing time under standardized conditions and is expressed in minutes. Arrival time is the time when the top of the curve touches the 500-BU line. This indicates the rate of flour hydration (the rate at which the water is taken up by the four). Departure time is the time when the top of the curve leaves the 500-BU line. This indicates the time when the dough begins to break down and is an indication of dough consistency during processing. Mixing tolerance index is the difference in BU value at the top of curve at peak time and the value at the top of the curve 5 minutes after the peak. This indicates the degree of softening during mixing. Based on the shape of the curve obtained, Farinograms are divided into 7 different types from very weak to very strong types.

1. Short peak time and short stability
2. Short peak time and long stability
3. Medium peak time and short stability
4. Medium peak time and long stability
5. long peak time and short stability
6. Long peak time and long stability
7. Double peak, swayback or dip in the early part of the curve

There are many intermediate types of curves that are variations or combinations of these general types, but most curves can be placed in one of the seven general categories. In addition Farinograph curve scan be classified into 8 different types in increasing order of strength. These are given in the (Fig.4).

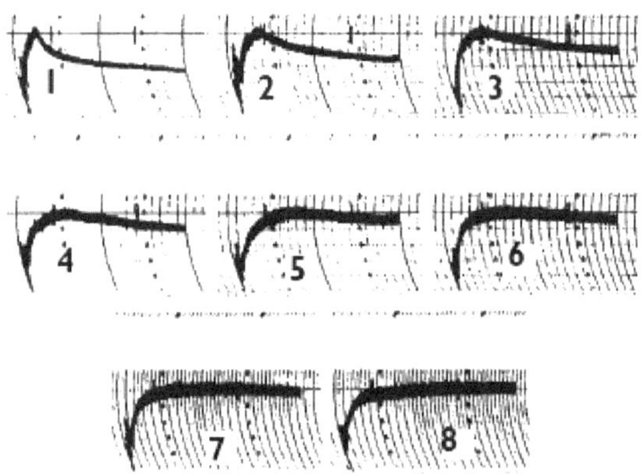

Fig. 4. Reference Farinograms for Hard Red Spring Wheat

The Farinograph analysis of Indian wheats exhibited wide range in Farinograph peak time and stability. Very low elasticity was observed in Nap Hal, an Indian land race of wheat, with Brabender tolerance of 140 BU. Farinograph analysis of other wheat varieties exhibited higher elasticity with tolerance of <100 BU. The peak time to attain the maximum dough consistency was lower (2.0 min.) as compared to other wheat varieties (>2.0 min.). Thus Farinograph curve showed very weak gluten (Lower elasticity) of Nap Hal not observed earlier in other bread wheats (Fig.5). Since, lower elasticity of the dough is preferred for biscuit making quality; Nap Hal can be an excellent source for improving biscuit making quality.

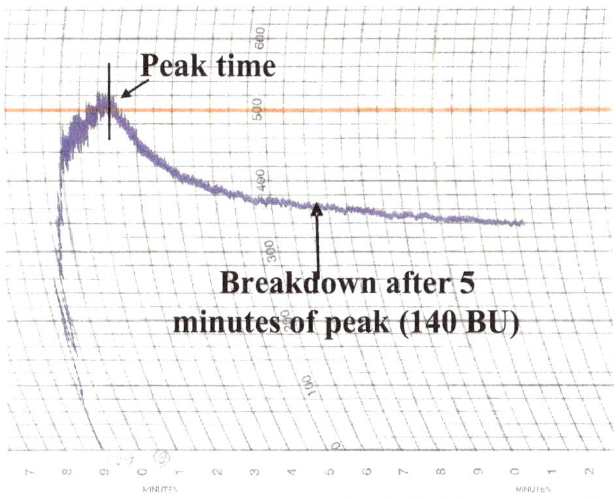

Fig. 5. Farinograph developed using flour of Nap Hal

Alveograph

Alveograph is different from Farinograph and Mixograph in measuring dough strength. In Farinograph and Mixograph, there is uniaxial extension of the dough during mixing while in Alveograph it is biaxial extension: Bubble inflation and lubricated squeezing flow. Bubble inflation is obtained with the Chopin Alveograph (Fig.6). A circular sheet of dough, clamped at its circumference, is inflated by air flowing through a hole in the base plate into an expanding, nearly spherical dough bubble and eventually the bubble ruptures. The excess pressure of the air in the bubble is recorded as a function of time. An accessory, relaxometer makes it possible to stop the flow of air at a chosen volume of the bubble and to record the decay of pressure. Time is a measure of the volume of the bubble and therefore extension of its wall. The excess air pressure in the bubble is a measure of the tensile strength in the wall. A typical alveograph record or alveogram is developed, the maximum height of which is the measure of the resistance and its length is a measure of extensibility. Unlike Brabender extensigraph, the alveogram does not show the effect of flour improvers.

Fig. 6. Alveograph showing the two dimensional expansion of the dough in the form of baloon.

Conventional rheological tests usually operate at small strains in the order of up to 1%, whilst strain in gas cell expansion during proof is expected to be in the region of several hundred percent. Furthermore, many rheological tests are carried out in shear, whilst most large strain deformations in dough are extensional in nature. The alveograph is considered to approximate these conditions best since it can, with certain modifications, be operated at rates and strain close to those observed during baking expansion, and can be used to obtain fundamental tensile rheological properties under conditions close to those of baking expansion. The alveograph consists of three parts: mixer, bubble blower, and a recording manometer. The procedure involves using air pressure to blow a bubble from a disc of flour-water-salt dough. The bubble is expanded to the breaking point, and the recording manometer records a curve from which three basic measurements are taken: distance (in mm) that the dough stretches before it ruptures, resistance to stretching at peak height (associated with dough stability), and area of curve. The area indicates bread making potential and is substantially higher in strong than in weak wheat flours.

Alveograph parameters have been widely used to assess the bread making quality of wheats. Highly significant correlation between W value and loaf volume has been reported for flours. Generally correlations are stronger when samples of same varieties were compared than different varieties of varying quality are used. However, Schoeggl (1998) obtained positive correlations between deformation value of the alveogram and parameters linked to the protein complex across 500 samples of wheat and further concluded that quality of Australian wheat could be satisfactorily assessed by Alveogram data. It is well known that grain hardness has a strong influence on flour granulometry, starch damage, dough hydration for a given consistency and finally rheological properties. Because the alveograph test was performed at constant hydration, grain hardness strongly influenced the dough tenacity (P) and dough strength (W).

Different values are taken from the analysis of the alveogram. To simplify, the values are:

- P — Stands for maximum pressure and shows the strength of the dough
- L — Stands for the height of the bubble achieved, measured from where the slope of the bubble started to the top of the bubble. It measures the extensibility of the dough
- P/L — Stands for the ratio of P to L;
- G — Stands for the square root of the volume of air in the bubble at the time of it's burst. It shows the flexible the dough;
- W — Stands for the energy needed to make the bubble as big as it could go before it burst. It measures the strength of the flour.

Extensograph

The extensigraph/extensograph records a force-time curve for a test piece of dough stretched until it breaks. It measures the stretching properties of wheat flour doughs, in particular the resistance to extension and the extensibility, and for gaining reliable information about the baking behavior of the dough. Characteristics of force-time curves, or extensigrams, are used to assess not only the general quality of flour but also its response to improving agents. The test is conducted as per the AACC Method

54-10 in which the doughs for testing the extensigraph are first prepared in Farinograph mixer, followed by rounding of the dough and subsequently load extension. Water addition is adjusted to make the final consistency of 500FU. The typical extensograph curve is shown (Fig.7).

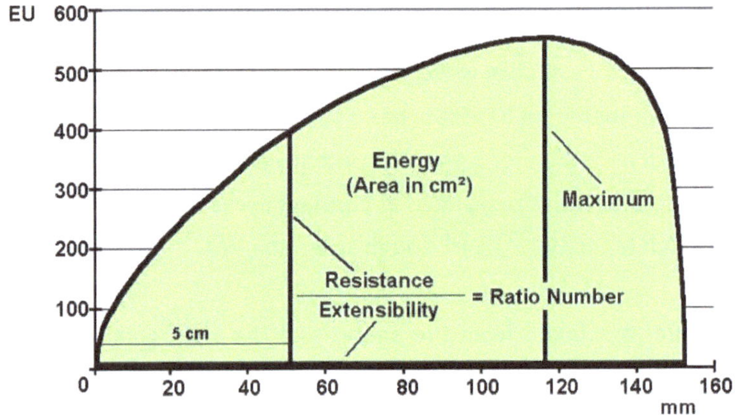

Fig. 7. Brabander[R] Measurements and Control System.

The four most common measurements made on force-time charts or extensigrams, are:

1. Resistance to extension (Rm): height of curve in BU or in cm either at maximum or at 5 cm from where test starts on chart.
2. Extensibility (E): total length of curve in cm. There is some "noise" in curve after dough breaks. This should be ignored.
3. Evaluate area under curve with planimeter and report in square cm. This area is proportional to the energy that is required to bring about rupture of the test piece along the predetermined path. Thus it is single most convenient figure for characterizing dough strength which is associated with large resistance and large extensibility.
4. Ratio between resistance and extensibility characterising the shape of the extensogram and the extension at the maximum force.

Based on the extensograph profiles developed from different wheat varieties; three main groups of profiles can be identified. The one group (A) includes varieties with strong extensible, elastic dough suited for long fermentation process, large proving tolerance, light voluminous baking product with higher volume. The second group (B) includes rigid and tough dough structure with poor extensibility and poor rising during proving and has lower baking volume. The third group (C) includes wet elastic and soft dough with narrow fermentation tolerance and small baking volume. The three main extensograph profiles are presented in the (Fig.8). (Brabander^{R} Measurements and Control System). Recently computerized extensograph has been developed where data is recorded and displayed.

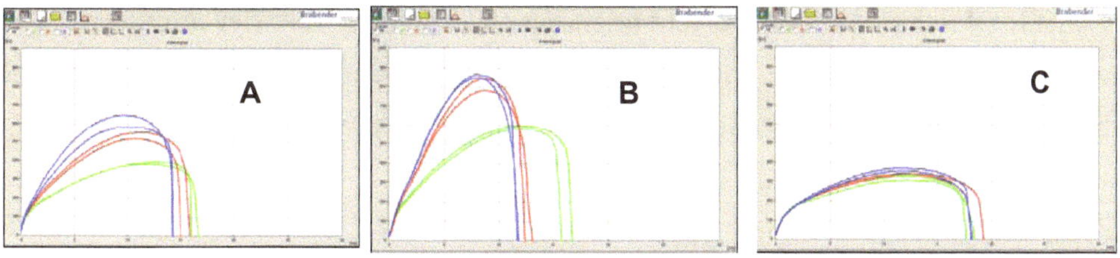

Fig. 8. Extensograph profiles developed from different wheat varieties A, B and C

Application of Rheological Tests

The tests with the instruments described above describe information about flour strength, balance of properties and consistency. Strong wheats are generally characterized by high water absorption, long dough development time and lower mixing tolerance index in dough mixing tests. In load extension tests, strong wheats are characterised by large resistance and large extensibility. Weak flours lack these characteristics. Strong flours are used in bread industry for making good quality breads with higher volume. Strong flours are tolerant to over mixing and can improve quality of weak flour by blending. Load extension tests with untreated flours give information on the inherent balance of or imbalance of the dough properties. They guide the millers in determining how to compose grist in which defects of one wheat lot are compensated by the attributes of another lot and how to exploit fully the potential in available raw materials. Moreover, physical dough tests are used to predict quality of baking products without baking tests. For example loaf volume can be predicted by rheological tests.

Chemical Tests

The development of fast prediction methods for end use quality continues to be a major focus of wheat breeding programs. Many small scales, predictive tests for wheat quality have been adapted from large-scale tests. Those methods share some of the basic attributes: they require relatively small amounts of grain, wheat meal, or flour; are rapid; and provide information that can be used to improve the end use quality potential of breeding material. In early segregating generations when small quantity of materials is available and short time is left in between next sowing, chemical tests are very useful in predicting the quality. Depending upon the chemical constituents of the grain, large numbers of tests have been developed to predict the quality traits. For example sedimentation tests were developed for gluten strength and flour swelling volume for starch properties. Solvent retention capacity tests were developed to predict the quality of biscuit flours. Later these tests were extended to other products as well. The details of important chemical tests are described below.

Sedimentation Test

Large numbers of rheological tests are conducted to know mixing properties of the dough as well as predicting gluten strength and end product quality. However these tests are time consuming and require larger quantity of material, their utility in predicting quality of early generation line in breeding programmes are limited. Therefore, microlevel tests developed based on the chemical constituents of the grain determining end product quality have larger utility in breeding. In the early stage chemical tests were developed based on swelling property of gluten in different solutions and showed strong relations with bread making quality. Later these tests were also used in durum quality improvement for pasta products. The Zeleny and SDS sedimentation tests measure the sedimentation value of a suspension of flour in dilute lactic acid and in SDS-lactic acid solvents, respectively. The sedimentation value depends mainly on the amount and the swelling characteristics of the glutenin, since other proteins such as gliadin are soluble in the Zeleny and SDS test solutions. The SDS sedimentation test correlates better with loaf volume than the Zeleny sedimentation test Axford *et al.* (1979). Therefore a number of modified SDS sedimentation tests have been developed and are widely used in predicting dough properties and bread making qualities in the early stages of wheat breeding programmes Weegels *et al.* (1996). Cultivars with different protein quality, as expressed by their gluten characteristics, should be differentiated by the SDS sedimentation test.

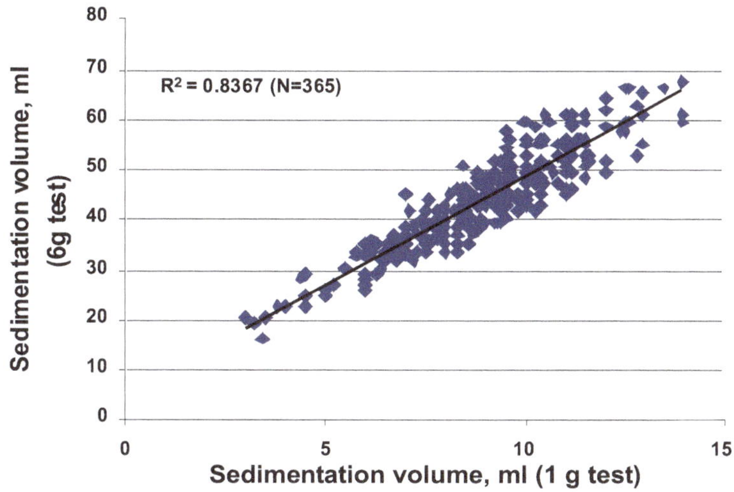

Fig. 9. Relationship between 1g and 6g tests for sedimentation volume

Sedimentation test using 1 g of the whole meal flour has strong correlation with sedimentation test using 6 gram whole meal four (Fig. 9). Though SDS sedimentation test assesses protein quality, protein content also influences its value. The effect of protein concentration on SDS sedimentation volume may be eliminated by multiplying SDS sedimentation volume by the grain protein concentration of the sample divided by 10 Baik *et al.* (1994). SDS sedimentation volumes also have been influenced by environment, crop year, and their interaction with cultivar. Nevertheless, SDS sedimentation volumes are highly heritable and can be used for selecting among early generation progeny Matuz, (1998). Estimates of different quality components including heritability are shown in the (Table 1). High SDS sedimentation values have been associated with stronger gluten and superior pasta making quality in durum wheat. Since 6 g test requires larger sample weight and comparatively fewer samples can be analyzed in day, 1 gram test is more useful in screening recombinant lines in breeding programme.

Table 1: Estimates of different grain quality components. sg^2 = Genotypic variance; se^2 = Environmental variance, h^2 = Heritability.

Traits	Mean ± SE	óg^2	óe^2	h^2
SKCS hardness index	78.7 ±0.3.20	209.19	20.44	0.91
Sedimentation value (ml)	9.5 ±0.66	31.95	8.63	0.79
Grain protein content (%)	11.9 ±0.56	0.70	0.63	0.53

Swelling Index of Glutenin (SIG)

Recently a new method has been developed to estimate insoluble glutenin using 40 mg flour sample, called as swelling index of glutenin (SIG) by Wang and Kovacs (2002a). The SIG values are correlated significantly with SDS sedimentation value and insoluble glutenin content. Sedimentation and SIG value were reported to have strong correlation with loaf volume as well. Strong positive relationships between insoluble glutenin content and loaf volume have also been found by many workers. The results also indicated that the SIG test, similar to SDS sedimentation test, strongly correlated to Mixograph dough development time and alveograph W index. Analysis of 400 wheat genotypes in India demonstrated strong positive correlation between SIG and sedimentation value (Fig 10). Because only 40 mg of the flour is used and large number of samples can be analyzed in a short time with excellent reproducibility, SIG can be very useful in screening breeding lines, predicting the quantity and quality of insoluble glutenin.

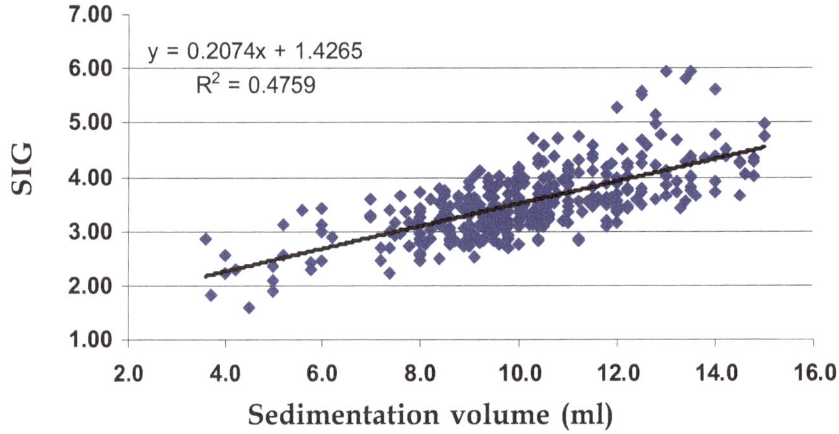

Fig. 10. Correlation between Sedimentation Volume (ml) using 1 gram test and Swelling Index of glutenin (SIG)

Wet Gluten and Gluten Index

When flour is mixed with the water, a viscoelastic complex called gluten is developed. Strong and extensible gluten is suitable for bread making while week and highly extensible for soft wheat products such as biscuits, white noodles, cakes and pretzels. Strength of the gluten is genetic trait and depends on the polymer forming abilities of HMW and

LMW glutenin subunits. Wet gluten is washed from whole-grain wheat meal or flour by an automatic gluten washing apparatus (Glutomatic) and centrifuged on an especially constructed sieve under standardized conditions. The weight of wet gluten forced through the sieve and the total weight of wet gluten (passed through and remaining on the sieve) are weighed. The total wet gluten is then dried under standardized conditions and weighed. The difference between the weights of total wet gluten and total dry gluten is calculated, which gives the water bound in the wet gluten, referred to as water-binding capacity. Total wet gluten and total dry gluten contents are expressed as percentages of the sample. The gluten index is the ratio of the wet gluten remaining on the sieve (after centrifugation) to the total wet gluten. This method is applicable to wheat meal (ground whole grain) and flour.

Calculations

Calculate total wet gluten, gluten index, dry gluten, and water binding in wet gluten as follows:

Wet gluten content (%) (14% moisture basis)

$$= \frac{\text{Total wet gluten (in g)} \times 860}{100 - \%\ \text{sample moisture}}$$

Gluten index

$$= \frac{\text{Wet gluten remaining on the sieve (g)} \times 100}{\text{Total wet gluten}}$$

Dry gluten content (%) (14% moisture basis)

$$= \frac{\text{Total dry gluten (g)} \times 860}{100 - \%\ \text{sample moisture}}$$

Water binding capacity (Water binding in wet gluten) % = Wet gluten content % - dry gluten content %

Solvent Retention Capacity Test

The solvent retention capacity (SRC) test is a relatively new method for evaluating wheat quality (Slade and Levin 1994, Gaines 2000). The SRC test is essentially the same as the Alkaline Water Retention Capacity test; however different aqueous solutions are utilized. In this test, four different solutions as lactic acid, sucrose, water and sodium carbonate are utilized for providing information on chemical and physical aspects of wheat samples. Each solvent provides information on a different chemical or physical aspect of the sample. Generally lactic acid SRC is associated with gluten strength, sodium carbonate SRC is associated with levels of damaged starch and, indirectly, hardness, sucrose SRC is associated with pentosan and gliadin content, and water SRC is influenced by all of these flour constituents Gaines (2000). Damaged starch absorbs much more water than undamaged starch and increased damaged starch is detrimental for cookie quality. Membrane associated pentosans were found to be most influential in accounting for variation (69%) in alkaline water retention capacity. Total pentosans, together with flour protein, accounted for 87% of the variation in cookie diameter for soft wheat samples, with the total pentosan fraction being the more influential Bettge and Morris 2000, Ram and Singh (2004). Therefore, low starch damage as well as low sodium carbonate and sucrose SRC are desirable for confectionery products. SRC tests can effectively differentiate soft wheat genotypes into groups of similar end use quality with very limited genotype into environment interaction Guttieri *et al.* (2001).

Flour extraction and sodium carbonate SRC were negatively correlated, and whole grain hardness measurements were negatively correlated with sucrose and sodium carbonate SRC. This was unexpected, as harder grains should produce flours with higher starch damage. Earlier Bettge and Morris (2000) also observed a negative correlation of grain hardness with membrane bound pentosan content among soft wheat cultivars. Authors suggested that this inverse relationship might be due to differences in amyloplast membranes, such as composition of membrane lipids. Guttieri *et al.* (2004) studied sources of variation in SRC test of wheat flour. Inbred lines from three soft spring wheat populations were assessed. Results indicated significant variance among genotypes for water, sodium carbonate, sucrose, and lactic acid SRC values for all the three populations. A small-scale version of SRC test using wheat meal was developed by Bettge *et al.* (2002). Significant correlations were observed between 1g and 5g wheat meal SRC.

The authors also tested a 0.2g method with wheat meal in a micro-centrifuge tube. Results indicated that advancing promising germplasm or eliminating germplasm with inferior end use quality potential was possible using the small scale SRC technique to evaluate early generation wheat breeding material. Ram and Singh (2004) reported significant correlations between whole meal flour and flour SRCs using 1 gram material. Whole meal flour test can be used more effectively as large number of samples can be ground using small mills as Cyclotec Mills. SRC profile of Indian wheat varieties was developed and three clusters were made representing different set of genotypes and each cluster having distinct profile. This information has been useful in developing breeding strategy for the improvement if Indian wheat cultivars for different products.

SRC tests can also be used in estimating Farinograph water absorption and Mixograph properties. Mixograph and Farinograph are used to measure functional properties of flour including water absorption and to differentiate wheat flours of good and poor baking quality. The determination of optimum water needed to obtain a certain dough consistency is essential to give dough with optimal handling characteristics suitable for rheological testing as well as product quality. Different flour constituents such as polysaccharides, pentosans, starch damage and protein content and kernel size have been reported associated with water absorption. Since rheological tests such as Farinograph and Mixograph tests are time consuming, labour intensive and expensive, small-scale tests are required to predict the end-use quality of grain from early generation breeding material. Further, there is a major problem in finding optimum water absorption of flour for use in Mixograph and Farinograph. The limited quantity of grain available from early generation of segregating materials and the longer time required to mill thousands of samples are other constraints for analyzing breeding materials with ease. Based on the SRC tests and using multiple regression analysis, Ram *et al.* (2005) developed an equation to predict Farinograph water absorption utilizing flours from 192 wheat varieties released in India (Fig. 11). The equation found to be very useful in predicting water absorption. Development of prediction models can be very valuable due to objective testing of flour quality, especially when the sample amount is limited and the numbers of samples to be tested are excessive as in early generation lines in wheat breeding programme.

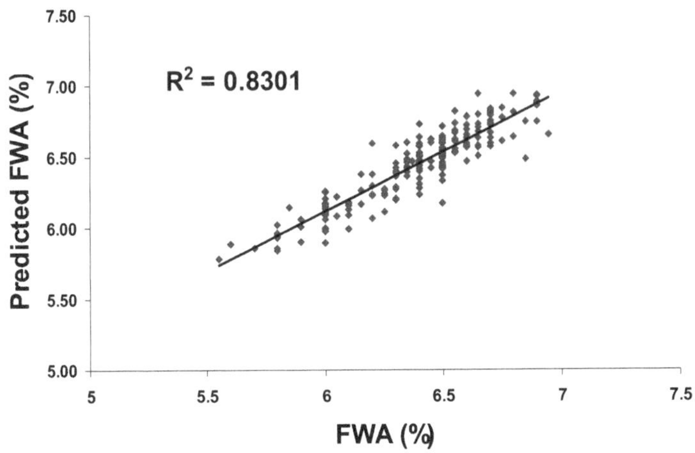

Fig. 11. Equal to predict farinograph water absorption.

Starch Gelatinization, Swelling and Hot Paste Viscosity Properties

Starch granule gelatinization involves the heat driven transformation of granules in excess of water from an ordered state to a disordered state. The initial stages of gelatinization involve the uptake of water, lowering the glass transition temperature in the amorphous regions and causing a crystalline region to begin to melt. Further heating and water uptake results in the mobilization of amylose and the initiation of its leaching from the granule and the complete melting of the crystalline regions, marked by the loss of birefringence. Further swelling of the granules occurs, typically producing greatly enhanced viscosity in the starch water paste. The overall process of gelatinisation is similar in both normal and waxy starch granules, although by definition, amylose is not present in a waxy starch and cannot be lost by leaching. These observations suggest that the properties of the amylopectin fraction of starch are most critical in controlling gelatinisation and granule swelling in cereal starches. Amylose content has a major influence on starch granule swelling, mediated through the reduced mobility of amylose/ lipid complexes in the granule that restrict water movement and swelling.

The pasting of starch is typically analysed in an instrument such as the Visco-amylograph or Rapid Visco Analyzer (RVA) which measures the resistance of starch to shearing forces under defined hydration and temperature regimes Kornik *et al.* (1992). Following complete gelatinization at high temperature, peak viscosity is reached; the starch is then cooled while continuing to be stirred. Further cooling of starch paste to ambient temperature results in the formation of either a gel in the case of starches containing amylose or a very week gel in the case of waxy starches. However, the analysis time and sample size required for these instruments is too restrictive for them to be used routinely in a wheat breeding programme for early generation testing. Therefore, simple test for measuring starch paste property is required. Flour swelling volume test has been developed for assessing the quality wheat flours of different varieties. Flour swelling volume (FSV) test measures the cumulative effect of starch quality, specially the amylose/amylopectin ration as reflected by the volume of gel produced when flour is heated in an excess of water. Starch composition has the greatest effect on FSV. FSV is related to the eating quality (texture) of white salted noodles as well as to the gene dosage of granule bound starch synthase that controls the amylose/amylopectin ratio. This method is applicable to wheat meal, flour, or starch.

Flour Swelling Volume

Gelatinization and swelling of wheat starch has an important effect on the quality of end-use products, especially on the quality of white salted noodles. The flour swelling volume (FSV) test measures the cumulative effect of starch quality, specifically the amylose-amylopectin (AM-AP) ratio as reflected by the volume of gel produced when flour is heated with an excess of water. Starch composition (*i.e.*, the AM-AP ratio) has the greatest impact on FSV. FSV is related to the eating quality (texture) of white salted noodles as well as to the gene dosage of granule-bound starch synthase that controls the AM-AP ratio. FSV is not much affected by mild to moderate preharvest sprouting. This method is applicable to wheat meal, flour, or starch. It may be useful in monitoring the impact of starch additives as well. Starch with high pasting viscosity in wheat flour is beneficial in the production of noodles with a smooth surface and soft texture. As the swelling ability of starch granules is largely responsible for the degree of paste viscosity observed when a starch suspension is heated, it is suggested to use flour swelling power as a simple test to assess the inter-cultivar differences in starch properties. Application of the swelling power test to starch or flour derived from samples of individual cultivars indicates a strong association between measurements of swelling power and noodles texture Crosbie *et al.* (1992). High swelling degree is positively related to the softness of Japanese white salted noodles and negatively correlated with the firmness of yellow alkaline noodles.

Flour swelling Volume is simple test and is determined by AACC Method 56-21. In this test 0.40 g of flour is mixed with water two times of 6.25ml each and vortexed. Tubes are equilibrated at 25°C and immersed in hot water bath at 92.5°C. Tubes are mixed continuously for the first 2 min, and then inverted twice after a total immersion time of 3, 4, 5, 10, 15, 20, and 25 min. The slurry should be kept in constant motion for the first 2 min to ensure even heating. After 30 min tubes are placed in ice-water bath for 1 min, followed by immersion in a 25° water bath for 5 min to bring the sample back to room temperature. Then samples are centrifuged at 1,000 × *g* for 15 min, using a swinging bucket rotor and the height of the gel is measured in mm to the top of the gel-water interface. Flour swelling volume is calculated from the height of the gel by using the regression equation derived from the calibration of tubes.

Rapid visco-analuzer measures the viscosity of flours or starch pastes during gelatinization in a manner similar to amylograoph. The behavior of starch in water during heating has been related to textural properties of certain types of noodles, especially soft textured Japanese noodles. Starch with high pasting viscosity in wheat flour is beneficial in the production of noodles with a smooth surface and soft texture. As the swelling ability of starch granules is largely responsible for the degree of paste viscosity observed when a starch suspension is heated Crosbie *et al.* (1992). During this process a graph is developed containing peak viscosity, trough and final viscosity reflecting quality of starch. Large variations exist in starch pasting properties in Indian wheats and thus flour from different sources can be useful in improving noodle quality. However, the analysis time and sample size required for these instruments is too restrictive for them to be use routinely in a wheat breeding programme for early generation testing. Therefore, simple test for measuring starch paste property is required. Flour swelling volume test has been developed for assessing the quality wheat flours of different varieties. Flour swelling volume (FSV) test measures the cumulative effect of starch quality, specially the amylose/amylopectin ration as reflected by the volume of gel produced when flour is heated in an excess of water. Correlations were studied among different parameters of RVA and FSP.

Strong significant positive correlations were observed between peak viscosity in RVA and FSP ($R^2=0.374$; $p<0.001$). Since FSP needs only 40 mg of the flour and easy to do as well as large number of samples can be done in a day, it is very useful in breeding for improving noodle making quality. Earlier reports indicated positive correlation between FSP and noodle quality (Kornik 1992, Crosbie *et al.* 1992, Fu *et al.* 1998). Strong correlations were also observed between peak viscosity and final viscosity (R^2=0.4811; <0.001). Moreover, large variability has been found in both FSP and peak viscosity which can be a source for enhancing noodle making quality. FSP was measured using both whole meal flour and flour to identify any relationship between them. There was highly significant positive correlation between flour FSP and whole meal FSP (R^2=0.479; $p<0.001$) (Fig. 12). This has additional advantage of predicting noodle starch quality for wheat products using simple whole meal flour test. This is because whole meal can be obtained with very small amount and several hundred whole meal samples can be ground in a day. The data thus demonstrated the utility of whole meal tests in predicting wheat quality based on starch swelling.

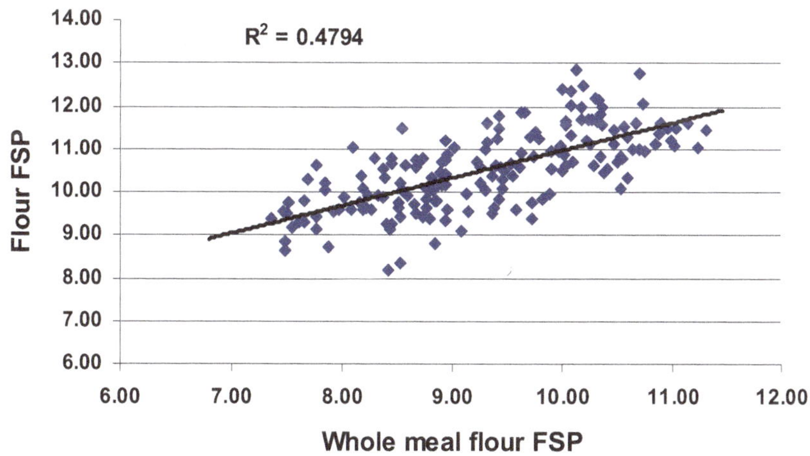

Fig. 12. Relationship between whole meal flour FSP and flour FSP

Summary

Rheological properties of dough determine the behaviour of dough during mixing, rounding and molding used in making different products and also effect on the quality of finished product. Therefore, understanding the various features of dough rheology and their relations with product quality is important. Recently progress has bee made in reducing the quantity of the flour used in measuring rheological properties and also the computerization of the data recording and analysis. 2 g mixograph has shown promise in differentiating strong and weak flour and hence useful in evaluating early segregating lines. The small scale tests based on the chemical constituents of the grain have been developed to generate profile of different classes of wheat and understand their potential for different products. Solvent retention capacity test, micro-sedimentation test, swelling index of glutenin (SIG), gluten index (GI) and flour swelling power (FSP) explain large variation in different traits and hence useful in predicting quality of early segregating lines in breeding programme. These methods share some of the basic attributes: they require relatively small amounts of grain, wheat meal, or flour; are rapid; and provide information that can be used to improve the end use quality potential of breeding material. This further enhances the utility of these tests in quality improvement.

References

Axford, D.W.E., McDermott, E.E. and Rehman, D.G. (1979). Note on the sodium dodesyl sulphate test of bread making quality. Comparison with Pelshenke and Zeleny tests. *Cereal Chem.* 56: 582-584.

Baik, B.K., Czuchajowska, Z. and Pomeranz, Y. (1994). An SDS-FY test to evaluate quality of wheat for oriental noodles. *J. Cereal Sci.* 19: 191-201.

Barnes, H.A., Hulton, J.F. and Walters, K. (1989). An Introduction to Rheology. Elsevier, New York.

Bekes, F., Grass, P.W., Anderssen, R.S. and Appels, R. (2001). Quality Traits of Wheat Determined by small scale dough testing methods. *Aust. J. Agric. Res.* 52: 1325-1338.

Bettge, A.D., Morris, C.F., DeMacon, V.L. and Kidwell, K.K. (2002). Adaptation of AACC method 56-11, solvent retention capacity, for use as an early generation selection tool for cultivar development. *Cereal Chem.* 79(5), 670-674.

Bloksma, A.H. and Bushuk, W. (1988). Rheology and chemistry of dough pp 131-218. In Wheat Chemistry and Technology, 3rd edition, Ed. Y. Pomeranz, American Association of Cereal Chemists, St. Paul, M.N.

Corbellini, M., Empilli, S., Vaccino, P., Brandolini, A., Borghi, B., Heun, M. and Salamini, F. (1999). Einkorn characterization for bread and cookie production in relation to protein subunit composition. *Cereal Chem.* 76 (5): 737-733.

Crosbie, G.B., Lambe, W.J., Tsutsui, H. and Gilmour, R.F. (1992). Further evaluation of the flour swelling volume test for identifying wheats potentially suitable for Japanese noodles. *J. Cereal Sci.* 15: 271-280.

Dobraszczyk, B.J. (2001). Wheat and Flour. In Cereals and Cereal Products : Chemistry and Technology pp 100-139. Eds. B.J. Dobraszczyk and D.A.V. Dendy, Aspen Publishers, Inc. Maryland.

Dobraszczyk, B.J. and Morgenstern, M. P. (2003). Rheology and The Breadmaking Process. *J. Cereal Sci.* 38(3): 229-245.

Ferry, J.D. (1980). Viscoelastic Properties of Polymers. John Wiley and Sons, New York.

Fido, R.J., Bekes, F., Gras, P.W. and Tatham, A.S. (1997). Effect of α-, β-, γ- and ω- gliadins on the dough mixing properties of wheat flour. *J. Cereal Sci.* 26: 271-277.

Fu, B.X., Kovacs M.I.P. and Wang C. (1998). A Simple Wheat Flour Swelling Test. *Cereal Chem.* 75: 566-567.

Gaines, C.S. (2000). Collaborative Study of Methods for Solvent Retention Capacity Profiles (AACC Method 56-11). *Cereal Foods World* 45(7): 303-306.

Guttieri, M.J., Becker, C. and Souza, E., (2004). Application of Wheat Meal Solvent Retention Capacity Tests Within Soft Wheat Breeding Populations. *Cereal Chem.* 81(2), 261-266.

Guttieri, M.J., Bowen, D., Gannon, D., O'Brien, K. and Souza, E. (2001). Solvent Retention Capacities of Irrigated Soft White Spring Wheat Flours. *Crop Sci.* 41,1054-1061.

Hou, G., Yamamoto, H. and NG, P.K.W. (1996). Relationship of quantity of glutenin subunits of selected U.S. soft wheat flours to rheological and baking properties. *Cereal Chem.* 73: 358-363.

Khatkar, B.S., Fido, R.J., Taham, A.S. and Schofield, J.D. (2002). Functional properties of wheat gliadins. 1. Effecs on mixing properties and bread making quality. *J. Cereal Sci.* 35: 299-306.

Kornik, C.M., Miskelly, D.M. and Gras, P.W (1992). Contribution of starch and non-starch parameters to the earing quality of Japanese white salted noodles. *J. Sci. Food Agri.* 58: 403-406.

Matuz, J. (1998). Inheritance of SDS sedimentation volume of flour in crosses of winter wheat (*Trticum aestivum* L.) *Cereal Res. Commun.* 26: 203-210.

Pomeranz, Y. (1987). Grain quality. In: Modern Cereal Science and Technology pp: 96-105. (Ed. Y. Pomeranz). VCH Publishers Inc.

Ram, S. and Singh, R.P. (2004). Solvent retention capacities of Indian wheats and their relationship with biscuit making quality. *Cereal Chem.* 81(1): 128-133.

Ram, S., Dawar, V., Singh, R.P. and Jag Shoran (2005). Application of solvent retention capacity tests for the prediction of mixing properties of wheat flour. *J. Cereal Sci.* 42 (2): 261-166.

Schoeggl, G. (1998). Investigation of the alveogram method for the evaluation of wheat quality. *Getreide Mehl und Brot.* 52: 218-223.

Shuey, W.C. (1975). Practical instruments for rheological measurements on wheat products. *Cereal Chem.* 52: 42-81.

Slade, L. and Levine, H., (1994). Structure-function relationships of cookies and cracker ingredients. in: The science of cookies and cracker production pp 23-141. Ed. H. Faridi Chapman and Hall: New York.

The Mixograph Handdbook, (1997). Edited by C.E. walker, J.L. Hazelton and M.D. Shogren. pp 70.

Wang, C. and Kovacs, M.I.P. (2002a). Swelling index of glutenin test. I. Method and comparison with sedimentation, gel-protein, and insoluble glutenin tests. *Cereal Chem.* 79: 183-189.

Weegles, P.L., Hamer, R.J. and Schofield, J.D. (1996). Functional properties of wheat glutenin. *J. Cereal Sci.* 23: 1-18.

Yamamoto, H., Worthington, S.T., Hou, G. and NG, P.K.W. (1996). Rheological properties and baking qualities of selected soft wheats grown in the United States. *Cereal Chem.* 73: 215-221.

❏❏❏

Chapter 6

Gluten Structure, Genetics and Functionality

Introduction

Wheat is the most important cereal crop to the human kind. It is the main source of energy and nutrition in human diet. Wheat is consumed in the form of large number of end products as chapati, bread, biscuit, cakes, pretzels, noodles and pasta products. Millions of people are involved in milling, baking and pasta products industry with huge turnover per annum. However, each end-use product of wheat has it's own specific requirement for example basically hard wheats along with strong gluten and high protein content are used for bead making. Soft wheat with weak gluten and low protein content are required for biscuit making Souza *et al.* (1994). Durum (tetraploid) wheat is used for pasta products as macaroni, vermicelli and spheghatti and Dicoccum for therapeautic purposes. Often chemical improvers are used to improve the dough properties for these products. Most of these chemicals have been banned because of their toxic effects. Therefore, only alternative is to develop varieties suited to meet the requirements of these end-use products. This needs better understanding of molecular basis of grain quality to accelerate breeding for producing wheat varieties for each end-use product separately. Storage proteins are the major component of grain which determines end-product quality. In this chapter detailed structure, genetics and functionality of gluten proteins are described.

Storage Proteins

Wheat endosperm proteins were among the first proteins to be studied when Beccari in 1745 described the separation of wheat flour into starch and gluten. At the beginning of 20th Century, a systematic study was conducted by Osborne (1907) to develop a classification of cereal seed proteins based on their sequential extraction and differential solubility. He classified the proteins into different solubility groups i.e. albumins (soluble in water and dilute buffers), globulins (not soluble in water but soluble in saline solutions), prolamins (which are soluble in 70-90% alcohol), and glutenins (which are soluble in dilute acids or alkalis). Although Osborne's fractionation was a major milestone in the development of cereal chemistry, there was overlapping in the solubility of these fractions. This is particularly true for the gliadin and glutenin proteins. For example, Beckwith *et al.* (1966) detected by gel filtration a small fraction of the ethanol-soluble proteins as an excluded peak, which he called gliadin of high molecular weight. Nielsen *et al.* (1968) found that these proteins behaved like glutenins in showing a drastic decrease in viscosity after the reduction of disulfide bonds and therefore called them glutenins of low molecular weight. Later, Jackson *et al.* (1983) using 2D electrophoresis showed that the constituent subunits of this fraction were identical to low molecular weight glutenins (LMW-GS) and different from the gliadin polypeptides. Field *et al.* (1983), using 50% 1-propanol instead of 70% ethanol, also found some HMW-GS along with the LMW-GS within the alcohol soluble fraction. After the reduction of disulfide bonds, all gluten proteins are soluble in 70% ethanol or other alcohol such as n-propanol as individual polypeptide chains (Kreis *et al.* 1985). They have thus been classified as prolamins Shewry *et al.* (1986).

The existence of close similarity in structure between LMW-GS and gliadins is another reason to consider them within the same group (prolamins). Within this group, further differences between them are based on biochemical characteristics (sulfur-rich, sulfur poor and HMW prolamins). Ewart (1990) differed from Shewry's classification, and argued that the polymeric glutenins are fundamentally different from the monomeric gliadins because of their intermolecular disulfide bonding capacity MacRitchie, (1992). While gliadins are single polypeptides chains (monomeric proteins), the glutenins are multichained structures that are held together by disulfide bonds. The very high molecular weight of these polymeric structures is responsible for their partial insolubility and for their distinct contribution to functionality compared with that of the gliadins. Therefore, the classification of these proteins into monomeric and polymeric forms is a good indicator of dough functional properties Payne and Lawrence, (1983). Nevertheless, there is consensus in that all gluten proteins are high in proline and glutamine contents and that should be the predominant basis for calling them prolamins, the name being derived from the combined names of these amino acids. Structure, genetics and functionality of glutenins and gliadins are described below.

High Molecular Weight Glutenin Subunits (HMW-GS) Structure and Genetics

Glutenins (acid soluble) are polymeric proteins whose monomeric units are divided into high (HMW) and low (LMW) molecular weight glutenin subunits and their molecular weight ranges between 67-130kDa and 35-45kDa, respectively. HMW-GSs represent 5-10% of total seed proteins depending upon the number of expressed genes present. The genes for HMW-GS are present at the complex Glu-1 loci on the long arm of the homoeologous group 1 chromosomes Payne *et al.* (1987). Each locus contains two tightly linked genes encoding two subunits designated as x and y type according to their decreasing molecular weights (Fig. 1) and each gene accounting for up to 2% of total proteins. The x-type subunits generally have a slower electrophoretic mobility in SDS- PAGE and higher molecular weight than the y-type subunits. They differ in their N-terminal domains and their repeat motifs, Shewry *et al.* (1992). HMW-GS consist of three domains; a central domain varying in length from 440-680 residues flanked by non-repetitive N and C terminal domains of 81-104 and 42 residues respectively (Fig. 2). The allelic variation in HMW-GS size is caused by deletions or duplications within the repetitive domain by recombination events. Both contain hexapeptides (consensus PGQGQQ) and nonapeptides (x-type consensus GYYPTSLQQ and y-type consensus GYYPTSPQQ) and in x- types only a tripeptide (consensus GQQ). Further, the repetitive domain in the y-type subunit is shorter than the x-type and the N-termimal domain of y-type is larger. C-terminal domains in both the x and y type subunits are of same size.

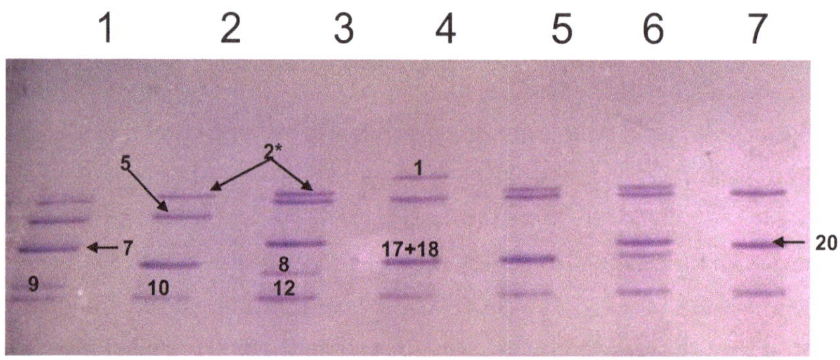

1=PBW443, 2=GW273, 3=NW1014, 4=K9465, 5=K9644, 6=NW1012, 7=C306

Fig. 1. SDS-PAGE separation of high molecular weight glutenin subunits from some of the wheat varieties released in India.

Electrophoretic studies have revealed appreciable polymorphism in number and mobility of HMW-GS in both bread wheats Lawrence *et al.* (1988) and pasta wheats Kovacs *et al.* (1993). Payne and Lawrence (1983) summarized the range of alleles at the Glu-1 loci as three allelic forms at the Glu-A1 and 11 alleles at the GluB1 and six alleles at the Glu-D1. In hexaploid wheat six HMW glutenin genes are present, but only those coding subunits 1Bx, 1Dx are always expressed whereas the 1Ax and 1By are not always expressed, and the 1Ay gene is not expressed. Non-expression of 1Ay gene may be caused by nucleotide changes in the promoter region Ford *et al.* (1985), by the presence of transposon like insertion in the coding region Harberd *et al.* (1987) or presence of stop codon in down stream region Bustos *et al.* (2000). Therefore, introduction of both Ax and Ay subunits from other sources as tetraploids or diploids wheats into hexaploid wheats can lead to enhanced genetic variability and improved technological properties Li *et al.* (2006).

In general there is little variation in the amount of individual subunits synthesized when alleles of the same locus are compared. However, genotypes from Israel (Line TAA36), Canada (Wheat cultivars such as Glenlea, Roblin and Bluesky) and the United States (Cultivar Red River 68) over express 1Bx subunit 7 when compared with other HMW-GS. In Red River 68, large proportion of subunit 7 is due to gene duplication. Israeli land race TAA36 showed similar results. On the other hand subunit 7 from Glenlea is encoded by a single gene dose. In this case either a more efficient transcription or more efficient translation is the cause of over-expression Anjum *et al.* (2007).

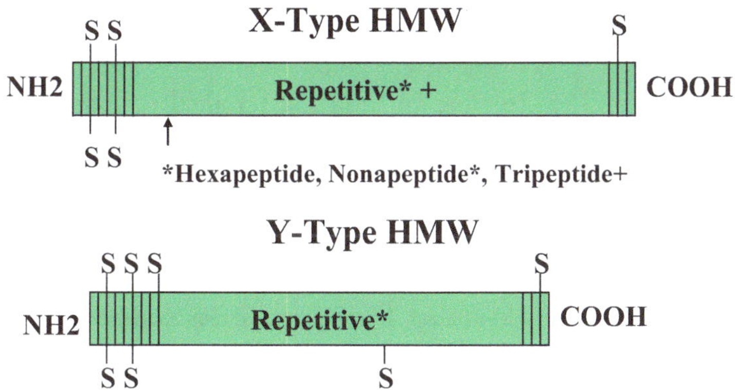

Fig. 2. General features of x and y type high molecular weight glutenin subunits showing repetitive domain of hexapaptide and nonapeptide in y subunit and tripeptide along with hexapeptide and nonapeptide repeat motifs in x type subunits in the center and N and C terminal regions having cysteine residues.

Recent developments in molecular techniques such as cloning and sequencing have boosted the isolation of cloned cDNA and genes coding for glutenins. The PCR, a technique that allows the specific amplification of a target DNA segment using a pair of flanking oligonucleotides as primers, has been used for genetic analysis. PCR has also been used for isolation and characterization of promoter sequences to understand the molecular basis of control of gene expression in endosperm Anderson *et al.* (1998) producing different levels of protein subunits. The direct cloning of genomic sequences amplified by PCR Saiki *et al.* (1988) opens a new opportunity for the characterisation of different allelic forms and detailed study of genetic polymorphism (D'Ovidio *et al.* 1991; Van Campenhout *et al.* 1995). The direct and specific analysis of genomic sequences amplified by PCR can infect allow the isolation of different alleles without constructing specific cDNA or genomic DNA libraries. Silent genes have also been amplified and compared with the allelic active ones. This approach offers the possibility of analyzing for the presence of cysteine codons.

Further the PCR technique can be used in screening of recombinant lines (D'Ovidio & Anderson, 1994; Varghese *et al.* 1996; Ahmad, 2000) as an alternative method. This easy, quick and non-destructive PCR based approach is an efficient alternative to standard procedures for selecting bread wheat genotypes with good bread making characteristic. Further this alleviates the problem of anomalous migration of protein subunits in SDS-PAGE. This has allowed the complete amino acid sequences of the proteins encoded by these DNA to be deduced, providing a basis of modeling and biophysical studies. Such biochemical and biophysical studies gave detained picture of HMW-GS structure. The structure of both the terminal regions showed a helical structure and the central region with repetitive domain. The elastic nature of glutenin was proposed to be due to reverse turns similar to elastin protein present in mammalian connective tissue, Tatham *et al.* (1990). Belton 1999 postulated that intra and inter molecular hydrogen bond formation by glutamine residues are accounted for elasticity and compared it with elasticity in rubber. During dough development these bonds break and reform on stretching and when the stress is removed, respectively.

Functionality

The role of HMW-GS has been studied extensively in bread making (Payne *et al.* 1987, Harberd *et al.* 1986, Kasarda *et al.* 1989, Shewry *et al.* 1992) and biscuit making (Souza *et al.* 1994; Hou *et al.* 1996, Czuchajowska *et al.* 1996) quality. The analysis of a large number of wheat cultivars by means of SDS-PAGE has proved that all loci display a large number of allelic variations. Payne & Lawrence (1983) reported 3 alleles at *Glu-A1* locus, 11 alleles in the locus *Glu-B1* and 6 alleles at *Glu-D1* locus and devised a *Glu-1* quality score for the HMW glutenin subunits to predict the quality of British grown wheats based on the scores of individual alleles. The score was subsequently used in breeding programmes in UK which led to enhanced quality for bread making within a short period of time. Similarly, Bushuk (1998) devised an equation to predict unit loaf volume in Canadian wheats from the HMW glutenin subunit composition. Report in India showed that combination of HMW glutenin subunits 2*, 17+18 and 5+10 coded by *Glu-A1*, *Glu-B1* and *Glu-D1* respectively exhibited highest gluten strength as measured by sedimentation volume, Ram, S. (2003). Though HMW glutenins explained larger variability in gluten strength, the synergistic role of LMW glutenins was observed in later studies which will be dealt in greater detail in subsequent section in this chapter. Electrophoresis testing for HMW glutenin subunit composition is now used routinely in many breeding programs around the world.

Table 1. Quality scores assigned to individual high molecular weight glutenin subunits (Payne *et al.* 1987)

	Locus		
Score	Glu-A1	Glu-B1	Glu-D1
4	-	-	5+10
3	1	17+18	-
3	2*	7+8	-
3	-	13+16	-
2	-	7+9	2+12
2	-	-	3+12
1	Null	7	4+12
1	-	6+8	-
1	-	20	-

The mechanism by which certain HMW-GS confer better bread making performance than others remains to be defined. The differences in the dough strength are considered due to differences in molecular size distribution of gluten polymers Gupta & MacRitchie, (1994). Three types of glutenin polymers are formed by the interactions of HMW-GS and LMW-GS Vasil & Anderson, (1997) (Fig.3). Larger polymers have mixture of HMW-GS and LMW-GS and can reach into tens of millions of Daltons. The shorter polymers have both the HMW-GS and LMW-GS and are soluble in either salt solutions or acid SDS extractions. The third types of polymers have LMW-GS only and are shorter in size. Increasing proportion of larger polymers correlates with higher dough strength and bread making quality. The polymer size depends upon the number and position of cysteine residues and the size of the repetitive domain in the glutenins (Tatham *et al.* 1994, Belton 1999). For example formation of larger or more branched polymers by 5Dx subunit has been attributed to the presence of additional cysteine residues. Certain allelic subunits impart different effect on gluten quality. Gliadins and LMW-GS with one available cysteine residue function as chain terminators and therefore, the polymer is reduced by the introduction of these during polymer formation. Recently this information has been used in the development of transgenics for enhanced dough strength.

Low Molecular Weight Glutenin Subunits (LMW-GS)

Structure and Genetics

LMW-GSs constitute around 40% of gluten proteins and are separated into three groups called B-type, C-type and D-type by SDS-PAGE (Jackson *et al.* 1983). LMW-GS are encoded at the Glu-3 loci (Singh & Shephard, 1988) on the short arm of chromosome-1 which are closely linked to *gli-1* loci for gliadins. The general features of LMW-GS have tetrapeptide motiff PQQP interspersed with heptapeptide motiff PQPQPFP and QQQQPUL in the repetitive domain as in s-rich prolamins (Fig. 3). The C-type subunits are related to α and γ gliadins and D-type to ω gliadins and are considered as the mutated form of these gliadins respectively (Okita *et al,* 1985; Masci *et al.* 1991). Additionally LMW-GSs contain a cysteine residue in the unrepetitive N-terminal domain which could be related to the ability of these polypeptides to form polymers. Based on the distribution of cysteine residues, the LMW-GS proteins can be classified into three types: i) those with one cysteine in n-terminal domain; ii) those with a cysteine residue in the repetitive domain; and iii) those with 8 cysteines in the c-terminal part of the protein. All the LMW glutenins were reported having 1st cysteine residue in the N-terminal part of the sequence until D'Ovidio *et al.* (1997) and Masci *et al.* (1998) detected 1st cysteine residue in the repetitive domain in durum and bread wheat respectively and subsequently reported by Ram, S. *et al.* (2006) in Indian wheat variety NP4.

Several total and partial gene sequences of LMW-GS genes have been reported (Okita 1985, Pitts 1988, D'Ovidio 1995, Van Campenhout 1995, Benmoussa 2000, Ram *S. et al.* 2006). The sequence studies have indicated the absence of introns in LMW-GS genes (Kreiss *et al.* 1985, Ciaffi *et al.* 1999, D'Ovidio 1999,

Lee 1999). All the complete LMWm or LMWs sequences that have been reported have 8 cysteins and considered as chain extenders. Chain extenders allow the formation of stronger doughs and chain terminators have the opposite effects (Greenfield 1998, Masci 1998, Tamas 1998). Reports of N-terminal amino acid sequences of LMW glutenin fractions revealed that they have either methionine or serine or isoleucine residue at the first position of the mature polypeptide. These subunits are therefore called LMW-m, LMW-s and LMW-I type glutenins respectively (Masci *et al.* 1995, D'Ovidio *et al.* 1999, Cloutier *et al.* 2001). Lkeda *et al.* (2006) classified LMW - GS sequences into 12 groups based on the alignment of the N and C terminal domains of the deduced amino acid sequences from 60 gene sequences available. Recently PCR techniques have been used to clone and characterise LMW-GS genes (D'Ovidio *et al.* 1997, Van Campenhout *et al.* 1995, Benmoussa *et al.* 2000, Cloutier *et al.* 2001, Zhang *et al.* 2004, Hai *et al.* 2005, Ikeda *et al.* 2006 and Zhao *et al.* 2007). These studies identified genome specific and allele specific primers for amplification and thus useful in breeding programme.

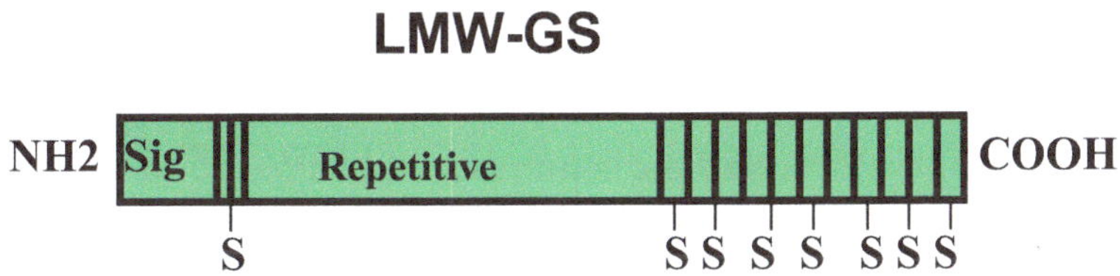

Fig. 3. General features of Low molecular weight glutenin subunits showing tetrapeptide motif PQQP interspersed with heptapeptide motif PQPQPFP and QQQQPUL in the repetitive domain and 8 cysteine reaidue of which 1 in tho N-terminal region and the rest 7 in the C-terminal region.

The deduced protein sequence coded by *Glu-B3* locus of LMW-GS genes from NP4 showed the presence of regular structure of the repetitive domain with high proportions of glutamine (Q) residues. The presence of 1st cysteine residue within the repetitive domain at 40th position exhibited it's availability for intermolecular disulphide bond formation. Although the octapeptide repeat PPFSQQQQ was most common, hexapeptide, nonapeptide and pepta peptide repeats were also present with variations in the number of glutamine residues. The irregular amount of Gln in the repeats ranging from 2 to 5 residues may prevent the formation of highly regular intra and intermolecular interactions. Highly regular repeats might lead to strongly interacting aggregates having excessive insolubility and enzyme inaccessibility in the endosperm. Moreover, the Gln stretches in the somewhat extended repeated sequence domain are likely to interact intermolecularly with their counterparts in other protein molecules through side chain and main chain amide hydrogen bonding. Thus the repeat might increase the viscosity and elasticity of the dough through intermolecular interactions of the large numbers of glutamine side chains which are both good hydrogen donors and acceptors. The presence of cysteine residue in the repetitive domain might have arisen by the substitution of phenylalanine amino-acid residue which is present in majority of the LMW-GS genes reported so far. Such a substitution could have been arisen from a T to G transversion event. Since cysteine is encoded by a TGT triplet, whereas phenylalanine present in the repeating units of LMW-GS are usually coded by TTT triplets. The position of the cysteine residue is important because 1 and 7 cysteine residues are involved in intermolecular disulphide bond formation and thus act as chain extenders. Preponderance of chain extender types in glutenin should lead to strong gluten with good viscoelastic properties.

Functionality

Various studies have indicated the importance of LMW-GS to dough properties (Metakovski *et al,* 1991; Gupta *et al.* 1994). Though a number of studies revealed the variability in LMW-GS genes associated with significant differences in dough quality in bread and durum wheat, the clear cut relationship of individual protein subunits with gluten quality has not been established. On the basis of screening of a collection of 222 hexaploid wheats, Gupta & Shepherd (1990a) reported 20 different band patterns (LMW-GS blocks), 6 for *Glu-A3* locus (*Glu-A3a, Glu-A3b, Glu-A3c, Glu-A3d, Glu-A3e, Glu-A3f*), 9 for *Glu-B3* locus (*Glu-B3a, Glu-B3b, Glu-B3c, Glu-B3d, Glu-B3e, Glu-B3f, Glu-B3g, Glu-B3h, Glu-B3i*) and 5 for *Glu-D3 locus (Glu-D3a, Glu-D3b, Glu-D3c, Glu-D3d and Glu-D3e)*. Because the actual number of electrophoretic band combinations in LMW-GS is much lower than the total of randomly possible combinations, it is assumed that the genes controlling LMW-GS are closely linked. They form clusters that are inherited together, similar to those controlling gliadin blocks (Lagudah *et al.* 1991, Gupta & Shepherd, 1990).

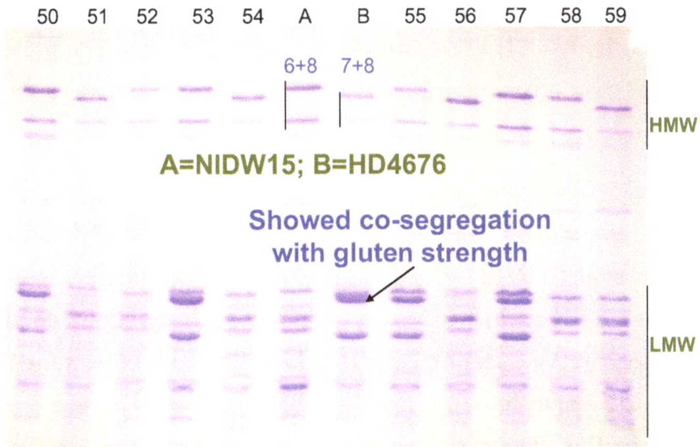

Fig. 4. LMW Glutenin II as Marker for Gluten Strength in Durum Wheat (Pasta Quality)

The ability of LMW-GS to form intermolecular disulphide bonds with each other and high molecular weight glutenins is considered important for gluten polymer formation. The presence of specific LMW-GS encoded at the *Glu B-3* locus has been associated with pasta making quality in durum wheat. The role of LMW subunits in bread making quality has been indicated by correlation studies of allelic variation and dough properties (Gupta *et al.* 1989, Metakovski 1990, Morel *et al.* 1993, Lukow *et al.* 1995). Several alleles have been ranked with respect to their effect on dough resistance (Gupta & Shepherd 1988, Gupta *et al.* 1994, Metakovski *et al.* 1990). In Australian wheat cultivars, LMW-GS provided better predictions than HMW-GS for Rmax (maximum dough resistance). For Rmax, the alleles of *Glu-A3* showed b>d>e>c: the alleles of *Glu-B3*: i>b=a=e=f=g=h>c: the alleles of *Glu-D3*: e>b>a>c>d. They concluded that HMW-GS are alone insufficient to account for differences in quality. LMW-GS must also be taken into consideration. Cornish *et al* (1993) catalogued information about the gluten alleles of about 600 wheats and summarized that null alleles were detrimental to the extensibility and *Glu-3* pattern b b b (at loci *Glu-A3*, *Glu-B3* and *Glu-D3* respectively) gave the best extensibility. Vasquez *et al.* (1996) reported the significant effect of allelic variation at *Glu-B3* locus on gluten strength in durum wheat. Luo *et al.* (2001) observed the significant effect of *Glu-B3* alleles on sedimentation value, Pelshenke value and various mixograph parameters. Vaccino *et al.* (2002) reported the positive effects of two fragments (Southern blot hybridisation) corresponding to LMW genes on several quality related parameters such SDS sedimentation volume, Farinograph found stability, Alveograph strength LMW-II (W), tenacity (P) and elasticity (L). In our studies we significant effect of glutenins on gluten strength in a set of recombinant inbored lines of a cross between NIDW-15 and HD4676 (Fig. 4).

Immunological techniques were used to assess the contribution of LMW-GS in bread making quality and dough extensibility (Brett *et al.* 1993, Andrews & Skerritt 1996). Manifesto *et al.* (1998) identified LMW-GS and gliadin loci affecting bread-making quality using PCR and RFLP analysis in a cross between Klen32 and Chinese spring having similar HMW-GS composition. Despite their importance only a limited number of LMW-GS genes have so far been characterised D' Ovidio *et al.* (1997). The functional properties such as dough mixing, extensibility and resistance of product of individual LMW genes were assayed using expression system in bacteria (Lee *et al.* 1999, Cloutier & Lukow 1998).

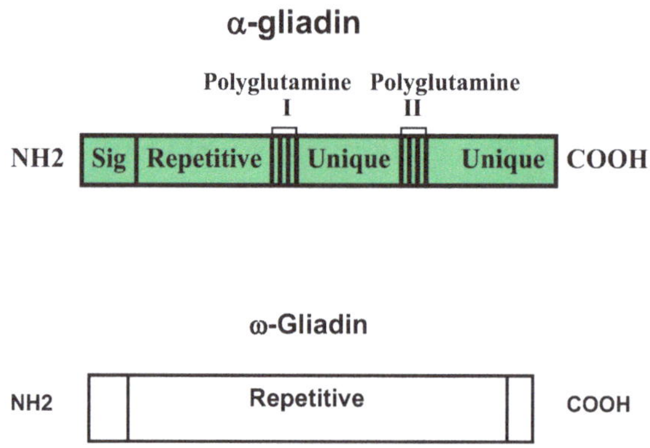

Fig. 5. General features of α and ω gliadins. Gliadins have a central repetitive domain flanked by unrepetitive N' terminal and C' terminal domain as in glutenins but the length of these domains varies from one type of polypeptide to other. The tetra peptide motif PQQP occurs in all the sulphur rich prolamins and interspersed with heptapeptide motif PQPQPFP in α-gliadins. The unrepetitive C- terminal domains have most of the cysteine residues conserved in α-gliadins. The main repeat motif in ω-gliadins is an octapeptide (PQQPFPQQ).

Gliadins

Structure and Genetics

Gliadins (alcohol soluble) represent around 50% of total gluten proteins in wheat. Gliadins are monomeric proteins and are divided into α, β, γ (30,000 daltons to

40,000 daltons) and ω (44000 to 80000 daltons) according to their mobility in acid-PAGE. ω gliadins are sulphur poor and α, β, γ gliadins are sulphur rich. The amino acid analysis of these proteins indicates the general similarity with glutenins. Each has a central repetitive domain flanked by unrepetitive N' terminal and C' terminal domain (Fig. 5). But the length of these domains varies from one type of polypeptide to other. Also the repeat motif in the central domain are based on different amino acid sequences in the different polypeptide types. These repeat motifs are clearly related to each other for the gliadins and LMW-GS Tatham *et al.* (1990). The tetra peptide motif PQQP occurs in all the sulphur rich prolamins and interspersed with heptapeptide motif PQPQPFP in α-gliadins. The main repeat motif in ω-gliadins is an octapeptide (PQQPFPQQ). The unrepetitive C-terminal domains have most of the cysteine residues conserved. In the sulphur rich prolamins, the C-terminal domains are large and comprise just over one half to two third of the polypeptides whereas the N' terminal domains are only 5 to 14 amino acids long.

Most gliadin alleles are controlled in common wheat by six main loci on the chromosome on the first (Gli-1) and sixth (Gli-2) homoeological groups (Payne *et al.* 1986). There are also some minor loci as *Gli-3, Gli-5* and *Gli-6* which control few minor gliadin bands (Pogna 1993, Metakovsky 1997). In a study, two new gliadin alleles *Gli-D4* and *Gli-D5* were mapped on the short arm of chromosome 1D (Rodriguez & Carrillo, 1996). *Gli-D4* controls γ type gliadins and is situated between centromere and the *Gli-d1* locus with a genetic distance of 10.1 ± 2.0cM from this locus. *Gli-D5* controls ω-type gliadin and located between *Gli-D1* and telomere. Two dimentional electrophoretic techniques have shown the 50 different gliadin polypeptides in a single wheat cultivar. There is a high degree of variability in the number of gliadin pattern and the restriction fragment length polymorphism among different wheat varieties (D'Ovidio *et al.* 1992; Ram *S. et al.* 2006) (Fig.6). increasing the complexity of these proteins. These differences may be due to duplications and deletions of chromosome segments generated by unequal crossing over and by gene conversion events. Gliadin pattern being a genetic characteristic, it has been used as a variety identification (Bushuk & Zillman 1978, Branlard *et al.* 1993, Ram S. *et al.* 2005).

Recent development in PCR-based methods can also be of value in differentiating between varieties and between different γ-gliadins groups; especially when applied to gene expression studies. Quantitative real time PCR (QRT-PCR) is a well-established technique, which allows the detection of a given sequence in a rapid, specific and very sensitive way Gachon *et al.* (2004). These features make the QRTPCR the best technology currently available to detect and quantify the expression of genes encoded by multigene families, such as γ-gliadin, which are expressed and accumulate during grain development. Based on these developments Piston *et al.* (2006) identified nine new full-length cDNA sequences for γ-gliadins from total mRNA from bread and durum wheat. The length of the open reading frames (ORFs) ranged from 720 to 936 bp, which corresponds to putative proteins of 239–311 amino acid residues, respectively. The γ-gliadin sequences fall into four major groups in the tree constructed by the neighbour-joining method. Primer pairs were designed to discriminate between these four γ-gliadin groups during grain development in bread wheat using quantitative real time polymerase chain reaction (QRT-PCR). Expression profiles of two lines of bread wheat showed higher levels of gene transcription for γ-gliadins from group II, whereas γ-gliadins from group I were expressed at the lowest level. Thus, the γ-gliadins from group II are expressed four- and 10-fold higher in 'Perico' and 'Anza', respectively, than the other γ-gliadins. The results reported may help to understand the order of synthesis and deposition of γ-gliadins in the grain of developing bread wheat. In addition, the differences found in the expression levels among the four groups could be used to select genotypes with specific profiles of γ-gliadins. These findings will also facilitate the up- and down-regulation of γ-gliadin gene expression in transgenic wheat.

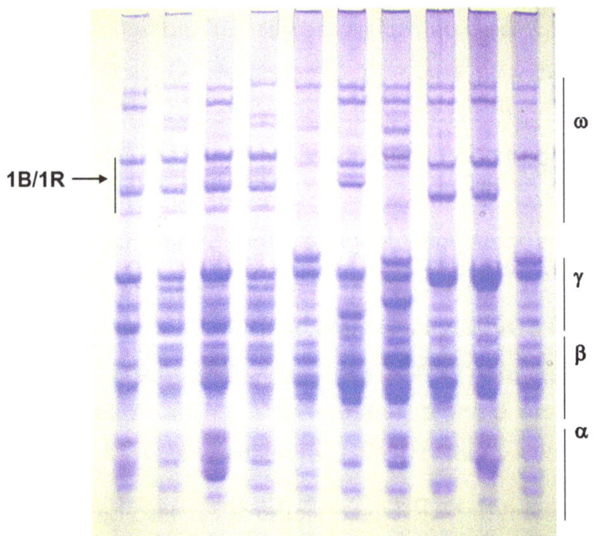

Fig. 6. Gliadin pattern of Indian wheat varieties. Gliadin pattern in ω (omega) region can be used in identification of IB/1R translocation.

Functionality

Although gliadins account for about 50% of the gluten proteins their role in determining the mixing properties of dough are not well understood. This is because correlations between gliadins patterns and functional properties are not as clear as for the HMW-GS. Unlike the HMW-GS, gliadins are encoded by large multigene families and inherited in blocks, thus effects of individual gliadins on dough properties are difficult to determine Fido *et al.* (1997). That is why association of gliadin alleles with dough quality has been somewhat less understood, though, the correlation between dough quality and certain gliadin polypeptides were reported by several workers (Pogna *et al.* 1982; Sozinov & Poperelya, 1980; Wrigley *et al.* 1982). In durum wheat γ gliadin 45 band showed strong correlation with gluten strength and good pasta making quality. However, Payne *et al.* (1987) suggested that the effect of allelic variation at the *Gli-1* loci on gluten strength was due to genes coding for low molecular weight glutenin sub-units at the *Glu-3* loci which are tightly linked to the *Gli-1* loci (Pogna *et al.* 1990; Singh & Shepherd, 1988). Ciaffi *et al.* (1996) reported the correlation of elasticity (strength) and extensibility (viscosity) of the dough with glutenins and gliadins respectively.

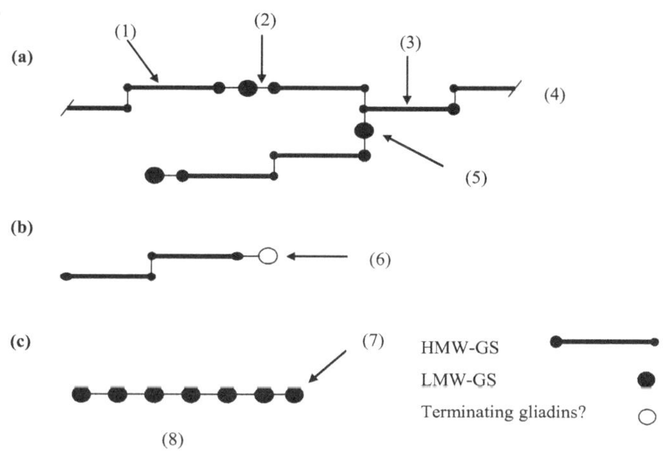

Fig. 7. Three types of glutenin polymers formed by the interactions of HMW-GS and LMW-GS, a: larger polymers having a mixture of HMW-GS and LMW-GS, b: shorter polymers with both the glutenins and c: polymers of LMW-GS only. Gliadins and LMW-GS with one available cysteine residue function as chain terminators. The size of the polymer determines the dough strength (Vasil and Anderson, 1997).

Generally gliadins are considered to have negative effects on dough strength. That may be due to their chain terminating effect on gluten polymer because of the presence of extra cysteine residue capable of forming only a single intramolecular disulfide bond (Fig. 7). Reconstitution studies of purified gliadins have shown weakening effect on dough strength (Fido *et al.* 1997, MacRitchie 1987). However Metakovsky *et al.* (1997) reported the significant positive relationship between three gliadin alleles (*Gli-B1b, Gli-B2c* and *Gli-A2b*) and gluten strength explaining about 41% of the observed variation in 54 cultivars used. The combination of this with the quality score based on Glu-1 alleles coding for HMW-GS could explain large variation in gluten strength. The effect could be related to interactions between particular gliadin and glutenin. Lonkhuisen *et al.* (1992) & Weegles *et al.* (1994) also reported the positive effect of γ-gliadin on bread making quality. The data for the effect of α and β gliadins is more ambiguous with certain bands showing negative effects and other positive effects on technological properties Buren *et al.* (2000). Besides their importance as a major protein component of the human diet and their relation to wheat quality, gliadins are major initiator of intestinal damage in celiac disease Shewry *et al.* (1992). There is also clear evidence that γ-gliadins are active in celiac disease Shan *et al.* (2002). The selection of genotypes with specific profiles of γ-gliadins may, therefore, be of interest in relation to determining the relative activities of individual γ-gliadins in provoking a response in celiac disease and developing cultivars with lower levels of the more toxic components.

Genetic Engineering of Gluten Proteins

The potential for the application of genetic engineering in wheat breeding is immense. Wheat transformation has progressed enormously during the last 15 years from the first transgenic wheat plants obtained in 1992. Transformation protocols have been developed in wheat using particle bombardment (Weeks *et al.* 1993, Becker *et al.* 1994; Altpeter *et al.* 1996, Barrow *et al.* 1997 & 2003) and agrobacterium mediated Cheng *et al.* (1997). The availability of transformation technology provides an opportunity to manipulate wheat to enhance its agronomic performance, resistance to abiotic/biotic stresses, yield and end-use quality. The quality traits that have got increasing attention are the manipulation of starch composition by the expression of various biosynthetic enzymes and manipulation of various storage proteins. Among the storage protein genes, high molecular weight glutenin sub-unit genes have been used for manipulating their number, composition and properties because the role of these proteins have been studied in greater detail in bread making quality. While there is well-documented correlation between bread making quality and the presence of specific HMW-GS, little is known about the precise organization and role of these in the formation of glutenin polymers. This is because of difficulty in purification of individual components and reconstruction of dough / gluten from their constituent parts. Further the complex multi-domain structure of HMW sub-unit makes its difficult to carry out detailed studies of the structure and functional properties of the β-spiral structure.

Genetic engineering techniques along with other molecular tools can help in providing solutions to these problems. The cDNAs or gene sequences provide complete amino acid sequences of the encoded proteins which in turn facilitate structural studies. These DNAs can also be expressed in heterologous systems as *E. coli* to provide wild type or modified proteins for structural studies or functional analysis. However such incorporation experiments are difficult to perform and it is not possible to ensure that the incorporated protein form the same molecular interactions as it would if expressed in the developing wheat grain. The cDNAs or genes can be used for wheat transformation, in order to explore the functionality of the encoded proteins when expressed in the developing grain. This will ultimately allow the development of new varieties in which the composition and properties of the gluten proteins have been tailored for specific end-use properties. The possibility to generate modified proteins enhances the capability to create new genetic variability in the laboratory for proteins. This approach, besides eliminating the need for natural variability of HMW-GS, offers an opportunity to generate variability at specific region. The details of genetic engineering techniques are provided in other chapter on molecular breeding.

Summary

The foregoing discussion clearly indicates that impressive advances have been made in understanding the role of certain protein subunits in imparting functional dough properties. However the role of individual protein subunits in polymeric structure of gluten is still to be elucidated. Direct cloning of genomic sequences using PCR has opened an opportunity for molecular genetic analysis of storage protein genes. This will accentuate understanding of molecular basis of gluten functionality and its manipulation by genetic engineering. Possibility of manipulating grain hardness property using genetic engineering has increased with the understanding of puroindoline genes and their endosperm specific promoters. Characterisation of starch biosynthetic enzymes and their corresponding genes along with improved understanding of structure and properties of starch will help in manipulating starch functionality for different end-use products. Recent advances in wheat transformation protocols will further elucidate the role of component genes and their manipulation.

References

Ahmad, M. (2000). Molecular marker assisted selection of HMW-glutenin alleles related to bread wheat quality by PCR-generated DNA markers. *Theo. Appl. Genet.* 101: 892-896.

Altpeter, F., Vasil, V., Srivastava, V. and Vasil, I.L. (1996). Integration and expression of the high molecular weight glutenin subunit 1Ax1 gene into wheat. *Nature Biotech.* 14: 1155-1159.

Alvarez, M.L., Guelman, S., Halford, M.G., Lustig, S., Reggiardo, M.I., Ryabushkina, N., Shewry, P.R., Stein, J. and Vallejos, R.H. (2000). Silencing of HMW glutenins in transgenic wheat expressing extra HMW subunits. *Theor. Appl. Genet.* 100: 319-327.

Anderson, O.D. and Green, F. (1989). The characterization and comparative analysis of high molecular-weight glutenin genes from genomes A and B of a hexaploid bread wheat. *Theor. Appl. Genet.* 77: 689-700.

Anderson, O.D., Litts, J.C. and Greene, F.C. (1997). The gliadin gene family. I. Characterisation of ten new wheat α-gliadin genomic clones, evidence for limited sequence conservation of flanking DNA an southern analysis of the gene family. *Theor. Appl. Genet.* 95: 50-58.

Anderson, O.D., Pierce, F.A.A. and Tam, A. (1998). Conservation in wheat high molecular weight glutenin gene. Promoter sequences: comparisons among loci and among alleles of the glu B-1-1 locus. *Threor. Appl. Genet.* 96: 568-576.

Andrews, J.L. and Skerritt, J.H. (1996). Wheat dough extensibility screening using a two sites enzyme-linked immunoabsorbent assay (ELISA) with antibodies to low molecular weight glutenin subunits. *Cereal Chem.* 73(5): 650-657.

Anjum, F.M., Khan, M.R., Din, A., Saeed M., Pasha, I. and Arshad, M.U. (2007). Wheat gluten: high molecular weight glutenin subunits-structure genetics and relation to dough elasticity. *J. Food Sci.* 72(3): 56-63.

Barcelo, P., Rasco-gaunt, S., Sparks, C., Cannell, M., Salgueiro, S., Rooke, L., He, G.Y., Lamacchia, C., Vina, G.D., Shewry, P.R. and Lazzeri, P.A. (1998). Transformation of wheat: State of the technology and example of application. In: *Proc. Int. Wheat Genetics Symposium*, Saskatoo.

Barrow, F. Rooke, I., Bekes, F., Gras, P., Tatham, A.S., Fido, R., Lazzeri, P.A. Shewry, P.R. and Barcelo, P. (1997). Transformation of wheat with high-molecular-weight glutenin subunit gene results in improved functional properties. *Nature Biotech.* 15: 1295-1299.

Barrow, F., Barcelo, P., Lazzeri, P.A., Shewry, P.R., Ballesteros, J. and Martin, A. (2003). Functional properties of flours from field grown transgenic wheat lines expressing the HMW glutenin subunit 1Ax1 and 1Dx5 genes. *Mol. Breed.* 12: 223-229.

Becker, D., Brettchneider, R. and Lorz, H. (1994). Fertile transgenic wheat from microprojectile bombardment of scutellar tissue. *Plant J.* 5: 299-307.

Beckwith, A.C., Nielsen, H.C., Wall, J.S. and Huebner, F.R. (1966). Isolation and characterization of a high-molcular-weight protein from wheat gliadin. *Cereal Chem.* 43: 14-28.

Belton, P.S. (1999). On the elasticity of wheat gluten. *J. Cereal Science.* 29: 103-107.

Benmoussa, M., Vezina, L.P., Page, M., Yelle, S. and Laberge, S. (2000). Genetic polymorphism in low molecular weight subunit genes from Triticum aestivum, variety Chinese Spring. *Theor. Appl. Genet.* 100: 789-793.

Blechl, A.E. and Anderson, O.D. (1996). Expresion of a novel high molecular weight glutenin subunit gene in transgenic wheat. *Nature Biotech.* 14: 875-879.

Branlard, G. and Dardevet, M. (1985). Diversity of grain proteins and bread wheat quality. I. Correlation between gliadin bands and flour quality characteristics. *J. Cereal Science* 5: 329-343.

Branlard, G., Dardevet, M. Nieto-Taeladriz, M.T. and Khelife, D. (1993). Allelic diversity of omega gliadins as revealed by SDS-PAGE. Their possible implication in quality variation. *Gluten* 93: 234-243.

Brett, G. M., Mills, E.N.C., Tatham, A.S., Fido, R.J., Shewry, P.R. and Morgan, M.R.A. (1993). Immunological identification of LMW subunits of glutenin associated with bread-making quality of wheat flours. *Theor. Appl. Genet.* 86: 442-448.

Bushuk, W and Zillman, R.R. (1978). Wheat cultivar identification by gliadin electrophoregrams. 1. Apparatus, method and nomenclature. *Can. J. Plant Science.* 58: 505-515.

Bushuk, W. (1998). Wheat breeding for end-product use. *Euphytica.* 100: 137-145.

Bustos, A.D., Rublo, P. and Jouve, N. (2000). Molecular characterisation of the inactive allele of the gene gluA-1 and the development of a set of AS-PCR markers for HMW glutenins of wheat. *Theor. Appl. Genet.* 100: 1085-1094

Campbell, K. G., Finney, P.L., Bergman, C. J., Gualberto, D.G., Andersan, J.A., Giroux, M.J., Sritunga, D., Jhu, J., Gender, F., Roue, C., Verel, A and Sorrells, M.E. (2001). Quantitative trait loci associated with milling and baking quality in a sift and hard wheat cross. *Crop Sci.* 41: 1275-1285.

Campenhout, S.V., Steppen, J.V., Sagi, L. and Volkaert, G. (1995). Locus specific primers for LMW glutenin genes on each of the group –1- chromosomes of hexaploid wheat. *Theor. Appl. Genet.* 91: 313-319.

Cheng, M., Fry, J.E., Zhou, H., Hironaka, C.M., Duncan, D.R. Conner,T.W. and Wan, Y. (1997). Genetic transformation of wheat mediated by Agrobacterium tumefaciens. *Plant Physiol.* 115: 971-980.

Ciaffi, M., Lee, Y.K., Tamas, L., Gupta, R., Skerritt, J. and Appels, R. (1999).The low molecular weight glutenin subunit proteins of primitive wheats. III. The genes from D-genome species. *Theor Appl Genet.* 98: 135-148.

Ciaffi, M., Tozzi, L. and Lafiandra, D. (1996). Relationship between flour protein composition determined by size exclusion high performance liquid chromatography and dough rheological parameters. *Cereal Chem.* 73: 346-351.

Cloutier, S and Lukow, D.M. (1998). Cloning and expression of a rare LMW glutenin. *Wheat Genet. Symposium.* 3(3): 2-4.

Cloutier, S, Rampitsch, C, Penner, G.A. and Lukow, O.M. (2001). Cloning and expression of a LMW-I glutenin gene. *J. Cereal Science.* 33: 143-154.

Cornish, G.B., Burridge, P.M., Palmer, G.A. and Wrigley, C.W. (1993). Mapping the origins of some HMW and LMW glutenin subunit alleles in Australian germplasm pp 255-260. Proc 42nd Aust Cereal Chem Conference, Sydney.

Czuchajowska, Z., Lin, P.Y. and Smolinski, S. (1996). Role in dough rheology of high molecular weight subunits of soft white winter and club wheats. *Cereal Chem.* 73(3): 338-345.

D'Ovidio, R. and Anderson, O.D. (1994). PCR analysis to distinguish between alleles of a member of multigene family correlated with wheat bread making quality. *Theor. Appl. Genet.* 88: 759-763.

D'Ovidio, R., Marchielli, C., Ercoli-Cardelli, L and Porceddu, E. (1999). Sequencing similarity between allele Glu-B3 genes related to quality parameters of durum wheat. *Threo. Appl. Genet.* 98: 455-461.

D'Ovidio, R., Simeone, M., Masci, S., Porcuddu, E. and Kasarda, D.D. (1995). Nucleotide sequence of a g-type glutenin gene from a durum wheat: Correlation with a γ-type subunit from the same biotype, *Cereal Chem.* 72: 443-449.

D'Ovidio, R., Tanzarella, O., Masci, S., Lafiandra, D. and Porceddu, E. (1992). RFLP and PCR analysis at *Gli-1, Gli-2, Glu-1* and *Glu-3* loci in cultivated and wild wheats. *Hereditas* 116: 79-85.

D'Ovidio, R., Tanzarella, O.A. and Porceddu, E. (1991). Cloning and sequencing of a PCR amplified gamma gliadin gene from durum wheat (*Triticum* turgidum L.) Thell. Conv durum (Desf.) MK). Plant Sci. 75: 229-236.

D'Ovidio, R., Simeon, M., Masci, S. and Porceddu, E. (1997). Molecular characterization of a LMW-GS gene located on chromosome 1B and the development of primers specific for the Glu-B3 complex locus in durum wheat. *Theor. Appl. Genet.* 95: 1119-1126.

Darlington, H., Fido, R., Tatham, A.S., Jones, H., Salmon, S.E. and Shewry, P.R. (2003). Milling and baking properties of field grown wheat expressing HMW subunit genes. *J. Cereal Sci.* 38: 301-306.

Fido, R.J., Bekes, F., Gras, P.W. and Tatham, A.S. (1997). Effects of α, β, γ and ω-gliadins on the dough mixing properties of wheat flour. *J. Cereal Sci.* 26: 271-277.

Ford, J., Malpica, M., Harford, N.G., Shewry, P.R., Anderson, O.D., Greene, F.C. and Miflin, B.J. (1985). The nucleotide sequence of a HMW glutenin subunit gene located on chromosome A of wheat (*Triticum aestivum* L.) *Nucleic Acids Res.* 13: 6817-6831.

Greenfield J J A (1998). Rheological properties of monomeric and polymeric forms of C hordeins, a sulphur-poor prolamin of barely. *J Cereal Sci.* 27: 233-236

Gupta R.B. Singh, N.K. and Shepherd , K.W. (1989). The cumulative effect of allelic variation in LMW and HMW glutenins subunits on dough properties in the progeny of the bread wheat. *Theor. Appl. Genet.* 77: 57-64.

Gupta, R.B. and MacRitchie, F. (1994). Allelic variation at glutenin subunit and gliadin loci, *Glu-1, Glu-3* and *Gli-1* of common wheats. II. Biochemical basis of the allelic effects on dough properties. *J. Cereal Sci.* 19: 19-29.

Gupta, R.B. and Shepherd, K.W. (1988). Low molecular weight glutenin subunits in wheat: their variation, inheritance and association with bread-making quality. In *Proceedings of the 7th International Wheat Genetic Symposium* pp. 943-949, edited by T E Miller & R M D Koebner. Cambridge, UK.

Gupta, R.B. and Shepherd, K.W. (1990). Two step one dimensional SDS-PAGE analysis of LMW subunits of glutenin I. Variation and genetic control of the subunits in hexaploid wheats. *Thoer Appl Genet.* 80: 65-74.

Gupta, R.B., Paul, J.G., Cornish, G.B., Palmer, G.A., Bekes, F. and Rathjan, A.J. (1994). Allelic variation at glutenin subunit and gliadin loci, *Glu-1, Glu-3* and *Gli-1* of common wheats. It's additive and interaction effects on dough properties. *J. Cereal Science.* 19: 9-17.

Hai, L., Yu, M.W., Ze, H.Y., Bernard, B., Eviatar, N. and You, L.Z. (2005). Classification of wheat low molecular weight glutenin subunit genes and its chromosome assignment by developing LMW-GS group specific primers. *Theor. Appl. Genet.* 111: 1251-1259.

Harberd, N.P., Bartels, T. and Thompson, R.D. (1986). DNA restriction fragment variation in the gene family encoding high-molecular weight (HMW) glutenin subunits of wheat. *Biochem. Genet.* 24: 579-592.

Harberd, N.P., Flavell, R.B. and Thompson, R.D. (1987). Identification of a transposon-like insertion in a Glu-1 allele of wheat. *Mol. Genet.* 209: 326-332.

Hou, G., Yamamoto, H. and NG, P.K.W. (1996). Relationship of quantity of glutenin subunits of selected U.S. soft wheat flours to rheological and baking properties. *Cereal Chem.* 73: 358-363.

Ikeda, T.M., Araki, E., Fujita, Y. and Yano, H. (2006). Classification of low molecular weight glutenin subunit genes and their protein products in common wheat. *Theor. Appl. Genet.* 112: 327-334.

Jackson, E.A., Holt, L.M. and Payne, P.I. (1983). Characterization of high-molecular-weight gliadin and low-molecular-weight glutenin subunits of wheat endosperm by two-dimensional electrophoresis and chromosomal localization of their controlling genes. *Theor. Appl. Genet.* 66: 29-37.

Jackson, E.A., Holt, L.M. and Payne, P.I. (1983). Characterization of high-molecular-weight gliadin and low-molecular-weight glutenin subunits of wheat endosperm by two-dimensional electrophoresis and chromosomal localization of their controlling genes. *Theor. Appl. Genet.* 66: 29-37.

Kasarda, D.D. (1989). In: Wheat is unique pp 277-301. (Ed. Y. Pomeranz). Am. Assoc. Cereal Chemist, St. Paul, Minnesota, USA.

Kovacs, M.I.P., Howes, N.K., Leisle, D. and Skerritt, J.H. (1993). The effect of high molecular weight glutenin subunit composition on the results from tests used to predict durum wheat quality. *J. Cereal Sci.* 18: 43-51.

Kreis, M., Shewry, P.R., Forde, B.G., Forde, J. and Miflin, B.J. (1985). Structure and evolution of seed storage proteins and their genes with particular references to those of wheat, barley and rye. *in* Oxford Surveys of Plant and Molecular Cell Biology pp. 253-317, edited by B J Miflin. Oxford University Press, London.

Lagudah, E.S., Appels, R., Brown, A.H.D. and McNeil, D. (1991). The molecular genetic analysis of Triticum tauschi, the D-genome donor to hexaploid wheat. *Genome.* 34: 375-386.

Lee, Y.K., Bekes, F., Gras, P., Ciaffi, M., Morell, M.K. and Appels, R. (1999). The low molecular weight glutenin subunit proteins of primitive wheats IV. Functional properties of products from individual genes. *Theor. Appl. Genet.* 98: 149-155.

Li, Q.Y., Yan, Y.M., Wang A.I., An X.L., Zhang, Y.Z., Hsam, S.L.K. and Zeller, F.J. (2006). Detection of HMW glutenin subunit variations among 205 cultivated emmer accessions (*Triticum turgidum* spp. Dicoccum). *Plant Breed.* 125:120.

Lonkhuijsen, H.J., Van Hammer, R.J. and Schreuder, C. (1992). Influence of specific gliadins on the bread making quality of wheat. *Cereal Chem.* 69: 174-177.

Lukow, O.M. and Townley-Smith, T.F. (1995). Association of extra strong dough mixing properties with HMW and LMW glutenin subunits via doubled haploid line comparison. *Cereal Foods World.* 40: 667.

Luo, C., Griffin, W.B., Branlard, G. and McNeil, D.L. (2001). Comparison of low- and high molecular weight wheat glutenin allele effects on flour quality. *Theor. Appl. Genet.* 102: 1088-1098.

MacRitchie, F. (1987). Evaluation of contributions from wheat protein fractions to dough mixing and bread making. *J. Cereal Sci.* 6: 259-268.

Manifesto, M.M., Feingold, S., Hopp, H.E., Schlatter, A.R. and Dubcovsky J. (1998). Molecular markers associated with differences in bread making quality in a cross between bread wheat cultivars with the same high molecular weight glutenins. *J. Cereal Sci.* 27: 217-227.

Masci S. Porceddu, E., Colaprico, G. and Lafiandra, D. (1991). Comparison of the B & D subunits of glutenin encoded in the *Glu-D3* locus in two biotypes of the common wheat cultivar Newton with different technological characteristics. *J Cereal Sci.* 14: 35-46.

Masci, L., Lew, E.J.L., Lafiandra, D., Porceddu, E and Kasarda, D.D. (1995). Characterisation of low molecular weight glutenin subunits in durum wheat by reversed phase high performance liquid chromatography and N-terminal sequencing. *Cereal Chem.* 72: 100-104.

Masci, S., D'Ovidio, R., Lafiandra, D. and Kasarda, D.D. (1998). Characterisation of a low-molecular weight glutenin subunit gene from bread wheat and the corresponding protein that represents a major subunit of the glutenin polymer. *Plant Physiol.* 118: 1147-1158.

Metakovski, E.V. (1991). Gliadin allele identification in common wheat II. Catalogue of gliadin allele in common wheats. *J. Genet. Breed.* 45: 325-344.

Metakovsky, E.V, Chernakov, V.M., Upelniek, V.P., Redaelli, R., Dardevet, M. Branlard, G. and Pogna, N.E. (1997). Minor ω-gliadin coding loci on chromosome 1A of common wheat. A revision. *J. Genet. Breed.* 50: 277-286.

Metakovsky, E.V. Wrigley, C.W., Bekes, F. and Gupta, R.B. (1990). Gluten polypeptides as useful genetic markers in Australian wheats. *Aust. J. Agri. Research.* 41: 289-306

Metakovsky, E.V., Annicchiarico, P., Boggini, G. and Pogna, N.E. (1997). Relationship between gliadin alleles and dough strength in Italian bread wheat cultivars. *J. Cereal Science.* 25: 229-236.

Morel, M.H., Bonicel, J., Melas, V and Autran, J.C. (1993). Multiple approach (IEF, SDS-PAGE and A-PAGE). Of the composition of LMW subunits of glutenin and it's effect on dough properties. *Gluten* 93: 244-253.

Nielsen, H.C., Beckwith, A.C. and Wall, J.S. (1968). Effect of disulphide bond cleavage on wheat gliadins fractions obtained by gel filtration. *Cereal Chem.* 45: 37-47.

Okita, T. W., Cheesbrough, V. and Reeves, C.D. (1985). Evolution and heterogeneity of the α-/β- type and γ-type gliadin DNA sequences. *J. Biol. Chem.* 13: 8203-8213.

Osborne, T.B. (1907). The protein of the wheat kernel. Publication No. 84. Carnegie Institute: Washington, DC.

Payne, P.I. and Lawrence, G.J. (1983). Catalogue of alleles for the complex gene loci, *Glu-A1*, *Glu-B1* and *Glu-D1* which code for the high molecular weight subunits of glutenin in hexaploid wheat. *Cereal Res Commun.* 11: 29-35.

Payne, P.I., Roberts, M.S. and Holt, L.M. (1986). Location of genes controlling the D-group of LMW glutenin subunits on chromosome 1D of bread wheat. *Genet. Res.* 47: 175-179.

Payne, P.I., Seekings, J.A., Worland, J.A., Jarvis, M.G. and Holt, L.M. (1987). Allelic variation of glutenin subunits and gliadins and it's effect on bread-making quality in wheat: analysis of F5 progeny from Chinese Spring Chinese Spring (Hope 1A). *J. Cereal Science.* 6: 103-118.

Pisto´n, F., Dorado, G., Martin, A. and Barro F. (2006). Cloning of nine γ-gliadin mRNAs (cDNAs) from wheat and the molecular characterization of comparative transcript levels of γ-gliadin subclasses. *J. Cereal Sci.* 43: 120–128.

Pitts, E.G., Rafalski, J.A. and Hedgcoth, C. (1988). Nucleotide sequence and encoded amino acid sequence of a genomic gene region for a low molecular weight glutenin. *Nucleic Acids.* 16: 11376.

Pogna, N.E., Autran, J.C., Mellini, F., Lafiandra, D. and Fiellet, P.C. (1990). Chromosome 1-B encoded gliadins and glutenin sub-units in durum wheat. Genetics and relationship to gluten strength. *J. Cereal Sci.* 11: 15-31.

Pogna, N.E., Boggini, G. Gorbellini, M., Cattanco, M and DalBelin Peruffo, A. (1982). Association between gliadin electrophoretic bands and quality in common wheat. *Canadian J. Plant Sci.* 62: 913-918.

Pogna, N.E., Metakovsky, E.V., Redaelli, R., Raeneri, F. and Dachkevitch, T. (1993). Recombination mapping of *Gli-5* a new gliadin-coding locus on chromosome 1-A and 1-β in common wheat. *Theor. Appl. Genet.* 87: 113-121.

Ram, S. (2003). High molecular weight glutenin subunit composition of Indian wheats and their relationships with gluten strength. *J.Plant Biochem. and Biotech.* 12: 151-155.

Ram, S., Bhatia, V., Jain, V. and Mishra, B. (2006). Characterization of low molecular weight glutenin subunit gene representing *Glu-B3* locus of Indian wheat variety NP4. *J. Plant Biochem. and Biotech.* 15: 79-83.

Ram, S., Jain, N., Dawar, V., Singh, R.P. and Shoran, J. (2005). Analyses of Acid-PAGE gliadin pattern of Indian wheats (*Triticum aestivum* L.) representing different environments and periods. *Crop Science.* 45:1256-1263.

Rodriguez, Q.M. and Carrillo, J.M. (1996). Linkage map of prolamin loci *Gli-D4* and *Gli-D5* in hexaploid wheat. *Plant breed.* 115: 189-191.

Rooke, L., Bekes, F, Fido, R., Barro, F. Gras, P., Tatham, A.S., Parcelo, P., Lazzeri, P. and Shewry, P.R. (1999). Over expression of a gluten protein in transgenic wheat results in greatly increased dough strength. *J. Cereal Science.* 30: 115-120.

Saiki, R.K., Gelfand,D.H., Stoffel, S., Scharf, S.J., Higuchi, R., Hom, G.T., Mullis, K.B. and Erlich, H.A. (1988). Primer directed enzymatic amplification of DNA with Thermostable DNA Polymerase. Science. 239: 487-491.

Shan, L., Molberg, O., Parrot, I., Hausch, F., Filiz, F., Gray, G.M., Sollid, L.M. and Khosla, C. (2002). Structural basis for gluten intolerance in celiac sprue. *Science* 297, 2275–2279.

Shewry, P.R., Halford, N.G. and Tatham, A.S. (1992). High-molecular weight subunits of wheat glutenin. *J. Cereal Science.* 15: 105-120.

Shewry, P.R., Tatham, A.S. and Kasarda, D.D. (1992). Cereal Proteins and coelic disease In: Marh, M.N. (ed) Coeliac disease pp. 305-348. Blackwell Scientific Publications, London.

Shewry, P.R., Tatham, A.S., Forde, J., Kreis, M. and Miflin, B.J. (1986). The classification and nomenclature of wheat gluten proteins: A reassessment. *J. Cereal Sci.* 4: 97-106.

Singh, N.K. and Shepherd, K.W. (1988). Linkage mapping of genes controlling of endosperm storage proteins in wheat. I. Genes on the short arm of group 1 chromosome. *Theor. Appl. Genet.* 75: 628-641.

Souza, E. Kruk, M. and Sundrman, D.W. (1994). Association of sugar-snap cookie quality with high molecular weight glutenin alleles in soft white spring wheats. *Cereal Chem.* 71: 601-605.

Sozinov, A.A. and Poperelya, F.A. (1980). Genetic classification of prolamins and its use for plant breeding. *Ann. Technol. Agric.* 29; 229-245.

Tamas, L., Bekes, F., Greenfield, J., Tatham, A.S., Gras, P.W., Shewry, P.R. and Appels, R. (1998). Heterologous expression and dough mixing studies of wild-type and mutant C hordeins. *J. Cereal Sci.* 27: 15-22.

Tatham, A.S., Shewry, P.R. and Belton, P.S. (1990). Structural studies of cereal prolamins including wheat gluten. *Adv. Cereal Sci. Techn.* 10: 1-78.

Vaccino, P., Redaelli, R., Metakovsky, E.V., Borghi, B., Corbellini, M. and Pogna, N.E. (2002). Identification of novel low M_r glutenin subunits in the high quality bread wheat cv Salmone and their effects on gluten quality. *Theor. Appl. Genet.* 105: 43-49.

Van Campenhout, S., Vander Steppen, A., Sagi, L. and Volckaert, G. (1995). Locus-specific primers for LMW specific genes on each of group-1 chromosomes of hexaploid wheat. *Theor. Appl. Genet.* 91: 313-319.

Varghese, J.P., Struss, D. and Kazman, M.E. (1996). Rapid screening of selected European winter wheat varieties ands segregating populations for the Glu-d1d allele using PCR. *Plant Breeding.* 115: 451-454.

Vasil, I.K. and Anderson, O.D. (1997). Genetic engineering of wheat gluten. *Trends in Plant Sci.* 2: 292-297.

Vasil, V., Srivastava, V. Castillo, A.M., Fromm, M.E. and Vasil, I.K. (1993). Rapid production ofntransgenic wheat plants by direct bombardment of cultured immature embryos. *Biotechnology.* 11: 1553-1558.

Vazquez, J. F., Ruiz, M., Nieto-Talatriz, M.T. and Albuquerque, M.M. (1996). Effects on gluten Strength of low Mr glutenin subunits coded by alleles at *Glu-A3* and *Glu-B3* loci in durum wheat. *J. Cereal Sci.* 24: 125-130.

VonBuren, M, Luthy, J. and Hubner, P. (2000). A spelt specific Y gliadin gene: discovery and detection. *Theor. Appl. Genet,* 100: 271-279.

Weegles, P.L., Hamer, R.J. and Schofield, J.D. (1996). Functional properties of wheat glutenins. *J. Cereal Science.* 23: 1-.

Weegles, P.L., Marseille, J.P., Bosveld, P. and Hamer, R.J. (1994). Large-scale separation of gliadins and their bread making quality. *J. Cereal Sci.* 20: 253-264.

Weeks, J.T., Anderson, O.D. and Blechl, A.E. (1993). Rapid production of multiple independent lines of fertile transgenic wheat (Tritucum aestivum.). *Plant Physio.* 102: 1077-1084.

Wrigley, C.W., Robinson, P.J. and Wiliams, W.T. (1982). Association between individual gliadin proteins and quality, agronomic and morphological attributes of wheat cultivars. Aust. *J. Agri. Resear* 33: 409-418.

Zhang, W., Gianibelli, M.C., Rampling, L. and Gale, K.R. (2004). Characterization and marker development for low molecular weight glutenin genes from *Glu-A3* alleles of bread wheat (*Triticum aestivum* L.). *Theor. Appl. Genet.* 108:1409-1419.

Zhao, X.L., Xia, X.C., He, Z.H., Lei, Z.S., Appels, R., Yang, Y., Sun, Q.X. and Ma, W. (2007). Novel DNA variations to characterize low molecular weight glutenin Glu-D3 genes and develop STS markers in common wheat. *Theor. Appl. Genet.* 114: 451-460.

❑❑❑

Chapter 7

WHEAT GRAIN HARDNESS

Introduction

Bread wheat (*Triticum aestivum* L.) is one of the most widely cultivated and consumed crops in the world. Wheat is perhaps best known for its different end-use baking properties as determined by grain texture, protein content and quality. Wheat is generally traded by and classified into "hard" and "soft" market classes on the basis of grain hardness or texture of the wheat kernel (caryopsis). Soft wheats have softer endosperm texture that fractures easily, therefore requiring less energy to mill, yields smaller particles, and has less starch damage after milling when compared to hard wheats. Hard and soft are used with slightly different connotations in different parts of the world. In Western Europe, hard is applied primarily to durum wheats and the soft applied to bread wheat. Elsewhere the term is applied to different cultivars within the species *Triticum aestivum*. In general, hard wheats are used for bread baking, while soft wheats are used for cookies and cake making. Hard wheat requires more force to crack the kernel, maintains a larger particle size, passes through sieves more easily and has more damaged starch MacRitchie (1980). Higher starch damage is required for good bread making quality Tipples *et al.* (1994) but is detrimental to the biscuit making quality.

Several theories were put to explain the mechanism of grain hardness. Using micro-penetrometer test, Barlow *et al.* 1973 found no difference in hardness between the hardness of either protein or starch from different cultivars and suggested that the differences in hardness among cultivars may be because of the bond between protein and starch molecules. Another evidence to support that hardness is related to the adhesion of starch and protein was provided by Simmonds (1971). The starch samples prepared by flotation in a non-aqueous solvent were quite different depending on whether the starting material was hard or soft flour. Similar conclusions concerning the strength of starch-protein interface were reached by Hoseney and Seib in (1973). Using scanning electron microscope (SEM), they showed that the wetting of protein on starch was much better with hard wheat than with soft wheat. The better wetting gives rise to stronger bond as evidenced by the number of starch granules that are broken when the kernel is fractured. Soft wheats have essentially no broken granules, hard wheats a moderate level and durum wheats a high level. This was interpreted as evidence of increasing strength of the protein-starch bond. Barlow *et al.* (1973) suggested that a water soluble material was responsible for the adhesion. Using fluorescence antibody technique they showed that water soluble protein is concentrated at the starch-protein interface. Later Greenwell and Schofield in (1986) implicated a low molecular weight starch surface protein called friabilin in determining grain softness. The friabilin was present abundantly in soft wheat water washed starch, low in hard wheat and absent in durum wheats. Subsequently molecular studies established the role of friabilin (having puroindolines) in explaining grain hardness.

Measurement of Grain Hardness

Generally, methods that measure the texture of wheat grain quantify the texture phenotype of bulk grain lots or individual kernels. As such, most methods provide a discrete, numerical separation of the qualitative classes of soft and hard. Although different methods provide slightly different phenotypes, the underlying utility of these methods is that they capture the physical manifestations of the hardness gene. As discussed below, the Hardness gene has two primary phenotypes (alleles), those being soft Ha) and hard (ha). Soft wheats are more friable, require less energy to mill and produce flours and meals with reduced particle- size distribution, including many free starch granules. Hard wheat meals are coarser, have more broken and damaged starch granules, but flow and bolt more easily. When the gene is absent, as in the case of durum, very hard phenotype is observed.

To facilitate information exchange, some standardization of texture methods has occurred. The three most commonly used texture methods are the particle size index (PSI), near- infrared reflectance (NIR) and the Single Kernel Characterization System (SKCS). As MacRitchie (1980) noted, methods of texture measurement might be grouped according to whether they grind, crush, abrade or indent the sample, but, none of the methods, however, measures a fundamental material property. One of the obstacles to accomplishing this is the difficulty which the sample geometry of the grain presents. Glenn *et al.* (1991) addressed this issue by devising a means of producing cylinders of uniform dimension from the center of the endosperm. However, this method is very labor-intensive and has not been widely adopted. Of the three common methods mentioned above, the two most wide-spread approaches to texture measurement rely on differences in granularity (particle size distribution) of meals or flours after grinding or milling, respectively. The first, PSI, quantifies granularity by sifting the ground or milled material and expressing the proportion of material that passes through a sieve of defined aperture. Consequently, higher numbers indicate softer texture (due to lower particle size distribution of soft wheat meals). The second, NIR provides an indirect assessment of particle size through the optical reflectance of ground samples. More recently, a desire in the U.S. system to identify mixtures of soft and hard wheats led to the development of Single-Kernel Characterization System (model 4100) (Perten Instruments, Springfield, IL). Of historical interest, apparently the first mechanical means of measuring wheat grain texture was developed around 1908 and determined the force required to crush individual kernels. In SKCS, grain hardness is determined on an approximately 300 kernels of each sample and by the weight, size and moisture of each grain is measured and then crushed. Hardness index (HI) is calculated by the software in the instrument taking into account the moisture, grain weight and size. The HI less than 50 is considered soft and more than 50 as hard.

SKCS Hardness Index of Indian Wheat Varieties

The SKCS hardness index of 285 released varieties in India showed harder texture of grains. HI varied from 40.2 to 101.3 with an average value of 82.0. Surprisingly only four varieties exhibited soft texture with HI around 40 and all other genotypes were hard with varying HI values. Since *chapati* has been the staple diet in India, the emphasis has been to develop wheat varieties suitable for this purpose. Harder grains and medium strong gluten are preferred for good chapati making quality while softer grains and weak gluten for good cookie-making quality. The data clearly demonstrated the limited number of soft varieties in India for their use in the development of cultivars for diverse products. This strongly supported the view that only within class hybridizations have been predominant in Indian wheat breeding programme with only few exceptions. Therefore to widen gene pool diversity and to harness the new recombinants through interclass hybridizations a large numbers of germplasm lines were evaluated for grain hardness. Seventy genotypes were identified with soft grain characteristics and batter biscuit making quality. However, most of the identified genotypes were tall in nature and with lodging susceptibility. Performance of all the wheat varieties and both hard and soft germplasm lines was assessed for two consecutive years for grain hardness (Fig. 1).

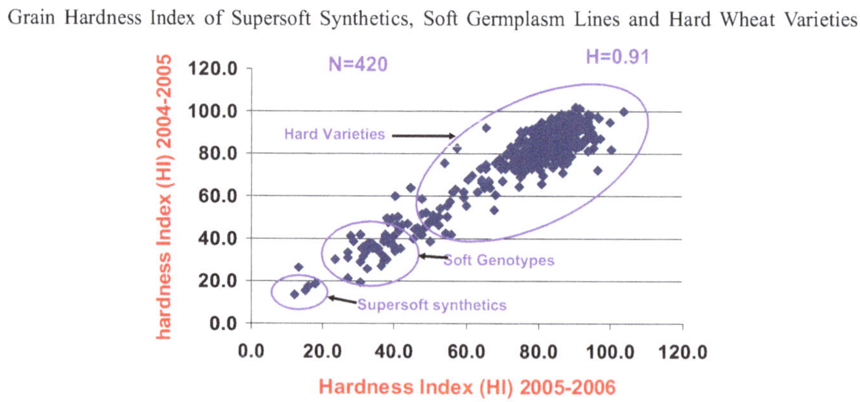

Fig. 1. Peformance of wheat varieties and germplasm lines over a period of two years for grain hardness showing high heritability

Genetics of Wheat Grain Texture

Although the physical difference among grain lots known as texture has been long recognized and attributed as a genetic factor, it was only in the second half of the 20th century that the means of measuring grain texture (see above) and the commercial interest in segregating and breeding soft and hard wheats combined to prompt careful genetic studies of the mode of texture inheritance. These reports indicated that a single major gene influenced grain texture. The gene was localized on 5D short arm and designated the gene with the soft allele as Ha, and the hard allele as ha. These studies were followed by physico chemical analysis of grains to identify chemical components responsible for hardness/softness. Using electron microscopic and staining techniques, Barlow *et al.*(1973) and Simmonds *et al.* (1973) indicated the presence of substances in between the starch granules and protein bodies in the endosperm. The amount of the material adhering to the surface of the starch granules differed between the hard and soft wheat grains. Subsequently Greenwell and Schofield (1986) demonstrated the presence of friabilin, a 15 kD starch surface protein, abundantly on water washed soft wheat starch, scarce on hard wheat starch and absent on durum wheat starch. They hypothesized that quantitative difference in friabilin might influence the degree of adhesion between starch granules and protein matrix surrounding the starch and thus variation in grain texture. This led to detailed biochemical and molecular studies of friabilin.

Molecular Basis of Grain Hardness

The report of Greenwell and Schofield prompted research in several laboratories across the world. The most significant result of this work was that friabilin was present in whole meal of hard and soft wheats. Although some hard wheat cultivars exhibited a reduced amount of friabilin, these data necessitated a re-evaluation of the then existing model which suggested that a quantitative difference in friabilin levels produced the difference between hard and soft wheats. Subsequent studies showed that two classes of bound polar lipid (glyco & phospholipids) follow the same pattern of occurrence as friabilin. This exhibited that friabilin might influence starch and protein interactions through polar lipids.

Molecular studies during 1990's exhibited that friabilin is composed of two components as puroindoline A (Pina) and puroindoline B (Pinb) whose transcripts are controlled by gene on 5D chromosome. To further localize puroindoline genes on sub-arm, deletion stocks of Chinese Spring for 5DS developed were utilized. There was no amplification of puroindoline genes in all the homozygous deletion lines. The heterozygous and normal lines exhibited PCR amplification using primers for both glycine type Pinb and Pina. This demonstrates that *Pina/Pinb* genes are tightly linked and present at the distal part of short arm of 5D chromosome where *Ha* gene is reported to be present. This supported the view that puroindolines may be the part of *Ha* gene itself. Transformation studies have also been conducted by some workers elucidating the role of puroindolines in grain softness Beecher *et al.* (2002). Krishnamurthy *et al.* (2001) reported that expression of wheat Pina and/or Pinb genes in transgenic rice reduced grain hardness. Flour prepared from these softer seeds had reduced starch damage and an increased percentage of fine flour particles. Campbell *et al.* (2001) observed high correlation between puroindoline with starch damage, cookie diameter and AWRC in a cross between soft and hard wheat.

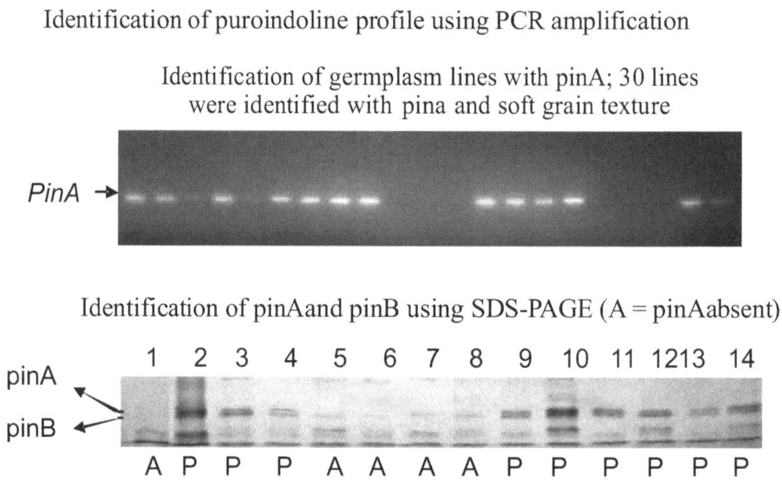

Fig. 2. Puroindoline profile can be known through either PCR based amplification of puroindoline genes or SDS-PAGE separation of puroindoline proteins.

Puroindolines are Triton X-114 soluble basic Cystine rich proteins with tryptophan rich domain. Amino acid sequence data from several sources shed new light on the friabilin phenomenon and supported the fact that friabilin exists in two forms. Giroux & Morris, 1997 showed that either glycine to serine change in Pinb at 46 position or null mutation in *Pina* were associated with hardness in wheat. However, subsequent studies by Ram, S *et al.* (2002 & 2005) in India & Turnbull *et al.* (2000) in Australia indicated exceptions where hard wheats do not exhibit either mutation reported above.

Puronidoline alleles can be identified by both PCR amplification as well as SDS-PAGE analysis (Fig. 2). A broad survey conducted in India for puroindolines indicated prevalence of null mutation in Pina in varieties released in India while glycine to serine changes in Pinb was prevalent wheat varieties from Kansas State, USA Ram, S *et al.* (2002). Recent studies by Lillemmo *et al.* 2006 confirmed that Pina null mutations are the predominant in wheat lines developed at CYMMIT and released throughout world including India.

Allelic Variation in Puroindolines

Glycine to serine change was due to presence of point mutation from adenine to guanine in the coding sequence. This information was used to design and synthesize glycine or serine specific primers. The full-length *Pina* and *Pinb* were amplified using *Pina* specific and *Pinb* specific primers, respectively, and produced 450 bp length PCR fragment. Partial *Pinb* sequence was amplified using glycine specific or serine specific forward primers and generated 250 bp long PCR fragment differentiating some soft and hard varieties. There was no amplification of both the puroindoline genes in durum wheats analyzed (Table 1) since durum wheats do not have D genome on which puroindoline genes are present. Protein analysis also demonstrated the absence of Pina band in null mutants of Pina. SKCS hardness analysis of Indian wheats showed that grains with null mutation in *Pina* were harder by5.3 points than those having serine type Pinb. The mean value of hardness was 77.6 in the varieties having null mutation in *Pina* and 72.3 in varieties with serine type of *Pinb*. This indicates that both the puroindolines together might influence the level of grain hardness.

In addition to null mutation in *Pina* and glycine to serine change in Pinb, large numbers of mutations were identified. The currently known mutations in *Pina* and *Pinb* conferring hard endosperm are listed in the (Table 1). These mutations appear to have occurred independently from each other and some can be traced back to specific geographic areas. Whereas the *Pinb-D1b* allele prevails among the spring and winter wheats of North America, Europe, China and Australia (Cane *et al.* 2004, Chen *et al.* 2006, Lillemo and Morris 2000, Morris *et al.* 2001 and Xia *et al.* 2005), *Pinb-D1c* and *Pinb-D1d* are mostly found in Western Europe (Lillemo and Morris 2000). The three rare alleles *Pinb-D1e, Pinb-D1f, Pinb-D1g* are confined to North American cultivars, Morris *et al.* (2001) and *Pinb-D1p* has only been found in Chinese wheat (Chen *et al.* 2006, Ikeda *et al.* 2005 and Xia *et al.* 2005). Likewise, the PINA null mutation (*Pina-D1b*) has been found to be the primary cause of hardness in Indian wheat Ram, S *et al.* (2002). Recent studies by Lillemmo *et al.* (2006) confirmed that Pina null mutations are predominant in wheat lines developed at CYMMIT and released throughout world including India. Additional variations in puroindoline were reported from diploids progenitors of hexaploids as Aegilops tauschii Massa *et al.* (2004). The synthetic hexaploids developed from these diploids exhibited softer grain characteristics (Gedye *et al.* 2004, Ram, S *et al.* 2008). Further details of allelic variation in puroindoline are provided in recent review by Bhave and Morris (2008).

Table 1: Molecular changes, phenotypes and references for the currently known *Pina* and *Pinb* hardness mutations in conventional bread wheat (Lillemmo *et al.* 2006)

Pina	Pinb	Molecular change	Phenotype
Pina-D1a	*Pinb-D1a*	–	Soft, wild-type
Pina-D1b	*Pinb-D1a*	Unknown	Hard, PINA null
Pina-D1l	*Pinb-D1a*	Base deletion in Gln-61→frame shift.	Hard, PINA null
Pina-D1m	*Pinb-D1a*	Pro-35 to Ser-35	Hard
Pina-D1n	*Pinb-D1a*	Trp-43 to stop codon	Hard, PINA null
Pina-D1a	*Pinb-D1b*	Gly-46 to Ser-46	Hard
Pina-D1a	*Pinb-D1c*	Leu-60 to Pro-60	Hard
Pina-D1a	*Pinb-D1d*	Trp-44 to Arg-44	Hard
Pina-D1a	*Pinb-D1e*	Trp-39 to stop codon	Hard, PINB null
Pina-D1a	*Pinb-D1f*	Trp-44 to stop codon	Hard, PINB null
Pina-D1a	*Pinb-D1g*	Cys-56 to stop codon	Hard, PINB null
Pina-D1a	*Pinb-D1p*	Base deletion in Lys-42→frame shift.	Hard, PINB null
Pina-D1a	*Pinb-D1q*	Trp-44 to Leu-44	Hard
Pina-D1a	*Pinb-D1r*	Base insertion in Glu-14→frame shift.	Hard, PINB null
Pina-D1a	*Pinb-D1s*	Base insertion in Glu-14→frame shift+ Ala-40 to Thr-40	Hard, PINB null
Pina-D1a	*Pinb-D1t*	Gly-47 to Arg 47	Hard

Modern cultivated wheats have functional *Pina-D1* and *Pinb-D1* genes only on chromosome 5D and absent in homeologous chromosomes 5A and 5B Tranquilli *et al.* (2002). Homeologous puroindoline genes on chromosome 5A and 5B have not been observed in cultivated wheats Bonafede *et al.* (2007). However functional pin homologs are present in wheat–related species such as *Aegilops tauschii.* Additional variation on the *Ha* locus has recently been introduced into bread wheat through the artificial hybridization of durum wheat (*T. turgidum* ssp. *durum*) with different accessions of *Aegilops tauschii* to produce synthetic wheat Mujeeb-Kazi *et al.* (1996). Seven additional alleles of *Pina* (*Pina-D1c, Pina-D1d, Pina-D1e, Pina-D1f, Pina-D1h, Pina-D1i, Pina-D1j*) and six different *Pinb* alleles (*Pinb-D1h, Pinb-D1i, Pinb-D1j, Pinb-D1m, Pinb-D1n, Pinb-D1o*) from *Ae. tauschii* were found to confer soft endosperm in synthetic wheat (Gedye *et al.* 2004 and Massa *et al.* 2004). Further studies showed that introgression of functional copies of puroindolines genes from diploid species into hexaploid wheats resulted into softer grains, See *et al.* (2004). However these studies used either complete chromosome substitution lines or large translocations involving most of the short arm of chromosome 5A, limiting their usefulness in commercial breeding programme. This is because of presence of undesirable genes in larger part of the alien chromosome. However, this can be overcome by eliminating undesirable segment by homeologous recombination using *Ph1* mutants and marker assisted selection. This was accomplished by Bonafede *et al.* (2007) by transferring short segment of *T. monococcum* chromatin proximal to *Ha* locus.

Puroindolines and Bread Making Quality

Generally the selection of bread wheat cultivars for bread making has been carried out using various technological tests without direct grain hardness measurements. However, recently association of puroindoline alleles with bread making quality has been reported. The reconstitution experiments using Pina free cultivars have shown that very low amounts of puroindolines (0.05 to 0.1%) have important effects on bread making Dubreil *et al.* (1998). A close relationship was found between fine grain crumb provided by reconstituted flours with puroindolines and the fine structure of corresponding dough liquor foams. Puroindolines were found to be essential for foaming properties of dough liquor. In addition lipid reconstitution experiments of defatted flours have shown that wheat polar lipids act as bread volume improvers, while non polar lipids are detrimental to bread volume. These results are the effects of complex mechanism involving the surface properties of polar lipids and proteins dispersed in the aqueous phase of the dough, the competitive adsorption between lipids and proteins which is essential for the formation and stability of the dough foam.

Puroindolines are basic proteins that contain five disulphide bridges and a unique tryptophan domain that is involved in lipid recognition. These proteins are capable of preventing the destabilization of protein foams by lipids. This foam controls gas retention and expansion during mixing, proving and the early stages of baking. Moreover, low amounts of puroindolines are also capable of inducing important changes in dough tenacity and extensibility. However, the relationship between good foaming and lipid binding properties of puroindolines and their effects on dough rheology remains obscure. Relationship of puroindolines to technological properties in French bread wheats was studied by Igrejas *et al.* (2001). They reported that in soft samples, pin (A+B) content was negatively correlated with loaf volume. Multiple regression analyses revealed that pinb content explained variation of dough strength (W) and loaf volume, both in hard and soft wheats, however it was more significant in soft wheats. The hardness allele *Pinb-D1b* has been found to improve milling quality and brightness of flour colour, desirable for making of white salty noodles, but negatively influenced viscoelastic properties of flour Nagamine *et al.* (2003). In addition milling yield experiments indicated that *Pina-D1b/Pinb-D1a* genotype had significantly higher water absorption and significantly lower milling yield than the *Pina-D1a/Pinb-D1b* genotype. This demonstrated that these genes will impede the development of hard-grained cultivars that combine high water absorption and high milling yield Cane *et al.* (2004).

Thus characterization and manipulation of puroindoline genes has a potential for improving grain quality. In addition, expression of puroindoline may give associated plant health benefits of anti-microbial activity. Puroindolines can be used as molecular markers in breeding programme to improve wheat quality. Reports indicate that interclass hybridizations where crosses among soft and hard wheats can be used in improving quality of soft and hard class separately. Molecular markers along with microlevel tests for grain hardness and gluten strength have proved useful in selecting desirable segregants for each class separately.

Puroindolines and Disease Resistance

Puroindoline not only affect grain texture but also have role in defense against microbes. Several studies have demonstrated that puroindolines have antimicrobial properties Bhave and Morris (2008). Transformation experiments by Krishnamurthy *et al.* (2001) in rice, Luo *et al.* (2007) in tetraploid wheat and Faiz *et al.* (2004) in apple have demonstrated that puroindoline impart fungal resistance. The effect of Pina and Pinb on diseases resistance individually is yet to be studies in detail. They may have synergistic effect or have different potency against pathogens. Ectopic applications have also shown the effect of puroindolines on different pathogens. Capparelli *et al.* (2007)

showed that purified native and recombinant Pina and Pinb had potent against *E.coli.* High concentrations of proteins were also effective against intracellular infections, indicating their potential for ectopic applications.

Summary

Based on grain texture wheat is classified into two distinct classes as hard and soft. Specific products are made from both the classes. Grain hardness is measured by different means such crushing, indenting etc. Recently single kernel hardness determination has been developed which is based on grain shape, size, moisture content and crushing force. The data is computerized and helps in differentiating hard and soft wheat as well as variation in hardness index among both the classes. This has made easier to select early generation line sin breeding programme. This along with molecular marker data has facilitated the improvement of wheat for different products. Molecular basis of grain hardness has been established and two closely linked major genes coding puroindolines have identified associated with grain hardness. Allelic diversity in puroindoline genes has been associated with some of the grain quality traits. However, the exact nature of the action of puroindolines in determining grain texture is yet to be established.

References

Barlow, K.K., Buttrose, M.S., Simmonds, D.H. and Vesk, M. (1973). The nature of the starch-protein interface in wheat endosperm. *Cereal Chem* 50: 443-454

Beecher, B., Bettge A., Smidansky, E. and Giroux, M.J. (2002). Expresion of wild type Pinb sequence in transgenic wheat complements a hard phenotype. *Theor. Appl. Genet.* 105: 870-877.

Bhave M. and Morris, C.F. (2008). Molecular genetics of puroindolines and related genes: allelic diversity in wheat and other grasses. *Plant Mol. Biol.* 66: 205-219.

Bhave, M. and Morris, C.F. (2008). Molecular genetics of puroindolines and related genes: regulation of expression, membrane binding properties and applications. *Plant Mol. Biol.* 66: 221-231.

Bonefede, M., Kong, L., Tranquilli, G., Ohm, H. and Dubcovsky, J. (2007). Reduction of Triticum monococcum chromosome segment carrying the softness genes Pina and Pinb translocated to bread wheat. *Crop Sci.* 4: 821-828.

Campbell K.G., Finney P.L., Bergman C.J., Gualberto D.G., Anderson J.A., Giroux, M. J., Siritunga D., Zhu,J., Gendre F., Roue C., Verel A. and Sorrells M.E. (2001). Quantitative trait loci associated with milling and baking quality in a soft x hard wheat cross. *Crop Sci* 41: 1275-1285.

Cane, K., Spackman, M. and Eagles, H.A. (2004). Puroindoline genes and their effects on grain quality traits in southern Australian wheat cultivars, *Aust. J. Agri. Res.* 55: 89–95.

Capparelli, R., Ventimiglia, I. and Palumbo, D. (2007). Expression of recombinant puroindolines for the treatment of staphylococcal skin infections (*Acne vulgaris*). *J. Biotechnology.* 128: 606-614.

Chen, F., He, Z.H., Xia, X.C., Lillemo, M. and Morris, C.F. (2005). A new puroindoline b mutation present in Chinese winter wheat cultivar Jingdong 11. *J.Cereal Sci.* 42: 267–269.

Chen, F., He, Z.H., Xia, X.C., Xia, L.Q., Zhang, X.Y., Lillemo, M. and Morris, C.F. (2006). Molecular and biochemical characterization of puroindoline a and b alleles in Chinese landraces and historical cultivars, *Theor. Appl.Genet.* 112: 400–409.

Dubreil, L., Meliande, S., Chiron, H., Compoint, J., Quillien, L., Branlard, G. and Marion, D. (1998). Effect of puroindolines on the bread making properties of wheat flour. *Cereal Chem.* 75: 222-229.

Faiz, M., Sourice, S. and Dupuis, F. *et al* (2004). Expression of puroindoline-β reduces scab susceptibility in transgenic apple (*Malus X domestica* Borkh). *Plant Science.* 167: 347-354.

Gazza, L., Nocente, F., NG, P.K.W. and Pogna, N.E. (2005). Genetic and biochemical analysis of common wheat cultivars lacking puroindoline-α. *Theor. Appl. Genet.* 110: 470–478.

Gedye, K.R., Morris C.F. and Bettge, A.D. (2004). Determination and evaluation of the sequence and textural effects of the puroindoline α and puroindoline β genes in a population of synthetic hexaploid wheat. *Theor. Appl. Genet.* 109: 1597–1603.

Giroux, M.J. and Morris, C.F. (1997). A glycine to serine change in puroindoline β is associated with wheat grain hardness and low levels of starch-surface friabilin. *Theor. Appl. Genet.* 95: 857–864.

Giroux, M.J. and Morris, C.F. (1998). Wheat grain hardness results from highly conserved mutations in the friabilin components puroindoline α and β, *Proceedings of the National Academy of Sciences of the United States of America* 95: 6262–6266.

Glenn, G.M, Younce, F.L. and Pitts, M.J. (1991). Fundamental physical properties characterizing the hardness of wheat endosperm. *J. Cereal Science.* 13: 179-194.

Greenwell, P. and Schofield, J.D. (1986). A starch granule protein associated with endosperm softness in wheat. *Cereal Chem.* 63: 379-380

Hoseney, R.C. and Seib, P.A. (1973). Structural differences in hard and soft wheat. *Bakers Digest.* 47(6): 26-28.

Igrejas, G., Gaborit, T., Oury, F., Chiron, H., Marion, D. and Branlard, G. (2001). Genetic and environmental effects on puroindoline-α and puroindoline-β content and their relationship to technological properties in French bread wheats. *J. Cereal Sci.* 34: 37-47.

Ikeda, T.M., Ohnishi, N., Nagamine, T., Oda, S., Hisatomi, T. and Yano, H. (2005). Identification of new puroindoline genotypes and their relationship to flour texture among wheat cultivars, *J.Cereal Sci.* 41: 1–6.

Krishnamurthy, K. Balconi, C and Sherwood, J.E. (2001). Wheat puroindolines enhance fungal disease resistance in transgenic rice. *Mol Plant Microbe Interact.* 14: 1255-1260.

Lillemo, M. and Morris, C.F. (2000). A leucine to proline mutation in puroindoline b is frequently present in hard wheats from Northern Europe. *Theor. Appl. Genet.*100: 1100–1107.

Lillemo, M., Chen, F., Xia, X., William, M., Peña, R.J., Trethowan, R. and He, Z. (2006). Puroindoline grain hardness alleles in CIMMYT bread wheat germplasm. *J. Cereal Sci.* 44(1): 86-92.

Luo L., Zhang J., Yang G., Li, Y., Li, K. and He, G. (2007). Expression of puroindoline a enhances leaf rust resistance in transgenic tetraploid wheat *Mol. Bio. Reports.* 35(2): 195-200.

MacRitchie, F. (1980). In *Advances in Cereal Science and Technology* pp 271-326. Vol III Y Pomeranz, Eds., American Association of Cereal Chemists, St. Paul, MN.

Massa, A.N., Morris. C.F. and Gill, B.S. (2004). Sequence diversity of puroindoline-α, puroindoline-b, and the grain softness protein genes in *Aegilops tauschii* Coss *Crop Sci.* 44: 1808–1816.

McIntosh, R.A., Devos, K.M., Dubcovsky, J., Rogers, W.J., Morris, C.F., Appels, R. and Anderson, O.D. (2005). Catalogue of gene symbols for wheat: 2005 supplement, published online at http;/wheat.pw.usda.gov/ggpages/wgc/2005upd.html.

Morris, C.F. (2002). Puroindolines: the molecular genetic basis of wheat grain hardness. *Plant Mol. Bio.* 48: 633–647.

Morris, C.F., Lillemo, M., Simeone, M.C., Giroux, M.J., Babb, S.L. and Kidwell, K.K. (2001). Prevalence of puroindoline grain hardness genotypes among historically significant North American spring and winter wheats. *Crop Sci.* 41: 218–228.

Mujeeb-Kazi, A., Rosas, V. and Roldan, S. (1996). Conservation of the genetic variation of *Triticum tauschii* (Coss.) Schmalh (*Aegilops squarrosa* auct. non L.) in synthetic hexaploid wheats (*T. turgidum* L.s. lat. x *T. tauschii*; $2n=6x=42$, AABBDD) and its potential utilization for wheat improvement, *Genetic Resources and Crop Evolution* 43: 129–134.

Nagamine, T., Ikeda, T., Yanagisawa, T., Yanaka, M. and Ishikawa, N. (2003). The effect of hardness allele pinB-D1b on the flour quality of wheat for Japanese white salty noodles. *J. Cereal Sci.* 37: 337-342

Ram, S., Boyko, E., Giroux, M.J. and Gill, B.S. (2002). Null mutation in puroindoline α is prevalent in Indian wheats: puroindoline genes are located in the distal part of 5DS. *J. Plant Biochem. and Biotech.* 11: 79–83.

Ram, S., Jain, N., Jag Shoran and Singh, R. (2005). New frame shift mutation in puroindoline b in Indian wheat cultivars Hyb65 and NI5439. *J. Plant Bioch. and Biotech.* 14: 45–48.

See, D.R., Giroux, M. and Gill, B.S. (2004). Effect of multiple copies of puroindolines genes on grain softness. *Crop Sci.* 44: 1248-1253.

Ram S., Tyagi B.S. and Mishra B. (2008). Utility of synthetic hexaploids in the improvement of wheat grain quality. *Indian J. Genet. Plant Breed.* 67(3): 243-247.

Simmonds, D.H., Barlow, K.K. and Wrigley, C.W. (1973). The biochemical basis of grain hardness in wheat. *Cereal Chem.* 50: 553-563.

Tipples, K.H., Kilborn, R.H. and Preston, K.R. (1994). In *Wheat Production, Properties and Quality* pp 25-36 (W Bushuk, VF Rasper, Editors), Blackie Academic & Professional, Glasgow.

Tranquilli, G., Heaton, J, Chicaiza, O. and Dubcovsky, J. (2002). Substitutions and deletions of genes related to grain hardness in wheat and their effect on grain texture. *Crop Sci.* 42: 1812-1817.

Turnbull KM, Gaborit T, Marion D and Rahman S (2000) Variation in puroindiline polypeptides in Australian wheat cultivars in relation to grain hardness. *Aust. J. Plant Physiol.* 27: 153-158

Xia, L.Q., Chen, F., He, Z.H., Chen, X.M. and Morris, C.F. (2005). Occurrence of puroindoline alleles in Chinese winter wheats, *Cereal Chem.* 82: 38–43.

□□□

Chapter 8

Wheat Starch Structure, Biosynthesis and Quality

Introduction

Carbohydrates constitute nearly 80% of the total dry matter of wheat kernel. Starch is the predominant form, with cellulose, hemicellulose and pentosans present in lesser amount. Starch is present in discrete granules within the cells of the endosperm. Starch content is available in the range of 65% to 75% and generally related negatively to the protein content. Starch structure is a crucial determinant of many end-product uses of wheat. While the amylose /amylopectin ratio is a useful descriptor of starch structure, a more complete analysis of the population of the molecules which constitute a particular starch is considerably more revealing. The molecules range from the high degree of polymerization (10^7 - 10^9 glucose units) and heavily branched amylopectin population, to the lower degree of polymerization (10^3 - 10^6 glucose units), and infrequently branched amylopectin population. Starches with widely differing structures are suited to particular end uses. For example, starches with no amylose, the so called waxy starches are particularly suited to use as thickeners or gel forming agents because they form a clear stable gel with little retrogradation. In contrast, starch with high amylose content form hydrogen bonded insoluble aggregations which are suited to use in adhesives, plastics and as a source of dietary-fibre starch. The size and branching pattern of starch are also important determinants of the suitability of starches for particular end-uses due to the effects these parameters have on gelatinization temperature, starch swelling and viscosity.

Starch occurs in granular forms as large lenticular (A type) ranging in size 15-40 μm and small spherical (B type) ranging in size 1-10 μm granules. Small starch granules have been reported to increase the extensibility of the dough whereas large granules increase the resistance to extension. Starch has two components as amylose and amylopectin. Amylopectin usually the most abundant of the two (~70-80% w/w) is a branch on branch polymer composed solely of α-glucopyranosyl residues. Amylose is also a polymer of (1, 4) linked α-glucopyranosyl residues but is sparsely branched, with long interchain segments (Fig.1). Amylose content of wheat starch generally varies only to a little extent though recently waxy wheats have been identified and new genotypes developed. In addition to starch, starch granule contains a range of starch biosynthetic enzymes and lipid complexed within the amylose fraction. Bound polar lipids are probably involved in the interaction between puroindolines and starch surface imparting grain softness Greenblatt *et al.* (1995).

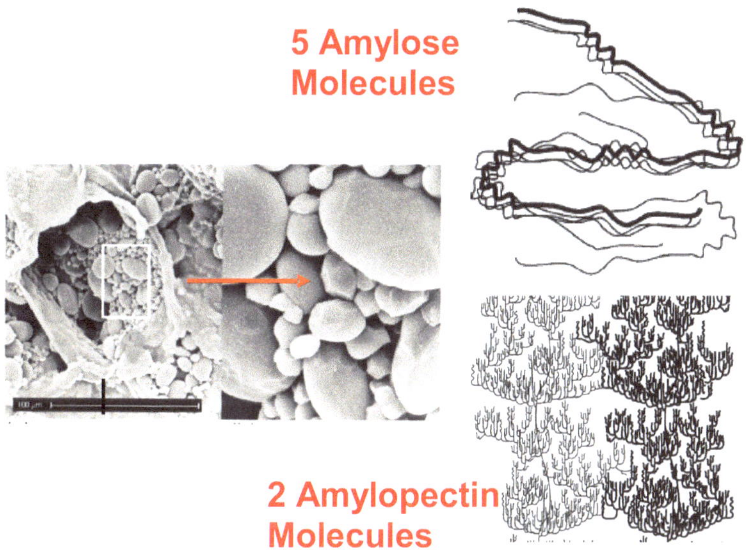

Fig. 1. Scanning electron micrographs of wheat starch granules showing a central endosperm region of a transverse section of a mature wheat grain which was prepared for scanning electron microscopy by briefly immersing the cut surface in water to remove soluble components. Magnified image of the area outlined by a white box in the Panel is also shown. Schematic diagram that illustrates the scale of the molecular size differences between amylose and amylopectin, and indicates the types of interactions involved in amylose-amylose intermolecular interactions, and amylopectin–amylopectin interactions. (figures are redrawn from Rahman *et al.* 2000)

Starch Biosynthesis

During the last two decades there has been dramatic expansion of knowledge of starch biosynthesis process in higher plants. Because of easy identification of mutants and transformation efficiency, rice, maize and arabidopsis have been used extensively for such studies. However, recently much attention has been paid towards wheat starch biosynthesis. Two issues have hindered the study of starch biosynthesis in wheat. One is the hexaploid nature of wheat genome which does not allow the identification of the wide range of revealing mutations because of the buffering effect of the presence of three genomes and second is the lower transformation efficiency as compared to diploid species such as rice. The probability of finding a spontaneous or induced triple null line is vanishingly small. However, recently these problems have been solved to an extent.

In cereals the analysis of starch biosynthesis has been dominated by the study of maize based on a large number of well-characterized mutants. However, in wheat there is relatively little information about starch biosynthesis. The identification of key genes involved in starch biosynthesis promotes the expectation of using metabolic engineering to specifically alter starch determinants to increase our understanding of biochemical interactions of isozyme activities. This knowledge is a prerequisite for the conception of plant products comprising novel or improved starches. The genes for each of these enzymes have been cloned in cereals. Attempts have been made to manipulate these genes to change the level as well as the structure of the starch. This will improve the starch functional properties not only for baking industry but also polymer industry. The detailed properties of these enzymes and the corresponding genes are given below (Fig.2).

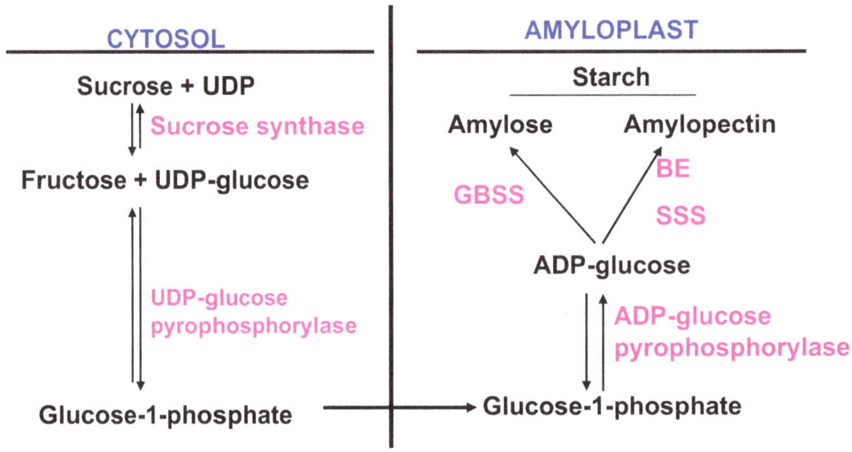

Fig. 2. Biochemical pathway from sucrose to starch in developing wheat endosperm. UDP = Uridine Diphosphate, GBSS = Granule Bound Starch Synthase, BE = Branching Enzyme, SSS = Soluble Starch Synthase

ADP-Glucose pyrophosphorylase

Starch content and its composition can be altered by manipulating starch biosynthetic enzymes. Starch biosynthesis in amyloplast is initiated with the formation of ADP-glucose (ADPG) from glucose–1 phosphate by the enzyme ADP-glucose pyrophosphorylase (AGP; EC 2.7.7.27). Subsequently other enzymes as granule bound starch synthase (GBSS), starch branching enzyme (BE) and starch debranching enzyme DBE) Preiss, (1992) are involved in starch biosynthesis during grain development. AGP is an allosterically regulated heterotetramer consisting of two large and two small subunits. The positive alloateric effector of AGP is 3-phosphoglycerate, whereas the negative allosteric effector is orthophosphate (Pi). Generally it is considered that AGP is located in amyloplast however recent evidence indicates that AGP is distributed between cytosol and amyloplast with the bulk of the activity in cytoplasm. The large and small subunit types share considerable homology and are likely to have arisen from gene duplication event. Immunological cross reactiveness and sequence comparison indicate that small subunit structure is highly conserved among plant groups and between photosynthetic and non-photosynthesis tissues and whereas the large subunits are more divergent.

In addition to the plastidial AGPase present in all starch-synthesizing tissues, biochemical evidence indicates the presence of at least two distinct AGPase enzymes in the endosperms of cereals which have been shown to correspond to plastidial and cytosolic isoforms of AGPase. In the developing endosperms of wheat, maize, barley, and rice the cytosolic isoform accounts for 65–95% of the total AGPase activity, implying that most of the storage starch biosynthesis in these tissues occurs through import of ADPGlc into amyloplasts. Plants possess multiple genes encoding either the AGP-L or AGP-S subunits, or both, and these are differentially expressed in different plant organs. This means that the AGPase subunit composition may vary in different parts of the same plant in tissues. The multiple genes encoding the AGP-L subunits show strong specificity in their expression, for example, being restricted to leaf, or root and endosperm in both barley and wheat or induced under specific conditions, such as increased sucrose or glucose levels. The differential expression of subunits in different tissues may produce AGPases with varying degrees of sensitivity to allosteric effectors which are suited to the particular metabolic demands of a given plant tissue/organ. In cereal endosperms, the subcellular localization of AGPase isoforms is thought to be regulated by differential splicing of AGPase genes Tetlow *et al.* (2004).

Starch Synthase

Starch synthase catalyses the synthesis of starch through the transfer of glucosyl moiety of ADP glucose to the non reducing end on an existing α-1,4 glucan chain. Two distinct classes of starch synthases are found in two locations within amyoplast; soluble starch synthase in stroma and starch synthase bound within starch granule. Kinetic analysis of these enzymes and mutants demonstrates that ADP glucose is the substrate for both soluble and granule bound starch synthases. The availability of waxy mutants from a number of species demonstrates that granule bound starch synthase is responsible for synthesis of amylose fraction and soluble starch synthase for amylopectin.

Analysis at the protein, cDNA and genomic DNA level shows the presence of 4 classes of starch synthase in wheat and maize endosperm, GBSS, SSI, SSII and SSIII (Li *et al.* 1999 and Cao *et al.* 2000). GBSS homeologous loci are present on 7A, 4A and 7D chromosomes. Loss of one of the three homeologous genes has large effect on noodle quality and starch swelling behaviour. While GBSS also contributes to amylopectin synthesis, the remaining components of starch synthases are thought to be primarily responsible for elongation of amylopectin chains.

In addition plants possess multiple isoforms of starch synthases (SSs), containing up to five isoforms that are categorized according to conserved sequence relationships. The isoforms within each of the major classes of SS genes are highly conserved in higher plants through the dicots and monocots. The major classes of SS genes can be broadly split into two groups, the first group primarily involved in amylose synthesis, and the second group principally confined to amylopectin biosynthesis. The first group of SS genes contains the granule-bound starch synthases (GBSS), and includes GBSSI and GBSSII isoforms. GBSSI is encoded by theWaxy locus in cereals, functioning specifically to elongate amylose and is found, essentially, completely within the granule matrix (one of the so-called granule-associated proteins). Expression of GBSSI appears to be mostly confined to storage tissues, and a second form of GBSS (GBSSII), which is encoded by a separate gene, is thought to be responsible for amylose synthesis in leaves and other non-storage tissues which accumulate transient starch.

The second group of SS genes (designated SSI, SSII, SSIII, and SSIV) are exclusively involved in amylopectin biosynthesis, and their distribution within the plastid between the stroma and starch granules varies between the species, tissue, and developmental stages. The individual SS isoforms from this group probably play unique roles in amylopectin biosynthesis. The study of SS mutants in a number of systems has been helpful in the assignment of in vivo functions/roles for the soluble and granule-associated SS isoforms in amylopectin synthesis. Valuable information about the roles of the SS isoforms *in vivo* is being derived from mutants lacking specific isoforms, and analysis of plants appears to show that each isoform performs a specific role in amylopectin synthesis. However, such data should be treated with caution, as in some cases there are pleiotropic effects of mutations on other enzymes of starch synthesis.

SSI is primarily responsible for the synthesis of the shortest glucan chains, and further extension of longer chains is achieved by the activities of SSII and SSIII isoforms, each of which act on progressively longer glucan chains. Two classes of SSII genes are found in monocots: SSIIa and SSIIb. The role of the latter in starch biosynthesis is unknown as no mutants have been identified to date. In vitro studies of the two SSII forms from maize reveal different substrate specificities and kinetic properties (Imparl-Radosevich *et al.* 2003). SSIIa predominates in cereal endosperms, whilst SSIIb is mostly confined to photosynthetic tissues. Loss of SSIIa (in monocots) and SSII (in dicots) results in reduced starch content, reduced amylopectin chain-length distribution, altered granule morphology, and reduced crystallinity, suggesting that the SSII forms have similar roles in starch biosynthesis across different species boundaries. In monocots, SSIIa plays a specific role in the synthesis of the intermediate-size glucan chains of DP 12–24 by elongating short chains of DP <10, and its loss/down-regulation has a dramatic impact on both the amount and composition of starch, despite the fact that SSIIa is a minor contributor to the total SS activities in cereal endosperms, as opposed to SSI and SSIII. Studies with plants lacking SSIII suggest that the primary role of this enzyme is amylopectin synthesis, although the impact of loss of SSIII appears to differ with the genetic background. Antisense suppression of SSIII in potato has a major impact on the synthesis of amylopectin, resulting in amylopectin with modified chainlength distribution and decreased starch synthesis. However, mutations in maize eliminating SSIII (du1) lead to a subtle phenotype which is only conspicuous in Waxy backgrounds Gao *et al.* (1998). Sequences for SSIV appear in a wide range of higher plants in EST databases although, to date, no mutants have been isolated with lesions in this gene and no role has been assigned for this class of SS in the process of starch biosynthesis.

Branching Enzyme

Starch branching enzymes (SBEs, EC 2.4.1.18) generate α-(1-6)-linkages by cleaving internal α-(1-4) bonds and transferring the released reducing ends to C6 hydroxyls to form the branched structure of the amylopectin molecule. This means that the enzyme is transglycosylase catalyzing first the hydrolysis of an α-1,4 glucosydic bond and the subsequent attachment of the severed chain (containing the original non-reducing end) to the hydroxyl group of C6 of a glucosyl unit of an α-1,4-glucan forming an α-1,6 branch point. Two major forms of branching enzyme (SBEI & SBEII) have been identified based on the order of elution of branching enzyme isoforms from DEAE–cellulose. Significant differences are observed in the activities of each isoform with different amylose and amylopectin substrates. The two classes of SBE differ in terms of the length of the glucan chain transferred in vitro and their substrate specificities; SBEII proteins transfer shorter chains and show a higher affinity towards amylopectin than their SBEI counterparts, which show higher rates of branching with amylose. Monocots and dicots differ in the isoforms of SBE they contain. In potato, two isoforms of SBE are present, SBEI and SBEII. Suppression of both SBEI and SBEII was required to generate starch with amylose content of >50%.

In monocots, SBEI and two forms of SBEII, SBEIIa and SBEIIb, are found. In maize, suppression of SBEIIb results in the "amylose-extender" phenotype, with amylose contents from 50% to 90%, whereas suppression of SBEI or SBEIIa had no impact on endosperm amylose content. Amylose-extender maize has been used to increase the RS content of processed foods. In contrast partial suppression by antisense of BE-I in potato tuber leads to subtle alterations in starch physico-chemical properties but not so alteration in the amylose/amylopectin ratio. In wheat, RNA interference used to down-regulate the two different isoforms of starch-branching enzyme (SBE) II (SBEIIa and SBEIIb) exhibited high amylose content. Suppression of SBEIIb expression alone had no effect on amylose content; however, suppression of both SBEIIa and SBEIIb expression resulted in starch containing >70% amylose Regina *et al.* (2006).

Debrancing Enzyme

In addition to SSs and SBEs, debranching enzymes (DBEs, EC 3.2.1.41, and EC 3.2.1.68) have been implicated in starch synthesis in plants. Two groups of DBEs exist in plants; isoamylase-type, and pullulanase-type (also known as limit-dextrinases) which efficiently hydrolyse (debranch) α-(1-6)-linkages in amylopectin and pullulan (a fungal polymer of maltotriose residues), respectively, and are part of the amylase 'super-family' of enzymes. The precise roles for the isoamylase-type and pullulanase-type DBEs in starch biosynthesis are not yet known. Two models have been proposed which could define a role for the DBEs in starch synthesis and phytoglycogen accumulation. The 'glucan-trimming' (pre-amylopectin trimming) model proposes that glucan trimming is required for amylopectin aggregation into an insoluble granular structure (Ball *et al.* 1996, Myers *et al.* 2000). DBE activity would be responsible for the removal of inappropriately positioned branches (pre-amylopectin) generated at the surface of the growing starch granules, which would otherwise prevent crystallization. As such, the debranched structure would favour the formation of parallel double helices, leading to polysaccharide aggregation. An alternative to the glucan-trimming model proposes that the DBEs function in starch synthesis indirectly in a 'clearing' role, removing soluble glucan that is not attached to the granule from the stroma. This removes a pool of substrates for the amylopectin synthesizing enzymes (SSs and SBEs), and thereby prevents the random/futile synthesis of glucan polymers by these enzymes which could cause the accumulation of phytoglycogen, ultimately leading to a reduction in the rate of starch synthesis. This model could also explain the accumulation of phytoglycogen at the expense of amylopectin observed in DBE mutants Zeeman *et al.* (1998b).

$$ATP + \alpha\text{-Glc-1-P} \xleftrightarrow{\text{ADP-Glc Pyrophosphorylase}} \text{ADP-Glc} + PPi$$

$$\text{ADP-Glc} + (\alpha\text{-1,4 Glc})_n \xrightarrow{\text{Starch Synthase}} \text{ADP} + (\alpha\text{-1,4 Glc})_{n+1}$$

$$\alpha\text{-1,4 Glucan} + \alpha\text{-1,4 Glucan} \xrightarrow{\text{Debranching Enzyme}} \alpha\text{-1,4 } \alpha\text{-1,6 Glucan} + \alpha\text{-1,4 Glucan}$$

Pathways of starch biosynthesis in amyloplast

Starch Gelatinization, Swelling & Hot Paste Viscosity Properties

Starch granule gelatinization involves the heat driven transformation of granules in excess of water from an ordered state to a disordered state. The initial stages of gelatinization involve the uptake of water lowering the glass transition temperature in the amorphous regions and causing a crystalline region to begin to melt. Further heating and water uptake results in the mobilization of amylose and the initiation of its leaching from the granule and the complete melting of the crystalline regions, marked by the loss of birefringence. Further swelling of the granules occurs, typically producing greatly enhanced viscosity in the starch water paste. The overall process of gelatinization is similar in both normal and waxy starch granules, although by definition, amylose is not present in a waxy starch and cannot be lost by leaching. These observations suggest that the properties of the amylopectin fraction of starch are most critical in controlling gelatinization and granule swelling in cereal starches. Amylose content has a major influence on starch granule swelling, mediated through the reduced mobility of amylose/lipid complexes in the granule that restrict water movement and swelling.

The pasting of starch is typically analysed in an instrument such as the Visco-amylograph or Rapid Visco Analuzer (RVA) which measures the resistance of starhse to shearing forces under defined hydration and temperature regimes. Following complete gelitinisation at high temperature, peak viscosity is reached, the starch is then cooled while continuing to be stirred. Further cooling of starch paste to ambient temperature results in the formation of either a gel, in the case of starches containing amylose, or a very week gel in the case of waxy starches. However the analysis time and sample size required for these instruments is too restrictive for them to be use routinely in a wheat breeding programme for early generation testing. Therefore simple test for measuring starch paste property is required. Flour swelling volume test has been developed for assessing the quality wheat flours of different varieties. Flour swelling volume (FSV) test measures the cumulative effect of starch quality, specially the amylose/amylopectin rationas reflected by the volume of gel produced when flour is heated in an excess of water. Starch composition has the greatest effect on FSV. FSV is related to the eating quality (texture) of white salted noodles as well as to the gene dosage of granule bound starch synthase that controls the amylose/amylopectin ratio. This method is applicable to wheat meal, flour, or starch.

Starch and Bread Quality

The key features of starch deposition in endosperm that control functionality are starch content, grain hardness, granule size distribution and shape, presence of endogenous lipids in the granule amylopection structure and the ratio of amylose to amylopectin. Each of these features may be amenable to modification by molecular/ genetic changes in genomic DNA. The starch serves as the source of carbon during yeast fermentation in bread making, dilutes gluten to desirable consistency, furnishes surface suitable for strong union with gluten, involves setting of bread loaf and plays key role in retrogradation during storage. The restricted type of starch gelatinization that occurs in bread, coupled with gluten denaturation, contribute in setting up of the structure of the final loaf of the bread. With cooling and ageing of bread, rearrangements in the starch fractions lead to a series of changes, including gelation and crystallization.

Starch retrogradation is thought to be the major cause of bread crumb firming commonly known as bread staling. It is during the baking stage that limited gelatinization occurs in the bread as a result of heat application. It is also during this stage that amylase action on starch will occur and that, if excess amounts of amylase enzyme is present, the bread crumb will become very sticky. In addition waxy wheats can increase shelf life of bread. However, optimum amylose/amylopectin proportions, in which the benefits of lower amylase levels overweigh the negative effects for specific bakery products needs to be identified. Bread prepared from 10-15% waxy wheats exhibits a crumb structure similar to that of bread from regular wheat but retained higher crumb moisture and lower crumb hardness. However when the blending ratio is increased to 25-50% waxy wheat, the loaves exhibited greater volume, softer crumb, more porous crumb structure and a higher level of starch degradation than loaves made with regular wheat Lee *et al.* (2001).

Starch and Noodle Quality

The behavior of starch in water during heating has been related to textural properties of certain types of noodles, especially soft textured Japanese noodles. Starch with high pasting viscosity in wheat flour is beneficial in the production of noodles with a smooth surface and soft texture. As the swelling ability of starch granules is largely responsible for the degree of paste viscosity observed when a starch suspension is heated, it is suggested to use flour swelling power as a simple test to assess the inter-cultivar differences in starch properties. Application of the swelling power test to starch or flour derived from samples of individual cultivars indicates a strong association between measurements of swelling power and noodles texture Crosbie *et al.* (1992). High swelling degree is positively related to the softness of Japanese white salted noodles and negatively correlated with the firmness of yellow alkaline noodles. Recently genetic variation for starch traits amongst different wheat genotypes responsible for variation in noodle quality have been recognized.

Dramatic differences occur in starch pasting and rheology with the small reduction in amylose content resulting from waxy allele at the three- homoeological loci found on chromosome 7A (wx-A1), 4A (wx-B1) and 7D (wx-D1) and affect eating quality of white salted Japanese noodles (Yamamori *et al.* 1994, Zeng *et al.* 1997). Detailed analysis of partial waxy lines revealed that a single null allele at the Wx–B1 locus demonstrated lower amylose contents and elevated flour swelling volume than null alleles at Wx-D1 and Wx-A1, suggesting a higher potential of Wx-B1 null for WSN. The study of wheat lines having null allele at GBSS locus indicated that starch viscosity as well as swelling volume increased without significant change in the relative amylose content Zhao *et al.* (1998). White salted noodles are typically produced from flour exhibiting high starch swelling volume and peak viscosity which confer a soft, elastic rexyture to the noodle. Reduced amylose content in partial waxy wheats increases flour swelling ability, decreases paste gel rigidity and enhanced deformation of gelatinized starch granules contribute to soft texture in WSN. Conversely partial waxy wheats are unsuitable for making yellow alkaline noodles which require a firm smooth and springy texture.

Starch and Soft Wheat Product Quality

Quality of soft wheat products depends on the fine granulation of flour particles and the extent of starch damage. Higher starch damage is detrimental to the biscuit making quality. Higher starch damage absorbs more water during mixing that leads to lower spread factor. In addition to starch damage, starch composition is also important in determining water absorption. Starch is composed of two components as amylose and amylopectin. Waxy wheats with reduced amylose content have higher water absorption. Limited literature exists on the application of waxy wheats in cakes and cookies. However, several characteristics of waxy flour, including higher ash content, starch damage and beta glucan in addition to amylopectin ratio affect cookie quality. Reports indicated that waxy wheat flour yielded cookies with smaller diameter that required greater force to snap the cookie than cookies made with non waxy flours. Waxy wheats have been reported suitable for improving noodle (Japanese) quality as well as keeping quality of pasta products and bread.

Other Industrial Uses of Wheat Starch

Bioplastics

The environmental impact of persistent plastic wastes is growing more global concern, and alternative disposal methods are limited. Incineration may generate toxic air pollution, and satisfactory landfill sites are limited. Also, the petroleum resources are finite and are becoming limited. It becomes important to find durable plastic substitutes, especially in short-term packaging and disposable applications. Recently, the continuously growing concern of the public for the problem has stimulated research interest in biodegradable polymers as alternatives to conventional nondegradable polymers such as polyethylene and polystyrene etc. Though, fully biodegradable synthetic polymers such as poly (lactic acids) (PLA), polycaprolactone (PCL), and polyhydroxybutyrate~valerate (PHBV) have been developed recently. Among these biopolymers, PLA was extensively studied in medical implants, suture, and drug delivery systems since 80s due to its biodegradability. PLA has been attractive for disposable and biodegradable plastic substitutes due to its better mechanical properties. However, PLA is still more expensive than conventional plastics. Also, the degradation rate is still slow as compared to the waste accumulation rate. Starch-based biodegradable plastics may have starch contents ranging from 10% to greater than 90%. Starch based polymers can be based on crops such as corn (maize), wheat or potatoes. Starch content needs to exceed 60% before significant material breakdown occurs. As the starch content is increased, the polymer composites become more biodegradable and leave less recalcitrant residues. Often, starch-based polymers are blended with high-performance polymers (e.g. aliphatic polyesters and polyvinyl alcohols) to achieve the necessary performance properties for different applications. At lower starch contents (less than 60%) the starch particles act as weak links in the plastic matrix and are sites for biological attack. This allows the polymer matrix to disintegrate into small fragments, but not for the entire polymer structure to actually bio-degrade. Biodegradation of starch based polymers is a result of enzymatic attack at the glucosidic linkages between the sugar groups leading to a reduction in chain length and the splitting off of sugar units (monosaccharides, disaccharides and oligosaccharides) that are readily utilised in biochemical pathways.

Therefore the potential for biodegradable plastics, particularly in single-use products (e.g., razors and golf tees) and packaging, has generated considerable interest for environmental considerations. However, the major drawback is that biodegradable plastics are currently 2-4 times more expensive then synthetic polymers. The resins used to make biodegradable plastics fall into two broad categories: natural and synthetic. Natural resins, or biopolymers, are largely based on renewable resources such as starch, cellulose, proteins and pectins, as well as polyhydroxyalcanoates (PHA) produced by microbes. Compared to other biopolymers, starch is inexpensive, available continuously, biodegradable in several environments and incinerable. Starch-based bioplastics can be extruded or formed through injection molding; however, an obstacle to product performance is the solubility of starch in water. Problems have been reported with swelling, deformation and cold weather brittleness. To improve its properties, starch is often blended with hydrophobic polymers. Current research and test-marketing are being conducted using wheat starch for fast-food containers, given better puffing performance than corn starch. Also, compared to PLA, wheat starch-based biodegradable products are claimed to withstand hot foods better, which may promote use in eating utensils, and to degrade more fully in either soil or seawater, which may allow use in a broader range of environments. Ultimately, adoption of biodegradable polymers from wheat starch will likely depend upon the ability of suppliers to achieve scale economies to bring costs nearly in line with petrochemical based products.

Paint Stripping

Generally toxic chemicals are used in removing paints from aircraft. Therefore finding environmentally safe alternative to chemical stripping has become important. Wheat starch blasting systems have attracted attention since their operating costs are estimated to be as much as 50% less than chemical paint stripping using methylene chloride. In this case abrasive wheat-starch particles propelled by compressed air are used. Wheat starch can be reused and actually becomes more effective as it breaks down. Finally, the product has a low potential for explosion. Applications include removing paint from aluminum alloys and composites like graphite and fiberglass. The wheat starch blasting process can also remove a variety of coatings. Coating types range from rain erosion-resistant coatings found on radomes and radar-absorbing materials, to tougher polyurethane and epoxy paint systems. However, high capital costs and requires complex systems for media recovery, recycling, and dust control ae associated with starch blasting. In addition wheat starch is highly sensitive to moisture and may require humidity control. Still, most of these drawbacks can be overcome, and the use of wheat starch blasting as an environmental safe alternative holds promise.

Summary

Starch being major storage form of carbohydrate in wheat grain, has great influence on quality of wheat products. Starch has important role in bread staling and noodle surface structure and quality. Starch is composed of linear chain (amylose) and branched chain (amylopectin) units of glucose. The enzymes of biosynthetic pathway and their corresponding genes of both the units have been characterized in greater detail. The information has been used in transformation experiments for manipulating ratio of amylose to amylopectin. This has demonstrated the utility of manipulation of starch structure and its synthesis in improving the processing and nutritional quality of wheat. This has also opened the opportunity to enhance industrial uses of wheat starch and consequently having beneficial effect on the environment.

References

Ball S, Guan H, James M, Myers A, Keeling P, Mouille G, Colonna P, Preiss J. 1996. From glycogen to amylopectin: a model for the biogenesis of the plant starch granule. *Cell* 86 : 349–352.

Cao, H., James M.G., Myers, A.M. (2000). Purification and characterization of soluble starch synthases from maize endosperm. *Arch Biochem Biophys.* 373:135-146.

Chibbar, R.N. and Chakraborty, M. (2005). Characteristics and uses of waxy wheat. *Cereal Foods World.* 50: 121-126.

Crosbie, G.B., Lambe, W.J., Tsutsui, H. and Gilmour, R.F. (1992). Further evaluation of the flour swelling volume test for identifying wheats potentially suitable for Japanese noodles. *J. Cereal Sci.* 15: 271-180.

Gao, M., Wanat, J., Stinard, P.S., James, M.G. and Myers, A.M. (1998). Characterization of dull1, a maize gene coding for a novel starch synthase. *Plant Cell.* 10: 399-412.

Greenblatt, G.A., Bettge, A.D. and Morris, C.F. (1995). Relationship between endosperm texture and the occurrence of friabilin and bound polar lipids on wheat starch. *Cereal Chem,* 72(2), 172-176.

Imparl-Radosevich, J.M., Gameon, J.R., McKean, A., Wetterberg, D., Keeling, P. and Guan, H. (2003). Understanding catalytic properties and functions of maize starch synthase isozymes. *J. Appl. Glycosci.* 50, 177–182.

Lee, M.R., Swanson, B.G. and Baik, B.K. (2001). Influence of Amylose Content on Properties of Wheat Starch and Bread Making Quality of Starch and Gluten Blends. *Cereal Chem.* 78(6): 701-706.

Li, Z., Rahman, S., Kosar-Hashemi, B., Mouille, G., Appels, R. and Morell, M.K. (1999a). Cloning and characterization of a gene encoding wheat starch synthase I. *Theor. Appl. Genet.* 98: 1208–1216.

Morrell, M.K., Rahman, S., Abrahams, S.L. and Appels, R. (1995). The biochemistry and molecular biology of strch synthesis in cereals. *Aust. J. Plant Physiol.* 22: 647-660.

Myers, A.M., Morell M.K., James, M.G., Ball, SG. (2000). Recent progress toward understanding the biosynthesis of the amylopectin crystal. *Plant Physiol.* 122 : 989–998.

Preiss J, 1992. Biology and molecular biology of starch synthesis and its regulation. *Oxford surveys Plant Mol. Cell Bio.* 7: 69-114.

Rahman, S., Li, Z., Batey, I., Cochrane, M.P., Appels, R. and Morell, M. (2000). Genetic Alteration of Starch Functionality in Wheat. *J. Cereal Sci.* 31: 91-110.

Regina, A., Bird, A., Topping D., Bowden, S., Freeman, J., Barsby, T., Kosar-Hashemi, B., Zhongyi Li, Rahman, S. and Morell, M. (2006). High-amylose wheat generated by RNA interference improves indices of large-bowel health in rats. *PNAS.* 103 (10) 3546-3551.

Tetlow, I.J., Morell, M.K. and Emes M.J. (2004). Recent developments in understanding the regulation of starch metabolism in higher plants. *J. Exp.Bot.* 55: 2131–2145.

Yamamori, M., Nakamura, T., Endo, R. and Nagamine, T. (1994). Waxy protein deficiency and chromosomal location of coding genes in common wheat. *Theor. Appl. Genet.* 89: 178-184.

Zeeman, S.C., Northrop, F., Smith A.M. and AP Rees T. (1998a). A starch- accumulating mutant of Arabidopsis thaliana deficient in a chloroplastic starch-hydrolysing enzyme. *The Plant Journal* 15: 357–365.

Zeng, M. Morris, C.F. Batey, I.L. and Wrigley, CW. (1997). Sources of variation for starch gelatinization, pasting and gelation properties in wheat. *Cereal Chem.* 74: 63-71.

Zhao, H.C., Batey, I.L., Sharp, P.J., Crosbie, G., Barclay, I., Wilson, R., Morell, M.K. and Appels, R. (1998). A single genetic locus associated with starch granule properties and noodle quality in wheat. *J. Cereal Sci.* 27: 7-13.

❑❑❑

Chapter 9

Nutritional Quality of Wheat

Introduction

Nutritional value of food stuff is defined by its ability to provide nutrients required by the body. Proteins, carbohydrates, fats, minerals and vitamins are main ingredients in food stuffs. Wheat is an important source of these nutrients. The distribution of nutrients within wheat kernel is typical of many cereals. Nutrients are generally found in the highest concentrations in the germ or embryo and in the aleurone cells surrounding the starchy endosperm followed by bran, whole wheat flour and white flour in decreasing order. Of special importance is that it contains the entire B complex, except for vitamin B12. B vitamins function as cofactors in many metabolic reactions involved in the release of energy. The germ, which includes the scutellum, is especially rich in vitamins B and E, high quality protein, unsaturated fats, minerals, and carbohydrates. The bran consists mostly of the insoluble carbohydrate cellulose, and contains incomplete protein, traces of B vitamins, and minerals including zinc and iron. The endosperm is the largest part of the grain, and consists mostly of the carbohydrate starch, incomplete protein, and trace amounts of vitamins and minerals. Significant variations in the content of grains occur because of variety, crop year, area, fertilizer, and soil type.

Because of its high content of vitamin E, wheat germ is promoted as a health food, and has been proposed as a cure for many diseases. Recent studies have shown that vitamin E increases the desirable HDL cholesterol in women, though in men only if they initially had low levels. Animal studies have also shown that vitamin E protects against free radicals released by the body when it is exposed to toxic chemicals. Vitamin E is used to treat intermittent claudication, which involves cramps in the calf muscles at night and extreme pain while walking. Vitamin E may be helpful for fibrocystic breast disease. Other vitamins and numerous minerals are found in the wheat kernels, though in small amounts. These include carotene, vitamin B6 or pyridoxine, pantothenic acid, biotin, and folic acid, vitamin C, and vitamin K. Other minerals are sodium, calcium, chlorine, manganese, zinc, copper, cobalt, nickel, chromium, molybdenium, fluoride, iodine, boron, selenium, lead, aluminum, and silicondioxide. The body is capable of converting the carotene to produce one sixth its amount as vitamin A (Health ~ Welfare, 1990).

Despite all its many nutritional qualities, wheat cannot meet all nutritional needs. It lacks adequate amounts of certain essential nutrients as vitamins A, B12, and vitamin C, fats and the amino acid lysine. These must come from other sources. There is a need for some animal products, since they are the only sources of vitamin B12, apart from intestinal bacteria capable of producing some. Large deficiencies of this vitamin may lead to anaemia. Fruits and vegetables are required to provide the missing vitamins A and C, and fats are needed to supply essential fatty acids, because wheat contains very little fat (about 2%). The quality of a protein is determined by the kind and composition of its constituent amino acids. When all essential amino acids are present in the proportions capable of promoting growth, the protein is complete. The complete protein has high biological value (BV), and would result in high net protein utilization (NPU) by the body. If a protein has a relatively small amount of one essential amino acid (called the limiting amino acid), body tissue repair will occur, but growth cannot be supported. Lysine is the limiting essential amino acid in cereals. A greater intake of lysine than that found in wheat is especially important for children. Wheat protein is adequate for adults, since they have been shown to maintain nitrogen equilibrium (intake of nitrogen from protein = loss), or to be in slightly positive nitrogen balance (intake > loss) when consuming bread diets Betschart *et al.* (1985). The requirements for lysine are about three times less for adults than for children. To assure an adequate supply of lysine, bread made solely from grain should be consumed in combination with milk products, meat, nuts, or legumes.

Though protein digestibility is higher in white flours than whole wheat flours, overall amino acid availability is higher in whole meal flours. Relative nutritional value (RNV) clearly differentiates protein quality, rating germ the highest followed by bran, whole wheat and white flour protein. However, if NPU is used as the criterion, protein quality is similar in flours with extraction rates of 66-100 %. Amino acid composition indicates protein quality when compared with a given amino acid profile. Aleurone cells are rich sources of minerals, many of the B vitamins, and protein. Significant quantities of the minerals and vitamins are lost when whole wheat is milled to produce white endosperm flour because outer layer of bran are removed along with aleurone cells and germ. Details of all these factors are given below.

Amino Acid Composition and Protein Quality of Wheat

Wheat is an important source of proteins in human diet. In wheat it varies from 6.0% to 21.0%. Though germ contains higher proportion of proteins in wheat, 72% of the total amount of proteins is present in endosperm. The quality and quantity of proteins in relation to processing quality are described in other chapter in the book.

Here an account of nutritional quality of wheat proteins will be provided. Based on solubility criteria wheat proteins are divided into albumins (water soluble), globulins (salt soluble), gliadins (ethanol soluble) and glutenins (acid soluble). Glutenins and gliadins are main storage proteins constituting around 80 % of total proteins. Albumins and globulins are mainly confined in embryo and aleurone layer while gliadins and glutenins in endosperm. Comparatively glutenins and gliadins are rich in glutamine and proline and poor in lysine while globulins have more lysine and lower content of glutamine and proline (Table 1). Because of higher content of glutenins and gliadins, wheat proteins are deficient in lysine which is one of the indispensable amino acids. Other indispensable amino acids are leucine, isoleucine, valine, lysine, threonine, methionine, phenylalanine and tryptophan. Recently transgenic approach has shown promise for the improvement of lysine content. The details are given in next chapter.

Table 1: Composition of amino acids in wheat, flour and protein fractions (g/16g N) (from Salunkhe *et al.* 1986).

Amino acid	Wheat	flour	albumin	globulin	gliadin	glutenin
Tryptophan	1.5	1.5	1.1	1.1	0.7	2.2
Lysine	2.3	1.9	3.2	5.9	0.5	1.5
Histidine	2.0	1.9	2.0	2.6	1.6	1.7
Arginine	4.0	3.1	5.8	7.0	1.9	2.7
Aspartic acid	4.7	3.7	5.8	7.0	1.9	2.7
Threonine	2.4	2.4	3.1	3.3	1.5	2.4
Serine	4.2	4.4	4.5	4.8	3.8	4.7
Glutamic acid	30.3	34.7	22.6	15.5	41.1	34.2
Proline	10.1	11.8	8.9	5.0	14.3	10.7
Glycine	3.8	3.4	3.6	4.9	1.5	4.2
Alanine	3.1	2.6	4.3	4.9	1.5	2.3
Cystine	2.8	2.8	6.2	5.4	2.7	2.2
Valine	3.6	3.4	4.7	4.6	2.7	3.2
Methionine	1.2	1.3	1.8	1.7	1.0	1.3
Isoleucine	3.0	3.1	3.0	3.2	3.2	2.7
Leucine	6.3	6.6	6.8	6.8	6.1	6.2
Tyrosine	2.7	2.8	3.4	2.9	2.2	3.4
Phenylalanine	4.6	4.8	4.0	3.5	6.0	4.1

Indispensable amino acids are not synthesized in the body and are obtained from the diet, while dispensable amino acids could be made within the body. However, even this nomenclature is incomplete and is not sufficient to cover all possibilities. Depending on the metabolic state of the body, certain dispensable amino acids may become indispensable. For example, glutamine, which is normally considered a dispensable amino acid, may be required in such high quantities under certain conditions, that enough can not be made in the body, and it becomes indispensable. One of these situations is following trauma or surgery, where glutamine requirements go up significantly. In this case, glutamine is termed a conditionally indispensable amino acid. Another example is cysteine, which reduces the requirements for methionine, and tyrosine which reduces the requirements for tryptophan. Under situations where insufficient methionine or tryptophan is consumed, cysteine and tyrosine would become indispensable.

Protein Quality Assessment

Protein quality relates to the efficiency with which various food proteins are used for synthesis and maintenance of tissue proteins. The composition of various proteins may be so unique that their influence on physiological function in the human body could be quite different. The quality of a protein is vital when considering the nutritional benefits that it can provide. The quality of a protein is determined by assessing its essential amino acid composition, digestibility and bioavailability of amino acids (FAO/ WHO, 1991). The capacity of a protein to meet the amino acid requirements of an organism depends on several factors such as amino acid composition, adequacy of a diet as a whole, physiological, nutritional and health status of an animal or human consuming it. There are several measurement scales and techniques based on chemical and biological evaluations.

Protein Rating Scales

Numerous methods exist to determine protein quality. These methods have been identified as protein efficiency ratio (PER), biological value (BV), net protein utilization (NPU), and protein digestibility corrected amino acid score (PDCAAS). Usually weanling rats, suckling pigs, mouse, chickens and adult and adolescent humans are used as experimental subjects.

Protein Efficiency Ratio

The protein efficiency ratio determines the effectiveness of a protein through the measurement of animal growth. This technique requires feeding rats a test protein and then measuring the weight gain in grams per gram of protein consumed. The computed value is then compared to a standard value of 2.7, which is the standard value of casein protein. Any value that exceeds 2.7 is considered to be an excellent protein source. PER values of wheat has been reported around 50% of casein control. However, this calculation provides a measure of growth in rats and does not provide a strong correlation to the growth needs of humans. The assumptions that weight increase is an index of protein synthesis and all proteins consumed result in tissue growth are not valid.

Biological Value

Biological value measures protein quality by calculating the nitrogen used for tissue formation divided by the nitrogen absorbed from food. This product is multiplied by 100 and expressed as a percentage of nitrogen utilized. The biological value provides a measurement of how efficient the body utilizes protein consumed in the diet. A food with a high value correlates to a high supply of the essential amino acids. Animal sources typically possess a higher biological value than vegetable sources due to the vegetable source's lack of one or more of the essential amino acids. There are, however, some inherent problems with this rating system. The biological value does not take into consideration several key factors that influence the digestion of protein and interaction with other foods before absorption. The biological value also measures a protein's maximal potential quality and not its estimate at requirement levels.

Net Protein Utilization

Net protein utilization is similar to the biological value except that it involves a direct measure of retention of absorbed nitrogen. Net protein utilization and biological value both measure the same parameter of nitrogen retention; however, the difference lies in that the biological value is calculated from nitrogen absorbed whereas net protein utilization is calculated from nitrogen ingested. NPU= (N intake-faecal N-Urine N)*100/N intake + (metabolic N + endogenous N)*100/N intake. NPU of wheat has been reported to be around 60%.

Protein Digestibility Corrected Amino Acid Score

Protein digestibility is calculated by the formula: True protein digestibility = PI – (FP – MFP)*100/PI; where PI = protein intake, FP = fecal protein and MFP = metabolic fecal protein. The amount of protein in the feces of rats fed the protein-free diet is used as the estimate for MFP. However, in 1989, the Food & Agriculture Organization and World Health Organization (FAO/WHO) in a joint position stand stated that protein quality could be determined by expressing the content of the first limiting essential amino acid of the test protein as a percentage of the content of the same amino acid content in a reference pattern of essential amino acids (FAO/WHO, 1991). The reference values used were based upon the essential amino acids requirements of preschool-age children. The recommendation of the joint FAO/WHO statement was to take this reference value and correct it for true fecal digestibility of the test protein. The value obtained was referred to as the protein digestibility corrected amino acid score (PDCAAS). PDCAAS is calculated by the following formula: PDCAAS (%) = True protein digestibility X amino acid score or the lowest amino acid ratio. This method has been adopted as the preferred method for measurement of the protein value in human nutrition.

Amino acid ratios [(mg of an essential amino acid in 1.0 g of test protein/mg of the same amino acid in 1.0 g of reference protein) × 100] for essential amino acids have been calculated using a rat growth pattern of amino acid requirements, Sarwar *et al.* (1985) and a human pattern of amino acid requirements (FAO/WHO/UNU 1985 suggested pattern of amino acid requirements for preschool children, 2-5 y) as the reference proteins.

The PDCAAS method has been considered to be a simple and scientifically sound approach for routine assessment of dietary protein quality for humans. The PDCAAS is now a federally approved alternative method to the protein efficiency ratio (PER) rat bioassay procedure. Although the PDCAAS is currently the most accepted and widely used method, limitations still exist relating to overestimation in the elderly (likely related to references values based on young individuals), influence of ileal digestibility, and antinutritional factors Sarwar, (1997). Antinutritional factors may occur naturally or may be formed during heat processing. Some examples of naturally occurring antinutritional factors include glucosinolates in mustard and rapeseed protein products, trypsin inhibitors and hemagglutinins in legumes, phytates in cereals and oilseeds and gossypol in cottonseed protein preparations, which could adversely affect nutrient utilization and may contribute to growth depression in animals. In addition to that, food processing conditions may influence protein digestibility.

Quality of Different Protein Sources

Protein is available in a variety of dietary sources. These include foods of animal and plant origins as well as the highly marketed sport supplement industry. In the following section proteins from both vegetable and animal sources, including whey, casein, and soy will be explored. Determining the effectiveness of a protein is accomplished by determining its quality and digestibility. Quality refers to the availability of amino acids that it supplies, and digestibility considering how the protein is best utilized. Typically, all dietary animal protein sources are considered to be complete proteins. That is, a protein that contains all of the essential amino acids. Proteins from vegetable sources are incomplete in that they are generally lacking one or two essential amino acids. Thus, someone who desires to get their protein from vegetable sources (i.e. vegetarian) will need to consume a variety of vegetables, fruits, grains, and legumes to ensure consumption of all essential amino acids. As such, individuals are able to achieve necessary protein requirements without consuming beef, poultry, or dairy. Protein digestibility ratings usually involve measuring how the body can efficiently utilize dietary sources of protein. Typically, vegetable protein sources do not score as high in ratings of biological value, net protein utilization, PDCAAS, and protein efficiency ratio as animal proteins as shown in the (Table 2).

Table 2: Protein Quality Rankings (Sarwar, 1997)

Protein Type	Protein Efficiency Ratio	Biological Value	Net Protein Utilization	Protein Digestibility Corrected Amino Acid Score
Whey protein	3.2	104	92	1.00
Egg	3.9	100	94	1.00
Milk	2.5	91	82	1.00
Soy protein	2.2	74	61	1.00
Casein	2.5	77	76	1.00
Beef	2.9	80	73	0.92
Black Beans	0	-	0	0.75
Peanuts	1.8	-	-	0.52
Wheat gluten	0.8	64	67	0.25

Micronutrient Density and Phytic Acid

Wheat is the main source of starch, fibre and also contain significant amount of proteins, minerals and vitamins. Minerals and trace elements are essential nutrients for good health. For example Calcium is required for formation and maintenance of skeleton, normal contraction of muscle, heart, nervous activity and blood clotting. Magnesium is implicated in the regulation of nervous system and prevents spasmophilia. Iron is incorporated in the haemoglobin molecule and plays a role in transport of oxygen. Zinc and copper act as constituents of a number of enzymes. The large part of the population in the world faces problem of micronutrient deficiency (especially Fe and Zn). This problem is more predominant in infants and pregnant women who require comparatively larger quantity of micronutrients. Though whole wheat is an important source of these minerals particularly Mg, Zn and Fe and contributes significantly to mineral consumptions; the daily requirements are not met by wheat based diets as shown in the (Table 3).

The recommended daily requirements of these micronutrients are not met by consuming wheat. Moreover, bioavailability of micronutrients in wheat based diet is reduced by the content of antinutritional factors such as phytic acid in the grain. Phytic acid inhibits the release of Fe and Zn in the intestine and thus their absorption is reduced. Phytate in wheat grains is found predominantly in aleurone which remains attached to the pericarp during milling and therefore is concentrated in the bran fraction. Therefore, Fe and Zn deficiencies are considered major micronutrient deficiency in people where cereals are consumed as major foods. The enhancement of their density and also their bioavailability by reduced antinutritional factors can led to improved nutritional status of vast sections of the human beings through out the world. Recent reports indicate that sufficient variability for the trait of Fe and Zn does not exist in cultivated hexaploids but wild species of wheat have wider variation of micronutrient content (upto 110 mg/kg) Chhuneja *et al.* (2006). Therefore wild progenitors of wheat can be used in biofortification for enhanced micronutrient density. In addition molecular tools are available to enhance both density and bioavailability of Fe and Zn in wheat, which are discussed in greater detail in other chapter of the book.

Table 3: Mineral content (mg/kg) of wheat and wheat fractions and RDA requirements

Wheat sample	Fe	Zn	Cu	Mn
Wheat	18-31	21-63	1.8-6.2	24-37
Bran	74-103	56-141	8.4-16.2	72-144
Germ	41-58	<100-144	7.2-11.8	101-129
Flour	3.5-9.1	3.4-10.5	0.62-0.63	2.1-3.5
Maximum adult RDA	15 mg	15 mg	1.5-3.0 mg	2-5 mg

RDA = Recommended dietary allowances are the daily levels of intake of essential nutrients judged to be adequate to meet the known nutrients needs of practically all healthy persons. RDA values are taken from Grusak and Penna 1999 and other values from Betschart 1998.

The bioavailability of these nutrients can be improved by reducing phytic acid content as well as increasing phytase activity in the grain. Therefore, identifying genotypes having low phytic acid content and high phytase activity along with higher Fe and Zn content is necessary before we proceed to improve nutritional quality of wheat. Phytic acid is the primary storage form of phosphorus (p) in seeds accounting for upto 85% of the total seed phosphorus. The negatively charged phosphates in phytic acid bind strongly to metallic cations as K, Mg, Mn, Fe , Ca and Zn to form mixed salt called phytin which reduces bioavailability to humans. Phytin accumulates in seed protein bodies either dispersed or dense inclusions called globoids. The stable complexes of these minerals with phytic acid can lead to micronutrient deficiencies (e.g., Fe and Zn) particularly in third world countries where diets are primarily seed based. Moreover, monogastric animals such as pig, poultry and fish can not digest phytic acid. As a result animal feeds are supplemented with P, and a large fraction of phytic acid is excreted leading to accumulation in soil and water causing pollution. Development of low phytate crops may improve phosphorus availability in animal feeds and reduce phytic acid excretion, thus lessening the negative effect of animal waste on the environment. Another potential advantage of such crops is in an increase in the availability of iron and zinc, which would significantly improve human nutrition. Initial studies have indicated that LPA products increase the absorption of several important minerals.

Resistant Starch

Carbohydrates constitute nearly 80% of the total dry matter of wheat kernel. Of the cereal grains cultivated widely, wheat supplies ~20% of the food calories for the world population and is the preferred base for most cereal-based processed products, such as bread, biscuit, pasta, and noodles. Starch is the predominant form, with cellulose, hemicellulose and pantosans present in lesser amount. Starch is present in discrete granules within the cells of the endosperm. Starch content is available in the range of 65% to 75% and generally related to the protein content. Starch structure is a crucial determinant of quality where amylose / amylopectin ratio is a useful descriptor of starch structure. The rate of starch digestion varies from source to source. The starch in potatoes, cereals, and baked goods is digested very rapidly. Other starchy foods, such as beans, barley, or long grained brown rice, are digested more slowly, and cause a much slower and lower blood sugar rise. Resistant starch goes all the way through the small intestine without being digested. In this way, it is more like fiber, and in some cases is classified and labeled as fiber.

When resistant starch reaches the colon, it is used for fuel by the bacteria there. This process, called fermentation, produces a certain type of fat called short-chain fatty acids (SCFAs). The fatty acids are the main products which produce most of the calories from resistant starch, and have many benefits. SCFAs are also produced by soluble fiber and oligosaccharides - this is the reason why on some food labels, some fiber is shown as having calories associated with it, but these calories do not raise blood glucose. Therefore, Modification of the starch composition of wheat by raising its RS content presents an opportunity for a potentially large-scale improvement in public health. Diet-related noninfectious chronic diseases, such as coronary heart disease, certain cancers (especially of the colon and rectum), and diabetes, are major causes of morbidity and mortality in both affluent industrialized countries and emerging nations. Increased consumption of whole-grain cereal foods can reduce the risk of these conditions substantially and also correlates negatively with obesity. Resistant starch (RS) and non-starch polysaccharides (NSP) are major components of dietary fiber and contribute to these benefits. NSP resist digestion by enzymes in the human small intestine completely, helping to explain their fecal bulking and laxative action, which may underpin the protection against colorectal cancer afforded by greater fiber consumption.

Although starch can be digested (theoretically to completion) in the human small intestine, amylolysis varies by rate and extent for many foods. Reduction of the rate of starch digestion in the small intestine has the potential to lower the rate of entry of glucose into the circulation and, thus, reduce demand for insulin. This is measured as the glycemic index (GI), and lowering the GI is emerging as an important mechanism for managing the incidence and severity of type II diabetes. The fraction of the ingested starch that escapes into the human large bowel is known as RS. Short-chain fatty acids (SCFA) are major end products of the fermentation of NSP and RS by the microflora, and they promote important aspects of large bowel function. One of the principal SCFA, butyrate, may also play a role in promoting a normal phenotype in colonocytes and lowering the risk of colorectal cancer. It is emerging that RS contributes substantially to some of the benefits that have been ascribed solely to dietary fiber. There is a case for increasing RS consumption as an effective means of improving nutrition for public health at the population level. The strategy of combining wheat bran (WB) with RS was effective at producing beneficial changes in fecal bulking, transit time, and fermentation-dependent indexes that persisted to the distal regions of the human colon. The WBRS combination clearly produced more benefits than did the WB diet alone (as shown here) or the RS diet alone (as shown in previous studies). Moreover, combining WB with RS may attenuate any potentially negative effects of RS. This work suggests that to maximize the health benefits of dietary fiber in the gastrointestinal tract, a combination of different fiber types may be required.

Benefits Accruing to Resistant Starch

It seems that the more it is studied, the more positive effects are being found. Many of these are common to oligosaccharides and fermentable fiber. Here are some of the benefits of resistant starch:

- Resistant starch is especially associated with one type of SCFA, called butyrate, which is protective of colon cells and associated with less genetic damage (which can lead to cancer). Butyrate also protects the cells in other ways. This is one of the real strengths of resistant starch over oligosaccharides and soluble fiber. Their fermentation does produce butyrate, but not at the levels of resistant starch.
- As with other fermentable fiber, resistant starch is associated with more mineral absorption, especially calcium and magnesium.
- Perhaps most exciting for people with sugar issues, resistant starch seems to improve insulin sensitivity. In the so-called "second meal effect", fermentable fiber and resistant starch are associated with improved glucose tolerance the next day. There is evidence that this is caused by the presence of the short chain fatty acids, and by a peptide produced in the fermentation process.

- Resistant starch produces more satiety, possibly partly through the release of a different peptide.
- Resistant starch consumption is associated with lower cholesterol and triglyceride levels.
- Promotes "good" bacteria, and suppresses "bad" bacteria and their toxic products.
- Promotes bowel regularity.
- Resistant starch in a meal is associated with less fat storage after that meal.

Amylose Content and Resistant Starch

Starches consist of glucose monomers polymerized through α-1,4 and α-1,6 linkages into two classes of polymers, amylose and amylopectin. Amylopectin is a large highly branched polysaccharide [degree of polymerization (DP) >5,000], whereas amylose has infrequent α-1,6 linkages and a lower DP (<2,000). In cooked foods, amylose molecules re-associate rapidly on cooling, forming complexes that resist digestion, whereas amylopectin molecules re-associate slowly and are more readily digested. This difference explains the higher RS content of high-amylose products. The synthesis of amylose and amylopectin occurs via two pathways. Amylose synthesis requires an active granule-bound starch synthase (GBSS). Amylopectin is synthesized by a complex pathway involving, among others, several isoforms of soluble starch synthase (SS), starch-branching enzymes (SBE), and starch-debranching enzymes. Monocots and dicots differ in the isoforms of SBE they contain. Details are given in other chapter on starch quality. In potato, two isoforms of SBE are present, SBEI and SBEII. Suppression of both SBEI and SBEII was required to generate starch with amylose content of >50%. In monocots, SBEI and two forms of SBEII, SBEIIa and SBEIIb, are found. In maize, suppression of SBEIIb results in the "amylose-extender" phenotype, with amylose contents from 50% to 90%, whereas suppression of SBEI or SBEIIa had no impact on endosperm amylose content. Amylose-extender maize has been used to increase the RS content of processed foods; however, additional sources of RS are needed.

Recently Regina *et al.* (2006) used RNA interference technique to down-regulate the two different isoforms of starch-branching enzyme (SBE) II (SBEIIa and SBEIIb) in wheat endosperm to raise its amylose content. Suppression of SBEIIb expression alone had no effect on amylose content; however, suppression of both SBEIIa and SBEIIb expression resulted in starch containing >70% amylose. When the >70% amylose wheat grain was fed to rats in a diet as a whole meal, several indices of large-bowel function, including short-chain fatty acids, were improved relative to standard whole meal wheat. Details are given in a chapter on molecular breeding. Earlier studies in barley also showed the beneficial effects of barley cultivar with high RS arising through altered starch synthesis. The results indicate that this high-amylose wheat has a significant potential to improve human health through its resistant starch content. The data justify further investigation of the health potential of high-amylose wheats, especially in processed foods, as an important additional mechanism to deliver significant health benefits to large numbers of consumers through their diet.

Wheat Gluten Allergy, Intolerance and Enteropathy

Gluten is the viscoelastic complex mainly formed of glutenins and gliadins. Adverse reactions to wheat, as to any food, can be allergic (in this case, wheat allergy), intolerance (in this case, wheat intolerance, gluten intolerance, and coeliac disease), or due to other naturally occurring constituents. Different mechanisms cause different adverse reactions. The resulting symptoms may be quite different from or confusingly similar to each other. The human body is able to mount a variety of defense mechanisms against proteins it regards as foreign or harmful. It is not clearly known why food proteins are regarded as harmful by the body, or how adverse affects occur. For example, wheat-sensitive allergic individuals typically produce IgE antibodies to the soluble grain proteins, but some develop gluten-specific IgE antibodies. Patients with coeliac disease develop gliadin-specific IgA and IgG antibodies. Wheat allergy refers specifically to adverse reactions involving immunoglobulin E (IgE) antibodies to one or more protein fractions of wheat, including albumin, globulin, gliadin and glutenin (gluten). The majority of IgE-mediated reactions to wheat involve the albumin and globulin fractions. Gliadin and gluten may also, rarely, induce IgE-mediated reactions. Allergy to wheat may occur in any individual, unlike Coeliac Disease, which is hereditary.

Allergic reactions to wheat may be caused by ingestion of wheat-containing foods or by inhalation of flour containing wheat (Baker's asthma). Clinical experience suggests that wheat allergy is relatively uncommon, but there are no accurate figures for prevalence. The allergy is more prevalent in certain groups: e.g., wheat allergy is responsible for occupational asthma in up to 30% of individuals in the baking industry. Allergic reactions to wheat (IgE-antibody mediated) usually begins within minutes or a few hours after eating or inhaling wheat. The more common symptoms involve the skin (urticaria [hives], eczema, angioedema [swelling due to allergy]), the gastrointestinal tract (abdominal cramps, nausea and vomiting, oral allergy syndrome) and the respiratory tract (asthma or allergic rhinitis). In association with exercise, reactions to gliadin or gluten can cause urticaria, angioedema or life-threatening anaphylaxis. As these proteins are present in other cereals, these symptoms may also occur in wheat-allergic individuals due to cross-reactivity.

The diagnosis may be easy if a person has the same reaction repeatedly after eating wheat-containing food. More often the diagnosis is difficult because wheat is usually consumed with other food. Diagnosis usually entails clinical evaluation (medical history, family history, food history) supported by appropriate laboratory tests (CAP® RAST blood tests, skin prick-testing). An elimination-challenge test may be employed to make the diagnosis. Medication is ineffective in treating this condition. Avoidance of wheat and wheat-containing foods is the only treatment. This may be difficult to maintain, particularly as wheat protein may be "hidden" in other foods. Rice or maize may be substituted as alternative cereals. A dietician must supervise treatment. Wheat-allergic patients who have sensitivity to gluten (or gliadin) should avoid other gluten-containing cereals.

Coeliac Disease (CD), also called Gluten Enteropathy, has until recently been known as Gluten Intolerance. CD is a hereditary disorder of the immune system in which eating gluten leads to damage of the mucosa (lining) of the small intestine (small gut). This results in malabsorption of nutrients and vitamins. CD is the result of IgA and IgG antibody responses to gluten. It is important to differentiate between CD, mediated by IgA and IgG antibodies, and wheat allergy, which is mediated by IgE antibodies. Coeliac Disease is one of the commonest life-long disorders in Western countries. CD is frequently under-diagnosed, particularly in adults, who may present with subtle symptoms. In some countries the incidence is as high as 1 in 200 (Sweden) or 1 in 10,000 (Denmark). The incidence in South Africa has not been ascertained, but is thought to be low, although the disease is most probably under-diagnosed. Recently cases of coeliac disease have been reported in India.

Typically CD presents at the age of 6-24 months with symptoms of intestinal malabsorption, impaired growth, abnormal stools, abdominal distension, muscle wasting, poor muscle tone (hypotonia), poor appetite or irritability, following the introduction of cereals into the diet. In adults, the symptoms of CD may be quite varied, from severe weight loss and diarrhoea and bulky, offensive stools to subtle complaints of cramps, abdominal bloating, flatulence and even constipation. These individuals are often mistakenly diagnosed as having Irritable Bowel Syndrome. Recent studies show that some individuals with CD present with no symptom but a form of ataxia. Recurrent oral aphthous ulcers are common and should arouse suspicion of the condition. Other symptoms may include persistent iron-deficiency anaemia, folate deficiency anaemia or a calcium metabolism disturbance. Dermatitis herpetiformis is a variant of Coeliac Disease in which clusters of itchy blisters occur, usually over the buttocks, knees and elbows. Doctors must have a low threshold of suspicion when seeing patients with symptoms such as those described above.

There are simple and fast blood tests that can be used to support the diagnosis of CD. They are particularly recommended for screening relatives of CD patients or patients who are affected by a related disease such as malabsorption or Diabetes Mellitus, and for monitoring the compliance to a gluten-free diet. None of these tests has shown 100% accuracy, and a small-intestinal mucosal biopsy remains the cornerstone for diagnosis. Any provisional diagnosis of Coeliac Disease must be confirmed by this biopsy. The procedure is safe and usually performed at the time of gastrointestinal endoscopy.

1. The anti-gliadin antibody (AGA) assay, which measures the amount of IgA and IgG antibody produced against the gliadin component of cereals
2. The anti-reticulin antibody (ARA) test, in which IgG antibodies are viewed in an immuno-fluorescent microscope examination
3. The anti-endomysial antibody (AEA) assay, which identifies IgA antibodies against the endomysium tissue.

Medication is ineffective in treating this condition. The only treatment available is the complete removal of gluten from the diet. This usually entails life-long avoidance of all cereals containing gluten, including wheat, oats, rye and barley. Individuals on any avoidance diet are at risk of developing deficiencies of micro-nutrients (e.g., thiamine, riboflavin, niacin, iron, selenium, chromium, magnesium, folacin, phosphorus and molybdenum). It is therefore essential that patients be managed in collaboration with a dietician. Information on gluten-free diets is becoming increasingly available worldwide. Gluten-free products are also becoming more abundant and more easily available. In conclusion, variety of reactions to proteins in cereals is possible. These include allergy and Coeliac disease. Adverse reactions may be mild to life-threatening, short-term to life-long. A correct diagnosis must be made as the treatment, dietary avoidance, is often very difficult and if incorrectly applied can lead to vitamin deficiencies or malnutrition.

Milling and Flour Nutritional Quality

The Egyptians were the first to use a selective milling system. With hand sieves, they separated the flour from large bran particles, dirt, and stone chips that had broken off their implements. Stone chips are not a problem with modern mills. Since Roman times, white flour and bread have been regarded as the foods of upper classes. Flour, however, was far from white compared to today's flour. It was not until the 19th century that major changes in the milling processes took place. The earliest version of today's iron roller mills were first used in Hungary in 1839. Between 1870 and 1890, they quickly replaced the stone mills throughout Europe and North America, and milling soon became completely automated. The roller mills were more economical and more efficient. The milling process could be controlled to produce as white a flour as the public demanded. However, the resulting flour was devoid of bran and germ, and consequently many nutrients were lacking.

Stone Grinding

In the third century B.C., rotary grindstones powered by animals, and small rotary hand mills called querns, replaced stone or wooden mortars and pestles for the grinding of grains. Querns are still used in rural areas of the Middle East, Far East, and parts of Africa. Earlier in India hand driven chakkies, made up of stones, were used to grind wheat grains for whole meal. Recently power driven chakkies have replaced hand driven chakkies. There are several advantages to Stone-ground wheat flour. The endosperm, bran, and germ remain in their natural, original proportions. Because the stones grind slowly, the wheat germ is not exposed to excessive temperatures. Heat causes the fat from the germ portion to oxidize and become rancid and much of the vitamins to be destroyed. Since only a small amount of grain is ground at once, the fat from the germ is well distributed which also minimizes spoilage. Nutritive losses due to oxygen exposure are also limited by the fact that stone-ground flour is usually coarser. Stone-ground flour is preferred by many bakers and natural food advocates because of its texture, its sweet and nutty flavour, and the beliefs that it is nutritionally superior and has a better baking quality than steel-roller-milled flour.

Milling Today

A very sophisticated process is currently employed for the milling of grain. Cleaning is accomplished by means of separators, aspirators, scourers, magnets, and washer-stoners. The wheat is tempered or conditioned in water to toughen the bran to reduce fragmentation when it is removed, and to obtain a moisture content resulting in particles of the desired size. The amount of water added to wheat varies, depending upon moisture content and the hardness of the grain. Soft wheat needs lesser water to temper wheat than hard wheat. The time required for tempering grains depends upon the grain hardness. Softer wheat requires much shorter time to temper than hard wheat. After wetting, water is concentrated in the bran and germ and then penetrates into dorsal region in the grain. The time required to reach an even distribution of moisture in grain varies from 6 hour for a soft, opaque kernel with low protein to over 24 hr for a hard, vitreous, high protein kernel. Conditioning implies the use of heat in conjunction with water to soften the endosperm. Higher temperature increases the rate of diffusion of water. However, high heat can damage gluten. During the milling process, steel rollers crush the grain, and the flour released from the endosperm is separated by sifters into different grades or streams, according to fineness. Each of these has different mineral and protein contents, and may be recombined later to form a variety of flours to be sold for diverse baking purposes.

In flour milling, bran and germ are separated by the break system of the mill. After breaking the grain, the endosperm is reduced to flour fineness by reduction system. The break system consists of four or five breaks. After each break, small particles are saved and large particles are further grinded by different rolls. This is accomplished by the sifting system. Break rolls are corrugated while in reduction system rolls are generally smooth rather than corrugated. In reduction system also after each grinding pass, stock is sifted, flour is removed and the coarser particles are sent to the appropriate reduction rolls. The sifting of soft wheat flour is more difficult than the sifting of hard wheat flour. This is because fine particles produced in soft wheat flour tend to aggregate and do not pass easily through flour cloth.

Different kinds of flours are produced in milling. *Straight-grade flour* is the composite of all the flours produced during milling. The flour produced in the first grinding step is called *patent flour*. The *patent flour* has lowest ash content and represents about 45% of the total products. Patent flour comes mainly from the central part of the kernel. Ash content shows the presence of bran in the flour. Ash content of whole wheat is around

1.5% and is not distributed evenly in the grain. The inner endosperm is relatively low in ash (~0.3%) whereas bran has ~6.0%.

The bran and germ, which make up about 28% of the wheat, are totally removed in this process. They are used in the production of animal feeds as well as by pharmaceutical laboratories for making diet supplements. Whole wheat flour is produced by recombining ground bran with endosperm flour, but the germ is usually left out, because it would go rancid. The resulting flour may represent only 95% to of the total grain (by weight), or in other words a 95% extraction. About 95% of the flour used in the USA is white and of only about 72% extraction. Only 20 to 30% of the grains original vitamins are retained, and the protein content is about 1.0 - 1.5% lower. However, since bran decreases protein digestibility, the available protein does not significantly change (Pomeranz, 1988). The NPU is similar in 66 to 100% extractions. Since bran and germ have different chemical composition than endosperm, the flour composition changes (Table 4). Overall flour nutritional quality is reduced and hence it needs enrichment with minerals and vitamins.

Table 4: Wheat grain and flour composition (Hoseney 1994)

	Wheat grain	70% Extraction flour
Ash, %	1.55	0.4
Fibre, %	2.17	Trace
Protein, %	13.9	12.9
Oil, %	2.52	1.17
Starch, %	63.7	70.9
Thiamin, µg/g	3.73	0.70
Ribiflavin, µg/g	1.70	0.70
Niacin, µg/g	55.6	8.50
Iron, mg/100g	3.08	1.42
Sodium, mg/100g	3.2	2.2
Potassium, mg/100g	316	83
Calcium, mg/100g	27.9	12.9
Magnesium, mg/100g	143.0	27.2
Copper, mg/100g	0.61	0.18
Zinc, mg/100g	3.77	1.17
Total Phosphorus, mg/100g	350	98
Phytate phosphorus, mg/100g	345	30.4
Chlorine, mg/100g	39.0	48.4

Flour Enrichment

In the 1940s, a flour enrichment program was instituted to compensate for wartime shortages of other foods. The decision to choose flour for enrichment was based on its commonality in the diets of those wartime populations, ranging from the rich to the poor. However, in the 'enriched' flour only the B vitamins - thiamin, riboflavin, and niacin - and the mineral, iron, were added, in amounts approximately equivalent to those removed from whole wheat. A healthy balanced diet consisting of enriched grain foods is a great way to ensure you receive the recommended daily intake of these B vitamins, without having to spend a lot of money on supplements. The addition of more nutrients to refined flour has been considered, but it is limited by, for example, the effect of some nutrients on sensitive individuals Pomeranz, (1988). Since research is incomplete concerning nutrient requirements, interactions, optimal ratios, and toxicities, many believe that the safest option is to consume flour containing the nutrients in their natural proportions.

Flour Storage

Because grains contain only about 12% water (or about 0.6 water activity), they are not predisposed to spoilage. However, grinding removes the protective layers and endangers the grain's biological stability. Deterioration of sensory and nutritional qualities depends on storage conditions, such as temperature, humidity, oxygen concentration, and light exposure. The lower the water activity, the lower is the loss of vitamins. For example, vitamin E loss of only about 23% occurred after a 13 months of storage at a 0.6 water activity. In order to reduce oxidation of essential compounds and the development of rancidity, many authors recommend storing ground flour for no more than two weeks. Antioxidants present naturally in grains (vitamin E and lecithin) help prevent oxidation of the fatty acids and the associated rancidity only for a limited time, and under 'favourable' conditions. Glutamic acid decarboxylase, the most sensitive enzyme in the grain, is used to indicate the health of the grain. When heated or exposed to increased humidity, even under 'favourable' conditions, it losses activity very quickly in wheat. It was found to be even more sensitive in rye. The B vitamins are liable to be destroyed by light and air, and it also seems that other substances, still unknown, are quickly destroyed. Other deteriorations include denaturation of lipoproteins, phospholipid hydrolysis, auto-oxidation of unsaturated fatty acids of phospholipids, polymerization within lipoproteins, browning, Maillard reaction of amino groups from phospholipids and aldehyde groups from sugars, and carotene and aroma losses.

Lipids in milled wheat are much more susceptible to enzymatic degradation, because enzymes are incorporated into the flour with fragments of bran and germ and with microorganisms from the surface of the grain. Associated with lipid deterioration are losses of carotenoids and vitamin E. The nutritional importance of using fresh stone-ground grains for bread-making was revealed in the results of feeding studies in Germany Bernasek, (1970). Rats were fed diets consisting of 50% flour or bread. Group 1 consumed fresh stone-ground flour. Group 2 was fed bread made with this flour. Group 3 consumed the same flour as group 1 but after 15 days of storage. Group 4 was fed bread made with the flour fed to group 3. A fifth group consumed white flour. After four generations, only the rats fed fresh stone-ground flour and those fed the bread made with it maintained their fertility. The rats in groups 3 to 5 had become infertile. Four generations for rats is believed to be equivalent to one hundred years in humans. Different ecological standards for flour storage set limits of 15 to 60 days, although rancidity has been detected as early as 2 to 14 days after milling. Nutrient analysis studies are required to determine the exact nutrient losses accompanying the development of rancidity and thereafter.

Summary

Wheat is the main source of energy and nutrition in human diet. It provides carbohydrates, proteins, vitamins, minerals and fats. Though wheat grain has many components of diet, it is deficient in certain essential nutrients as vitamins A, B12, and vitamin C, fats and the amino acid lysine. In addition it has certain anti-nutritional factors which reduce the availability of micronutrients especially Fe and Zn. Major part of the nutrients is lost during milling because white endosperm is preferred in baking industry. Most of the minerals and vitamins are concentrated in outer layers of grain. There are reports of allergy to wheat based products and also enteropathy to wheat gluten. This has led to considerable research in the area of improving wheat nutritional quality by improving lysine content and also reducing anti-nutritional factors. Recently genetic engineering techniques have opened new opportunities for manipulating starch content and structure, lysine content and phytic acid. This may led to enhanced nutritional quality of wheat and making it as complete food.

References

Betschart, A.A. (1988). Nutritional quality of wheat and wheat foods. In Wheat Chemistry and Technology pp 91-132. Ed. Y. Pomeranz. St. Paul, Minnesota, USA.

Chhuneja, P. Dhaliwal, H.S., Baines, N.S. and Singh, K. (2006). *Aegilpos kotschyi* and *Aegilops tauschii* as sources of higher levels of grain Iron and Zinc. *Plant Breeding*. 125: 529-531.

Food and Agriculture Organization-World Health Organization (1991) Protein quality evaluation; report of the joint FAO/WHO expert consultation. FAO Food and Nutrition Paper 51, Rome, Italy.

Food and Agriculture Organization-World Health Organization-United Nations University (1985) Energy and protein requirements; report of a joint FAO/WHO/UNU expert consultation. *WHO Tech. Rep. Ser.* 724, WHO, Geneva, Switzerland.

Greiner, R. and Konietzny, U. (2006). Phytase for Food Application, Food Technol. Biotechnol. 44 (2) 125–140 (2006).

Grusak, M.A. and Penna D.D. (1999). Improving the nutrient composition of plants to enhance human nutrition and health. *Annual Review Plant Physiology & Plant Molecular Biology*. 50: 133-161.

Guttieri, M., Bowen, D., Dorsch, J.A., Raboy, V. and Souza, E. (2004). Identification and Characterization of a Low Phytic Acid Wheat. *Crop Sci.* 44:418-424.

Hitz, W.D., Carlson, T.J., Kerr, P.S. and Sebastian, S.A. (2002). Biochemical and molecular characterization of a mutation that confers a decreased raffinosaccharide and phytic acid phenotype on soybean seeds. *Plant Physiol.* 128: 650–660.

Hoseney, R.C., (1994). *Principles of Cereal Science and Technology*. 2nd ed. *Am. Assoc. of Cereal Chem.* Inc., St. Paul, Minnesota, USA

Larson, S.R., J.N. Rutger, K.A. Young, and Raboy, V. (2000). Isolation and genetic mapping of a non-lethal rice (*Oryza sativa* L.) low phytic acid mutation. *Crop Sci.* 40:1397–1405.

Larson, S.R., Young, K.A., Cook, A., Blake, T.K. and Raboy, V. (1998). Linkage mapping of two mutations that reduce phytic acid content of barley grain. *Theor. Appl. Genet.* 97: 141–146.

Lucca, P., Hurrell, R.F. and Potrykus, I. (2001). Genetic engineering approaches to improve the bioavailability and the level of iron in rice grains. *Theor. Appl. Genet.* 102: 392-397.

Mattern, P.P.J., Schmidt, J.W. and Johnson, V.A. (1970). Screening for high lysine content in wheat. *Cereal Sci. Today*. 15: 409.

Pomeranz, Y. (1998). Wheat Chemistry and Technology. *American Associaion of Cereal Chemists*, St. Paul Minnesota, USA.

Raboy, V. (2002). Progress in breeding low phytate crops. *J. Nutr.* 132: 503S–505S

Raboy, V., Gerbasi, P.F., Young, K.A., Stoneberg, S.D., Pickett, S.G., Bauman, A.T., Murthy, P.P.N., Sheridan, W.F. and Ertl, D.S. (2000). Origin and seed phenotype of maize low phytic acid 1–1 and low phytic acid 2–1. *Plant Physiol.* 124: 355–368.

Rasmussen, S.K. and Hatzack, F. (1998). Identification of two low-phytate barley (*Hordeum vulgare* L.) grain mutants by TLC and genetic analysis. *Hereditas* 129: 107–112.

Salunkhe, D.K., Kadam, S.S and Austin, A. (1986). Quality of wheat and wheat products. Proceedings of symposium on quality of wheat and wheat products held at Pune, India.

Sarwar G., Peace R.W. and Botting H.G. (1985). Corrected relative net protein ratio (CRNPR) method based on differences in rat and human requirements for sulfur amino acids. *J. Assoc. Off. Anal. Chem.* 68: 689-693.

Sarwar, G. (1997). The Protein Digestibility-Corrected Amino Acid Score Method Overestimates Quality of Proteins Containing Antinutritional Factors and of Poorly Digestible Proteins Supplemented with Limiting Amino Acids in Rats. *The J. Nutr.* 127(5): 758-764.

Shukla, S., Vantoai, T.T. and Pratt, R.C. (2004). Expression and nucleotide sequencw of an INS(3)P1 synthase gene associated with low phytate kennels in maize (*Zea mays* L.). *J. Agric. Food Chem.* 52: 465-470.

❑❑❑

Chapter 10

MOLECULAR BREEDING & TRANSGENIC APPROACH FOR THE IMPROVEMENT OF WHEAT GRAIN QUALITY

Introduction

Improvement of wheat grain quality depends on the knowledge of quality requirements of different end-use products and the genetic components controlling different quality traits. In previous chapters we have dealt in-depth with quality requirements of different end-use products and their genetic components. Depending upon the goal of quality improvement, different strategies are adopted to improve wheat quality. It depends upon the number of genes controlling the traits as well as the availability of the simple methods for phenotyping the traits especially during early segregating generations. In addition stability analysis of the traits across the environments is essential before we go for breeding. Constraints in time, cost, and sample availability usually necessitate the use of predictive tests during early generations, followed by milling and baking tests on advanced lines that have met a number of agronomic, disease and predictive quality criteria. Knowledge of germplasm lines having specific quality traits is needed to initiate breeding for the improvement of wheat grain quality for different end-use products. The quality requirements of different major end-use products are provided in the following table (Table 1).

Table 1: Quality requirements of different products

	Grain texture	Protein content	Gluten strength	Starch
Chapati	Hard/Medium	10-13%	Medium, extensible	Waxy?
Biscuit, Cakes	Soft/very soft	8-10%	Weak, highly extensible	—
Pan Bread	Hard	>13%	Strong,	Waxy
Hearth Bread	Hard	11-14%	extensible	for self life
White Noodle	Soft	10-12	Medium	Waxy
Yellow Noodle	Soft/medium	10-13	Medium	—
Pasta, Durum	Hard	>13	Strong	Waxy

The table clearly indicates that each individual end-use product has specific quality requirements. For example, soft wheat with low protein content and weak and extensible gluten are preferred for biscuit making. Hard wheat with high protein content (>13%) and strong and extensible gluten are preferred for bread making while hard wheat with medium strong gluten and high protein content for chapati making. Waxy wheats have been found associated with improved shelf life of breads. White salted noodles have different requirements than yellow alkaline noodles. White salted noodles need soft wheat with medium protein content and medium gluten strength. In addition, starch properties have greater role in determining noodle quality. Very recently waxy wheats with comparatively lower amylose content have been found useful in white salted noodles. In contrast yellow alkaline noodles need comparatively harder wheat with medium strong gluten and without waxy trait. Waxy wheats have detrimental effect on alkaline noodle quality. Pasta made from durum wheats need high protein content and strong gluten. Waxy durums have been reported showing positive effect on pasta quality. In addition beta-carotene content imparting yellow colour to pasta products is another specific requirement of good pasta making quality. The above facts clearly demonstrate that there are specific requirements of each product and therefore, accordingly breeding strategy has to be developed. Conventional breeding along with molecular breeding and transgenic approaches can led to enhanced processing and nutritional quality of wheat.

Marker Assisted Selection (Molecular Breeding)

The recent developments in plant biotechnology including molecular mapping and marker assisted selection (MAS) offer a choice of options for targeted breeding for enhancing wheat grain quality. Molecular breeding is more effective if the molecular map is densely populated with markers. Molecular marker assisted selection is faster, more efficient and cost effective than conventional screening methods. Progress in molecular genetics has resulted in the development of DNA tags, which can be used in marker-assisted selection (MAS) strategies for cultivar development. These molecular markers can be used as chromosome landmarks to facilitate the selection of chromosome segments including useful agronomic traits during the breeding process. These markers are particularly useful for incorporating genes that are highly affected by the environment, genes for resistance to diseases that cannot be easily screened for, and to accumulate multiple genes for resistance to specific pathogens and pests within the same cultivar, a process called gene pyramiding. An additional advantage of the incorporation of MAS into breeding programs is that very different types of traits, e.g. a disease resistance gene or a gene to increase grain protein content, can be manipulated using the same technology.

Using conventional backcrossing methods, it typically takes 6-8 back crosses to fully recover the recurrent parent genome, while in molecular marker assisted breeding, it is possible to recover the recurrent parent genome in 3-4 back crossing. Therefore the number of cycles of hybridizations can be reduced to achieve the similar amount of recurrent parent genome. The theoretical proportion of the recurrent parent genome after n generations of backcrossing is given by: $(2^{2+1}-1)/2^{n+1}$ (*where* n = number of backcrosses; assuming an indefinite population size). The percentage of recurrent parent recovery after each backcross generation is given in table. Although, *average percentage* of the recurrent parent genome is 75% for the entire BC1 population, some individuals possess more of the recurrent parent genome than others. Therefore, if tightly linked markers flanking QTLs or availability of functional markers within the gene itself and evenly spaced markers from other chromosomes of the recurrent parent are used for selection, the introgression of QTL and recovery of the recurrent parent can be accelerated. This process is called marker assisted back crossing (Table 2).

Table 2: Percentage of recurrent parent genome after backcrossing

Generation	Recurrent parent genome (%)
BC1	75.0
BC2	87.5
BC3	93.8
BC4	96.9
BC5	98.4
BC6	99.2

Analysis of DNA markers with a good coverage of entire genome can be used to select for individuals with a high proportion of recurrent parent genome and, thus reduce the number of backcross generations required. This was demonstrated by Frisch and Melchinger (2001) in maize where a shortening of the backcross program from six to three generations resulted from applying marker-assisted background selection (Fig.1). In breeding programs with three backcross generations, the least marker data points were required when (i) applying selection strategies consisting of three or four selection steps on the basis of presence of the target genes and selection indices calculated from the marker genotype, (ii) increasing the population size from early to advanced generations, and (iii) merging the target genes in an early generation. These principles can be used for optimizing the design of marker-assisted backcross programs for the simultaneous introgression of two genes. indicate generations with © crossing, (b) backcrossing and (s) selfing respectively Frisch and Melchinger, (2001).

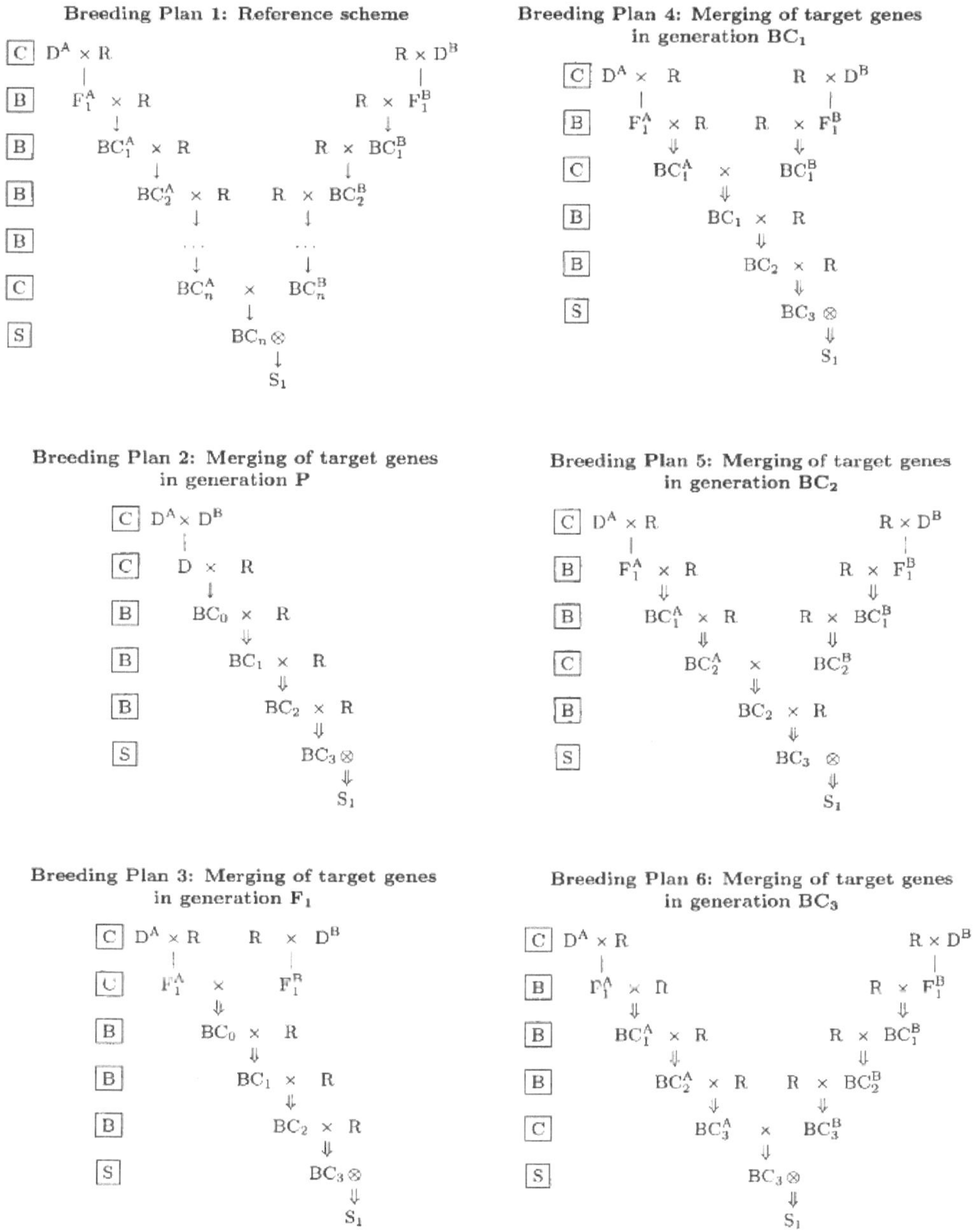

Fig. 1. Five breeding schemes for marker assisted introgression of two genes and a reference scheme (Breeding Plan 1) for introgression of two genes without back cross selection DA and Db are the donor lines of the target gene, R is the recipient line. The symbols denote: no selection, selection only for presence of target gene(s) and marker assisted background selection, x crossing, and selfing indicate generations with © crossing, (b) backcrossing and (s) selfing respectively. (Frisch and Melchinger, 2001).

The alleles that are incorporated by MAS are generally present within the gene pool of a particular crop and are transferred by meiotic chromosome recombination. One of the positive aspects of this approach is that these genes reside at their natural chromosomal locations, thereby minimizing the possibility of gene silencing. Another important aspect of cultivars developed by MAS is that they are not transgenic and therefore, do not face the public resistance against transgenic crops. The MAS strategy is a way to capitalize on available markers and to incorporate valuable traits into elite lines that are suitable for cultivar release. In addition, release of these MAS-improved cultivars is an efficient way of demonstrating the power of these technologies to the public. SSRs based PCR markers are the markers of choice which have the advantages of being single locus markers, co-dominant, multi-allelic, and widely dispersed over the genome. Recently high-density micro-satellite consensus maps representing different crops have been developed.

During the last 10 years, public researchers constructed detailed wheat genetic maps including more than 3000 molecular markers and physical maps including more than 16000 loci (http://wheat.pw.usda.gov/NSF/; verified 2 July 2004). In addition to mapping, U.S. federal agencies have funded the sequencing of more than 105000 wheat ESTs, the construction of wheat Bacterial Artificial Chromosome (BAC) libraries, the assembly of BACs into physical maps (http://wheat.pw.usda.gov/NSF.physical mapping; verified 2 July 2004), and the sequencing of large segments of wheat DNA. These powerful genomic resources have started to yield the first successful positional cloning efforts in wheat (Faris *et al.* 2003; Feuillet *et al.* 2003; Huang *et al.* 2003; Yahiaoui *et al.* 2004;Yan *et al.* 2003; Yan *et al.* 2004). Cloning of agronomically important genes has made possible to develop "perfect markers," based directly on the allelic variation responsible for the differences in the trait. Examples of perfect markers in wheat include the glutenin genes for gluten strength (Ahmad *et al.* 2000), the waxy genes for starch properties Briney *et al.* (1998), the puroindoline genes for hardness (Giroux and Morris 1997), the vernalization genes for vernalization requirement (Yan *et al.* 2003; Yan *et al.* 2004), the *Rht* genes for semi-dwarf habit Peng *et al.* (1999), and the *Lr10* and *Lr21* genes for leaf rust resistance (Feuillet *et al.* 2003, Huang *et al.* 2003). Wheat researchers have also developed closely linked molecular markers to yet unidentified genes with positive effects on quality characteristics and resistance to fungi, viruses, and insects. Dubkovsky (2004) provided more detailed account of molecular marker assisted selection giving wheat as an example.

The perfect markers for quality are being used in breeding programmes in many of the countries including India. The challenge to plant breeders is to implement these molecular markers into modern breeding programs in a cost-effective and efficient way. Essentially, the screening of a plant during marker-assisted selection (MAS) in a wheat breeding program is a four-step process: (1) collection of leaf samples from the plant to be screened in the field plot or glasshouse trial; (2) the extraction of DNA from the leaf sample; (3) the PCR-based marker assay and fragment analysis, and finally (4) the integration of marker results into breeder's database. The ability to automate each step is a highly desirable aim to enable the screening of thousands of plants required annually by plant breeders in large breeding populations. Streamlining of this process will considerably reduce the cost and increase the efficiency of marker-assisted breeding for wheat breeders, ultimately leading to the aim of the release of wheat cultivars with superior agronomic characteristics. High thorough techniques have been developed to isolate DNA from leaves and conducting PCR tests. This has resulted in the decrease of the total cost. Although molecular markers allow more accurate selection in early generations than conventional screen, large numbers can make selection impracticable while screening in later generations may provide little or no advantage over conventional breeding. Recently Bonnett *et al.* (2005) devised breeding strategy for efficient implementation of molecular markers in wheat breeding in which F_2 enrichment increased the frequency of selected alleles. This reduced minimum population size for recovery of target genotypes and selection at a greater number of loci. In addition backcrossing was also found to be a useful strategy to reduce population size compared with a bi-parental population.

Genetic Engineering Approach

The potential for the application of genetic engineering in wheat breeding is immense. The availability of transformation technology provides an opportunity to manipulate wheat to enhance its agronomic performance, resistance to abiotic/biotic stresses, yield and end-use quality. Wheat transformation has progressed enormously during the last 15 years from the first transgenic wheat plants obtained in 1992. Transformation protocols have been developed in wheat using particle bombardment (Weeks *et al.* 1993, Becker *et al.* 1994, Altpeter *et al.* 1996 and Barrow *et al.* 1997) and agrobacterium mediated (Chang *et al.* 1997). In earlier experiments Bob White was used for transformation experiments (Vasil *et al.* 1993, Weeks *et al.* 1993, Altpeter et al 1996, Blechl and Anderson 1996) because of easy transformation and regeneration. Subsequently other wheat lines were used for HMW glutenin manipulation (Barrow *et al.* 1997, Rooke *et al.* 1999 and Alvarez *et al.* 2000). These studies showed enhanced strength of gluten in transformed lines and hence improved functional properties of the dough.

New Developments

Transformation of cereal crops is a powerful research tool for gene discovery and function to investigate genetically controlled traits and is fast becoming a key element in the process of varietal improvement. For specific crops, it also enables the introduction of novel genes directly into locally-adapted germplasm and the creation of new genetically modified varieties. Wheat was among the last of the major crops to be transformed with the first fertile transgenic plants being reported using particle bombardment little over a decade ago. This is due to the inherent difficulties associated with gene delivery into regenerable explants and recovery of plantlets with the introduced gene. The last two decades have witnessed the widespread use of varied approaches for introduction of exogenous DNA into wheat, among which particle bombardment and co-cultivation with *Agrobacterium tumefaciense* are the two most widely used methods

Advances in the design of micro-projectile devices, choice of explant, media composition and selection systems have enabled the application of wheat transformation to study the role of specific genes in a wide range of agronomically important traits. Particle bombardment remains a robust, relatively efficient method for the genetic manipulation of wheat, however at the molecular level the DNA integration sites are often complex. The disadvantages of particle bombardment are multiple copy insertions of the transgene, low transformation efficiency and high tendency of gene silencing. The significant advantages to transferring DNA via *Agrobacterium,* including a reduction in transgene copy number, the stable integration with fewer re arrangements of long molecules of DNA with defined ends and the ability to generate lines free from selectable marker genes. This has been a driving force in the development of methods using *Agrobacterium tumefaciens* to deliver DNA although the ability to routinely transform wheat in this way is currently restricted to a few, well-resourced public and commercial laboratories world wide. This is partly due to the need for experienced personnel and expensive laboratory and plant growth infrastructure but also through a lack of clearly-written, complete, publicly-available protocols. There are several research papers and patents describing specific improvements to methodologies but these fail to provide a step-by-step guide to the transformation process as a whole.

Despite problems in wheat transformation, large information has been added in improving transformation efficiency particularly the stage and size of immature embryos and their osmotic and desiccation treatments. A marked effect of embryo size/age on T-DNA delivery and regeneration has been demonstrated, with large embryos (>2 mm) giving significantly higher transient expression levels but lower regeneration frequencies Wu *et al.* (2003) than smaller ones (<1.5 mm). Osmotic and desiccation treatments have been evaluated and incorporated into protocols based on particle bombardment, and have also been tested for *Agrobacterium* transformation of wheat. Air-drying pre-cultured immature embryos and embryogenic callus explants during *Agrobacterium* co-cultivation increased T-DNA-delivery and suppressed *Agrobacterium* cell growth which in turn facilitated better plant cell recovery Jones *et al.* (2005).

Agrobacterium Mediated Transformaton

Agrobacterium mediated transformation involves the immersion of explants in an *Agrobacterium* suspension followed by washing and co-cultivation for a further 1–3 days. One or both of these steps are carried out in darkness at approximately 25°C, although a two temperature co-cultivation step has also been tried with one day at 27°C then two days to 25°C. During the co-cultivation period, phenolic inducers such as acetosyringone work alongside other signaling factors such as temperature and an acid environment to promote the expression of *vir* genes. The presence of 200 μM acetosyringone in the *Agrobacterium* or co-cultivation medium markedly increased T-DNA delivery. Enhanced transient GFP expression was observed in wheat cell clusters with acetosyrigone at 400 μM in the co-cultivation but not the inoculation media. The need for acetosyringone been reported for a variety of wheat explants types but not for wheat cell suspension cultures where exogenous induction agents were not necessary for stable transformation Cheng *et al.* (1997). The use of surfactants during inoculation and co-cultivation significantly increases T-DNA delivery. Increasing concentrations of Silwet L-77 up to 0.04% had positive effects on T-DNA delivery as measured by the number of immature embryos with GUS foci and the number of GUS foci per embryo Wu *et al.* (2003). However, concentrations higher than 0.05% reduced survival and callus formation in freshly isolated immature embryos and an optimal concentration of 0.01% was chosen. Positive effects of surfactants were also reported in study Cheng *et al.* (1997) which used Silwet and pluronic acid F68 at 0.02%. Silwet has been used at concentrations as high as 0.05% for pre-cultured embryos and calli.

Incubation of suitable explants with *A. temefaciens* is needed in most of the experiments on Agrobacterium-mediated transformation. Therefore, the selection of suitable explants can have a significant impact on the transformation efficiency. Various explants have been attempted to find out the parameters necessary for successful, regeneration transgenic wheat. But it was not until 1997 that Chang *et al.* reported the stable transformation of wheat via cocultivation of Agrobacterium with freshly isolated immature embryos, precultured immature embryos, embryogenic calli and demonstrated the successful transmission of the exogenous gene to the subsequent generation. Later several workers reported the successful transformation of wheat using Agrobacterium-mediated transformation (Xia *et al.* 1999, Xue *et al.* 2004) using immature embryos. Recently Zhao *et al.* (2006) reported the transformation of wheat using basal portion of wheat seedlings (Shoot apical meristem) as explants. The exogenous gene (Beta-1, 3-glucanase) successfully integrated into the wheat genome, transmitted and expressed in the transgenic progeny. Since the method is very simple and efficient, it can be new technique for wheat genetics and breeding.

After the co-cultivation period, infected explants progress in a series of tissue culture steps on media designed to inhibit the growth of *Agrobacterium* cells and promote regeneration and selection of transformants. The antibiotics used to control the growth of *Agrobacterium* are added immediately after co-cultivation during the callus induction phase and are maintained in all subsequent media. Timentin or carbenicillin are commonly used but other compounds such as cefatoxin, cefotaxime, ticarcillium and vancomycin have also been reported. Plant selection agents complementary to the marker gene on the T-DNA are introduced to kill or compromise the growth of untransformed material. Selection for plant transformation is often initiated a few days after co-cultivation during the callus-induction phase and maintained during the latter regeneration steps.

Three selectable marker gene systems have been reported for *Agrobacterium* transformation of wheat. The first is based on antibiotic selection using either *hpt* (*aph4*-Ib) or *npt*II (*aph3'*II) which encode phosphotransferase enzymes that confer tolerance to the aminoglycoside antibiotics such as kanamycin, neomycin, paromomycin, G418 and hygromycin. A second system utilizes the bar gene which confers tolerance to glufosinate ammonium-based herbicides such as PPT, Basta, Bialaphos etc. A third system is based on the *aro*A:CP4 genes conferring tolerance to glyphosate-based herbicides such as Roundup. The use of 0.02 mM glyphosate on regenerating meristems has been reported to reduce the number of plants escaping selection to zero Hu *et al.* (2003). *NptII*, *bar* and *aro*A:CP4 have been successfully used by different groups to produce transgenic wheat plants but it is not possible to draw direct comparisons between selection systems because often a visual marker was also used in conjunction with chemical selection. For example, in several studies, the GUS reporter gene has been used in addition to the conventional selectable marker to help optimize the identification of transformants. Also, a T-DNA containing both *hpt* and GFP, along with hygromycin selection, has been used to identify early events in the transformation process.

The advantages arising from simple molecular integrations of single copy DNA fragments with defined ends have driven research into *Agrobacterium*-mediated plant transformation. Compared to rice and maize, progress with wheat has been slower but as described here, robust methods for the transformation of wheat using *Agrobacterium* now exist. There is scope to further optimise the media components and pH and to investigate the ideal virulence gene complement. Current bottlenecks limiting throughput include the labour-intensive steps of embryo isolation and transfers between media. Unlike biolistics, *Agrobacterium* suspensions can be manipulated by liquid handing robots and this combined with the use of callus cultures and the automation of transfer steps would enable a higher throughput which even at low efficiency would allow significantly more transgenic lines to be produced.

Genetic Engineering of Wheat Gluten

The fact that the bread making quality of wheat cultivars is mainly determined by specific alleles of HMW-GS genes and gene dosage has led to the suggestions that bread making quality could be improved by integration and expression of specific HMW-GS genes. Among the storage protein genes, high molecular weight glutenin sub-unit genes have been used for manipulating their number, composition and properties because the role of these proteins have been studied in greater detail in bread making quality. In addition, expression of limited number of LMW glutenin genes also showed enhanced gluten strength especially in durum wheat. This information can be used in the development of transgenics for individual glutenin subunits to improve quality. However, there is a lot of scope to improve further the transformation efficiency to make it routine as there is large variation between the efficiency achieved with different genotypes. Therefore, major bottleneck is to achieve higher transformation efficiency and to develop procedures applicable to a range of germplasm.

In earlier experiments Bob White was used for transformation experiments (Vasil *et al.* 1993, Weeks *et al.* 1993, Altpeter *et al.* 1996, Blechl & Anderson, 1996) because of easy transformation and regeneration. Subsequently other wheat lines were used for HMW glutenin manipulation (Barrow *et al.* 1997; Rooke *et al.* 1999; Alvarez *et al,* 2000). Field grown experiments of transgenic wheats indicated different effects of HMW glutenin subunits on milling and baking properties (Darlington *et al.* 2003, Barro *et al.* 2003). The transgene of subunit 1 (1Ax1) showed improved bread making quality in contrast transgene for subunit 5 (1Dx5) decreased loaf volume. The differences in bread making quality may be because of presence of an additional cysteine residue in subunit 5, which promotes more highly crossed linked gluten network. This was further validated by Rakszegi *et al.* (2005) by transforming wheat with 1Dx5 HMW glutenin subunit gene and over expressing it. Overexpresion of 1Dx5 led to decreased wet gluten content and lower SDS sedimentation volume. This demonstrated that a balance of elasticity and extensibility is essential for improving bread making quality.

Transgenic experiments in durum wheat have shown that low-molecular weight (LMW) glutenin subunits can be used in improving gluten strength and consequently pasta quality. Lines with better quality can be selected among the set of transgenic lines because transgene integration at different sites influence can have different effects. For example, the expression of the transgene in one line can led to an increase in the amount of large glutenin polymers resulting in stronger and more stable dough. In another line, however, the expression of the transgenic subunit can be accompanied by decreased expression of endogenous LMW subunits with consequent detrimental effects on glutenin polymers and dough visco-elasticity. These results demonstrate that the LMW glutenin subunits contribute to the functional properties of wheat by influencing the amount and the distribution of glutenin polymers and indicate that either plant breeding or GM technology can be used to 'fine tune' the properties of durum wheat for different end uses by manipulating the amount and structures of individual LMW subunit proteins.

These studies showed enhanced strength of gluten in transformed lines and hence improved functional properties of the dough. This information can be used in the development of transgenics for individual glutenin subunits to improve quality. However there is a lot of scope to improve further the transformation efficiency to make it routine as there is large variation between the efficiency achieved with different genotypes.

Therefore, major bottleneck is to achieve higher transformation efficiency and to develop procedures applicable to a range of germplasm. Moreover, there are other challenges also to bring genetically modified wheat in the market (Barcelo *et al.* 1998; Alvarez *et al.* 2000). The major challenges are i) identification of more of the key genes governing the traits of interest ii) the availability of the appropriate promoter sequences to derive transgene expression at desired levels at specific developmental stages in appropriate tissue iii) the identification of transgene instability and silencing and iv) to minimize the effect of transgene integration sites in the genome.

Genetic Engineering of Wheat Starch (Resistant Starch)

What is Resistant Starch?

The fraction of the ingested starch that escapes into the human large bowel is known as RS. Modification of the starch composition of wheat by raising its RS content presents an opportunity for a potentially large-scale improvement in public health. The increased incidence in many countries in lifestyle diseases such as colorectal cancer, cardiovascular disease, and diabetes has led to an enhanced interest in disease-prevention measures that can be delivered to target populations through diet. Resistant starch (RS) is emerging as an important dietary component that has the potential to reduce the incidence of bowel health disorders. However, the range of crop species that can serve as effective sources of RS is limited. Diet-related noninfectious chronic diseases, such as coronary heart disease, certain cancers (especially of the colon and rectum), and diabetes, are major causes of morbidity and mortality in both affluent industrialized countries and emerging nations. Increased consumption of whole-grain cereal foods can reduce the risk of these conditions substantially and also correlates negatively with obesity. Resistant starch (RS) and nonstarch polysaccharides (NSP) are major components of dietary fiber and contribute to these benefits. NSP resist digestion by enzymes in the human small intestine completely, helping to explain their fecal bulking and laxative action, which may underpin the protection against colorectal cancer afforded by greater fiber consumption. The details of resistant starch are given in other chapters on nutritional quality and starch structure and functionality.

Development of Wheat with Enhanced Amylose Content (RS)

Starches consist of linear chain (amylose) and branched chain (amylopectin) glucose monomers. In cooked foods, amylose molecules reassociate rapidly on cooling, forming complexes that resist digestion, whereas amylopectin molecules reassociate slowly and are more readily digested. This difference explains the higher RS content of high-amylose products. Therefore, altering the ratio of amylose to amylopectin in starch can led to higher resistant content having potential health benefits. This can be accomplished by altering the levels of different enzymes involved in starch biosynthesis. Amylose synthesis requires an active granule-bound starch synthase (GBSS). Amylopectin is synthesized by a complex pathway involving, among others, several isoforms of starch synthase (SSI, SSII, SSIII), starch-branching enzymes (SBEI, SBEIIa, SBEIIb), and starch-debranching enzymes (DBE). Details are given in other chapter on starch structure and biosynthesis. Theoretically, the suppression of the activities of starch synthase and starch branching enzymes can led to enhanced amylose content. Mixed results haven reported with respect the effect of alteration of gene expression in manipulation starch composition. In potato, two isoforms of SBE are present, SBEI and SBEII. Suppression of both SBEI and SBEII was required to generate starch with amylose content of >50%. In monocots, SBEI and two forms of SBEII, SBEIIa and SBEIIb, are found. In maize, suppression of SBEIIb results in the "amylose-extender" phenotype, with amylose contents from 50% to 90%, whereas suppression of SBEI or SBEIIa had no impact on endosperm amylose content. Amylose-extender maize has been used to increase the RS content of processed foods; however, additional sources of RS are needed. Similarly, RS content in wheat can be increased by manipulating genes of starch biosynthesis pathway. Recently RNAi technology has been developed to silence expression of genes in plants. This has the benefits of silencing genes present in all the homeologous chromosomes at a time with obvious advantage in wheat being hexaploid. Silencing homeologous genes in wheat by other methods as mutation are difficult. For the benefit of readers, brief account of RNAi technology is given below.

RNAi Technology

RNA interference (also called "RNA -mediated interference", abbreviated RNAi) is a mechanism for RNA-guided regulation of gene expression in which double-stranded ribonucleic acid inhibits the expression of genes with complementary nucleotide sequences Regina *et al.* (2006). Conserved in most eukaryotic organisms, the RNAi pathway is thought to have evolved as a form of innate immunity against viruses and also plays a major role in regulating development and genome maintenance. The RNAi pathway is initiated by the enzyme dicer, which cleaves double-stranded RNA (dsRNA) to short double-stranded fragments of 20–25 base pairs. One of the two strands of each fragment, known as the *guide strand*, is then incorporated into the RNA-induced silencing complex (RISC) and base-pairs with complementary sequences. The well-studied outcome of this recognition event is a form of post-transcriptional gene silencing. This occurs when the guide strand base pairs with a mRNA molecule and induces degradation of the mRNA by argonaute, the catalytic component of the RISC complex. The short RNA fragments are known as small interfering RNA (siRNA) when they derive from exogenous sources and microRNA (miRNA) when they are produced from RNA-coding genes in the cell's own genome (Fig. 2).

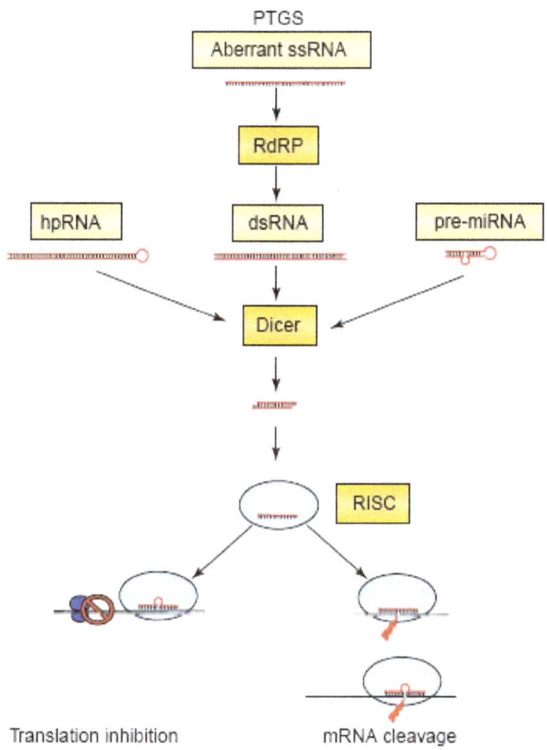

Fig. 2. The molecular mechanism of RNAi and RNAi-related phenomena in plants. PTGS involves the generation of dsRNA by RdRP. Micro RNA (miRNA) is an endogenous siRNA-like RNA known to be involved in the developmental regulation of gene expression in animals and plants. Its precursor (pre-miRNA) is a small hpRNA with 'bulges' in its stem region. All dsRNA, hpRNA and pre-miRNA are processed by Dicer into 21 nt RNA duplexes and the unwound ssRNA is then incorporated into RISC. In plants, like siRNA, miRNA cleaves mRNA despite a small number of mismatches with the target mRNA. It should be noted, however, that some miRNAs also inhibit translation in plants as well as in animals (Kusaba, 2004).

The selective and robust effect of RNAi on gene expression makes it a valuable research tool, both in cell culture and in living organisms; synthetic dsRNA introduced into cells can induce suppression of specific genes of interest. RNAi may also be used for large-scale screens that systematically shut down each gene in the cell, which can help identify the components necessary for a particular cellular process or an event such as cell division. Exploitation of the pathway is also a promising tool in biotechnology and medicine. Historically, RNA interference was known by other names, including post translational gene silencing, transgene silencing, and quelling. Only after these apparently-unrelated processes were fully understood did it become clear that they all described the RNAi phenomenon. RNAi has also been confused with antisense suppression of gene expression, which does not act catalytically to degrade mRNA but instead involves single-stranded RNA fragments physically binding to mRNA and blocking translation. In 2006, Andrew Fire and Craig C. Mello shared the Nobel Prize in Physiology and Medicine for their work on RNA interference in the nematode worm *C. elegans*, which they published in 1998.

Wheat is a hexaploid having three different genomes, A, B, and D. The hexaploid nature of the wheat genome makes finding and combining gene mutations in genes, such as the SBE genes, which are found in each of the three genomes, a challenging proposition. In contrast, gene silencing through RNA interference (RNAi) offers the potential to suppress simultaneously the expression of a target gene from each locus. Regina *et al.* 2006 used the RNAi technology to silence the expression of SBEIIa and SBEIIb in wheat, which resulted into a high-amylose (>70%) phenotype. DNA fragments containing sequence corresponding to exons 1, 2, and 3 and intron 3 of the wheat SBEIIa gene, *wSBEII-DA1* (GenBank accession no. AF338431), and the wheat SBEIIb gene, *wSBEII-DB1*, were ligated to generate SBEIIa and SBEIIb RNAi constructs, respectively. The respective ligated sequence was introduced into an intermediate vector pDV03000 containing a high-molecular-weight glutenin promoter sequence from wheat and the terminator sequence of nopaline synthase gene from *Agrobacterium* to generate SBEIIa (pDV03-IIa) and SBEIIb (pDV03-IIb) intermediate vectors. Each of the expression cassettes from pDV03-IIa and pDV03-IIb was then inserted into binary transformation vectors derived from pSB1 and pSB11, to generate hp-SBEIIa and hp-SBEIIb constructs. Hp-SBEIIa and hp-SBEIIb constructs were transformed into wheat by using an *Agrobacterium tumefaciens* transformation system. The transformed plants showed enhanced (>70%) amylose content in grain starch.

The experiment demonstrated that suppression of SBEIIa, not SBEIIb, is critical for high amylose in wheat. There are three main mechanisms for increasing amylose content in plants. The first is to increase granule-bound starch synthase (GBSS) activity. Overexpression of GBSS has recently been reported to yield a rice starch with slightly increased amylose content; however, the lack of reports concerning this mechanism precludes making an assessment of its general utility at this time. The second mechanism is to decrease amylopectin synthesis by suppression of the activity of starch synthases and/or isoamylases leading to a net increase in amylose content. Generally, this approach yields only modest increases in amylose content. For example, amylose contents of 35–45% have been in reported in maize sugary-2, su1, du-1 and wheat Sgp-1 mutants. However, the one exception is the much higher levels of amylose (>70%) present in a barley lacking SSIIa activity.

Rat nutritional studies were conducted by using high amylose wheat generated and found positive effects on indices of gastrointestinal health. Indices of large-bowel fermentation were all significantly higher in rats fed the high-amylose wheat, consistent with more RS, compared with those fed the standard wheat. Thus large-bowel digesta wet weight and SCFA pools and fecal SCFA excretion were all ≈100% higher in rats fed the novel wheat compared with those fed the control diet. pH values were also significantly lower, again consistent with greater fermentation. These data in rats fed the high-amylose wheat were generally similar to those noted previously in rats fed *Himalaya 292*, a barley cultivar with high RS arising through altered starch synthesis. Interestingly, the molar ratios of the major SCFA in rats fed the novel wheat were rather different from *Himalaya 292* with considerably more butyrate in the investigation. This is of some interest in view of the apparent importance of butyrate in promoting large-bowel function. Collectively, the data support the potential of the new high-amylose wheat to produce foods high in RS and with a low glycemic index. The data justify further investigation of the health potential of high-amylose wheats, especially in processed foods, as an important additional mechanism to deliver significant health benefits to large numbers of consumers through their diet.

Improvement of Lysine Content

Glutenins and gliadins are main storage proteins in wheat and they are rich in glutamine and proline and poor in lysine. Because of higher content of glutenins and gliadins, wheat proteins are deficient in lysine which is one of the indispensable amino acids. It was demonstrated during 1960's that there is not much variability in lysine content in wheat. Hence, attempts were made to enhance lysine content through mutation in wheat but not much success. During the last two decades the biochemical pathway of lysine biosynthesis and degradation has been studied extensively. Aspartate kinase (AK), the first enzyme of the aspartic acid pathway, has been extracted, partially purified and well characterized in several higher plants. A bifunctional polypeptide with both threonine-sensitive AK and homoserine dehydrogenase (HSDH) activities has been observed in some plant species. The activity of the enzyme dihydrodipicolinate synthase (DHDPS), which is strongly inhibited by low concentration of lysine, has been shown to be a key regulatory point of the lysine biosynthesis branch of the pathway Azevedo *et al.* (1997). Although a great deal of information is available about the lysine biosynthetic enzymes and their regulation, little is known about lysine degradation. Success in obtaining high-lysine plants for human or animal consumption relies on the full understanding of the metabolism of lysine as well as for subsequent genetic manipulation. Recently transgenic approach has shown promise in improving lysine content in many crops. Similar approach can be applied for improving lysine content in wheat.

Recent studies have indicated that the lysine catabolism plays an important role for lysine accumulation in plants and the control of the lysine content, particularly in seeds Arruda *et al.* (2000). Initial studies with enzymes involved in the lysine degradation reinforced the main role of catabolism in the control of the soluble lysine concentration in the maize endosperm. The amount of lysine that was shown to be translocated from other tissues to the developing endosperm, for the synthesis of storage proteins, was 2 to 3-fold higher than what would be necessary. Due to excess lysine, an accumulation of lysine in the soluble form would be expected, however this does not occur, since the soluble lysine concentration was maintained at a low level during the development of the endosperm. These low lysine levels could be contributed by preventing the inhibition of AK activity and facilitate the subsequent biosynthesis of methionine. These results indicated that the soluble lysine concentration is mainly controlled by the catabolic rate, instead of by the feedback inhibition of its synthesis. Two main enzymes have been shown to be part of the lysine catabolic pathway in animals, micro-organisms and recently in plants. Lysine 2-oxoglutarate reductase (LOR; EC 1.5.1.8), (also known as lysine α-ketoglutarate reductase, LKR) is the first enzyme of the pathway, which condenses lysine and 2-oxoglutarate to form saccharopine, which is then hydrolyzed to α-amino adipic acid and glutamic acid in a reaction catalyzed by the enzyme saccharopine dehydrogenase (SDH; EC 1.5.1.9). The net result of these two reactions resembles a transaminase reaction in which the α-amino group of lysine is transferred to 2-oxoglutarate to form glutamate (Azevedo and Lea, 2001).

Mainly in the last four decades, research groups have focused attention on understanding the biochemical and genetic controls of the aspartate pathway Azevedo *et al.* (1997). The data allowed researchers to induce and select for lysine and threonine overproducing plants through genetic manipulation of key points of the pathway such as catalysis by the enzymes; AK, HSDH, DHDPS, threonine synthase (TS), LOR and SDH (Azevedo and Lea, 2001). Based on the information provided by the work with the biochemical mutants and on newly developed transformation techniques, a similar strategy has been used to obtain plants that accumulate lysine in the seeds though little success for enhanced lysine content. Transgenic tobacco plants expressing a lysine-insensitive AK from *E. coli* exhibited similar results to those observed for the biochemical mutants, with threonine accumulation, but without changes in the soluble lysine content of the seeds. However, recently transgenic experiments showed promise for enhancing lysine content by combining the genetic manipulation of the biosynthesis and lysine degradation mechanisms (Fig.3).

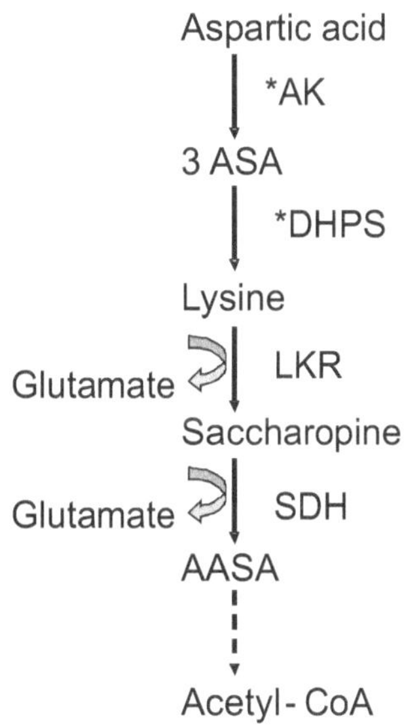

Fig. 3. Main steps involved in the biosynthesis and catabolism of lysine. ASA - Aspartic semialdehyde; AASA - Amino adipic semialdehyde; AK - Aspartate kinase; DHPS - Dihydrodipicolinate synthase; LKR - Lysine ketoglutarate reductase; and SDH - Saccharopine dehydrogenase. * - feedback inhibited by lysine. Dashed arrow in the last step indicates many enzymes.

Such a suggestion was supported mainly by the fact that the manipulation of enzymes involved in lysine biosynthesis did not produce lysine accumulation in cereal seeds. This could be explained by the fact that vegetables and the maize opaque-2 mutants, which exhibit higher concentration of soluble lysine in the seeds, exhibited a drastic reduction in the lysine catabolic rate in the endosperm, allowing excess lysine to be incorporated into storage proteins, as well as the accumulation in the soluble form Azevedo and Lea, (2001). The maize opaque-2 mutant has been extensively studied Gaziola *et al.* (1999). This mutation is characterized by an opaque phenotype with a farinaceous endosperm. The high lysine concentration observed in the endosperm is related to an increase in the concentration of soluble lysine and storage proteins with the simultaneous reduction of the prolamin fraction, which has only trace amounts of lysine Lefèvre *et al.* (2002). The introduction of the opaque phenotype modifier genes allowed the production of opaque-2 maize lines with good grain productivity, that also exhibit characteristics of high lysine and tryptophan contents, but with a translucent phenotype, which have been denominated as quality protein maize - (QPM) (Vasal, 1994, Gaziola *et al.* 1999). QPM inbred lines have been included in breeding programs with several hybrid of QPM been produced and agronomically tested that are now commercially available. Through transcriptome and proteome approaches, the regulatory role of the opaque-2 gene has been confirmed, since a 3' restriction site was shown to be associated with LOR-SDH mRNA abundance Lefèvre *et al.* (2002). Kemper *et al.* (1999) demonstrated that lysine levels in the endosperm are likely to be controlled at the transcriptional level by the Opaque2 transcription factor.

The functional role of Lys catabolism in balancing Lys levels in plants was demonstrated in developing seeds in Arabidopsis by Zhu and Gallili, (2004). Temporal and spatial expression of genes of lysine metabolic pathway can led to enhanced lysine in seeds. Genetic mutations in plant DHPS genes to render them insensitive to lysine cause the overproduction of lysine in all plant organs. But although an increase in lysine in seeds is beneficial, an increase in vegetative tissues is undesirable, because high concentrations of lysine cause abnormal vegetative growth and flower development which in turn reduces seed yield. Targeted expression of transgenic DHPS in seeds of several crop plants by using seed-specific promoters eliminates its undesirable effects in vegetative tissues, resulting in plants with good growth characteristics that accumulate high concentrations of lysine in their seeds (Falco *et al.* 1995 and Mazur *et al.* 1999).

Seed-specific expression of a bacterial feedback-insensitive dihydrodipicolinate synthase (DHPS) in an Arabidopsis knockout mutant of the AtLKR/SDH gene that regulates Lys catabolism synergistically boosted Lys accumulation in mature seeds, but it also severely reduced the growth of seedlings derived from them. Co-expression of a bacterial DHPS gene with an RNAi construct of AtLKR/SDH, both under control of the same seed-specific promoter, to restrict Lys synthesis and catabolism to the developing showed that post germination seedling growth was significantly improved when the reduction of Lys catabolism was restricted to seed development, suggesting that defective post germination Lys catabolism was responsible for inhibition of seedling growth in the AtLKR/SDH knockout plants expressing the bacterial DHPS gene in a seed-specific manner. Constitutive expression of the bacterial DHPS in the AtLKR/SDH knockout mutant boosted Lys levels in vegetative tissues in a similar manner to that observed in seeds, further demonstrating that Lys catabolism plays an important regulatory role in balancing Lys levels.

Another approach of enhancing lysine content may be to start with native plant proteins that are rich in essential amino acids and are naturally synthesized and accumulated in the tissue of interest. The content of essential amino acids in these proteins may be further improved by inserting them into desirable places that will interfere minimally with the structural conformation and stability of the proteins. An excellent example of such successful engineering is the modification of a native barley gene encoding a Lys-rich seed chymotrypsin inhibitor, termed 'barely high lysine' (BHL) protein, to increase its Lys codons (Roesler and Rao 1999, Forsyth *et al.* 2005). The most successful application of engineered native Lys-rich proteins so far has been achieved in maize. Among a variety of natural, modified or synthetic genes that were tested in maize, the most significant increase in seed Lys levels was obtained by the expression of genetically engineered BHL or hordothionine (HT12), containing 24% and 28% Lys, respectively (Jung and Falco, 2000). These proteins accumulated in transgenic maize to 3–6% of total grain proteins and, when introduced together with a bacterial Lys-insensitive DHPS, resulted in very high elevation of total Lys to over 0.7% of seed dry weight (Jung and Falco, 2000) compared to approximately 0.2% in wild-type maize. Combining these traits with seed-specific reduction of Lys catabolism offers an optimistic future for commercial application of high-Lys maize.

Phytic Acid and Phytase Levels

Phytic acid, which is an inhibitor of iron and zinc absorption in humans, is believed to be a major contributing factor to the world-wide problem of iron and zinc deficiency.

The negatively charged phosphates in phytic acid bind strongly to metallic cations as K, Mg, Mn, Fe, Ca and Zn to form mixed salt called phytin, and do not allow release of micronutrients in the intestine. Phytic acid is also the primary storage form of phosphorus (P) in seeds. Phytate in wheat grains is found predominantly in aleurone which remains attached to the pericarp during milling and therefore, is concentrated in the bran fraction. Monogastric animals such as pig, poultry and fish can not digest phytic acid. As a result animal feeds are supplemented with P, and a large fraction of phytic acid is excreted leading to accumulation in soil and water, causing pollution. Thus, any reduction in the content of phytic acid in staple foods is likely to result in improved iron and zinc bioavailablity. However, very low variations have been reported in phytic acid contents in wheat grains, therefore generation of *lpa* mutants and development of lpa transgenic lines may prove useful in wheat breeding.

Non-lethal recessive mutations that decrease seed phytic acid concentration have been isolated in maize, barley, rice, soybean and in wheat. Low phytic acid (*lpa*) mutants affect the partitioning of P into phytic acid, inorganic phosphorus (P_i), and inositol phosphorus (Ins P's) with five or fewer P esters. The *lpa* mutations are divided into two groups, designated *lpa*1 and *lpa*2. Phytic acid P reductions in *lpa*1 mutants are counterbalanced by molar-equivalent increases in P_i. In contrast, phytic acid P reductions in *lpa*2 mutants are counterbalanced by increases in both P_i and non-Ins P_6 and Ins P's of lower phosphorylation. Though, *lpa* mutants are not acceptable directly as cultivars because of undesirable changes and reduced yield, the deleterious effects can be mitigated by backcross breeding. Reducing phytate content through lpa mutants have been attempted through knock-out of genes involved in PA biosynthesis Bohn *et al.* (2008). With the only exception of one barley mutant accumulating Fe, the mineral composition does not dramatically change in lpa mutants, indicating that there is no direct link between mineral distribution and phytic acid biosynthesis. Due to the involvement of the lower inositol phosphates in plant cell metabolism, production of the perfect lpa mutant has however turned out to be quite a challenge. Since many kinases and tranferases are involved in the synthesis of phytic acid, the mutants often find ways to circumvent the knocked out pathway. Furthermore, the yield or germination ability is affected if PA content is reduced more than 50% Raboy, (2007), thereby making this approach unattractive from an economic perspective. However recently in wheat *lpa* mutants with perfect germination and good yield potential under field conditions have been identified. Moreover, maize *lpa*1 mutant have been identified, which is mutated in an embryo-specific ATP binding cassette (ABC)-tranporter and it is able to hold up to 90% reduction of PA without compromising seed viability. The effect on mineral distribution in this mutant is yet to be evaluated Shi *et al.* (2007).

In addition to low phytic acid, high phytase levels can be useful in enhancing nutritional quality. The phytases (*myo*-inositol (1,2,3,4,5,6) hexa*kis*phosphate phosphohydrolase) are a class of phosphatases with the in vitro capability to release at least one phosphate from PA, thereby releasing phosphate and lowering inositol phosphates and potentially chelated minerals. The number of enzymes described as phytases has increased rapidly over the years and details regarding the enzymatic properties can be found in several reviews (Haefner *et al.* 2005, Bohn *et al.* 2008). Most of the studies have been conducted in microbial systems. The largest group of phytases to date contains the 3-phytases (EC 3.1.3.8), which in general are found in fungi and bacteria. Structurally, most of the 3-phytases show homology to β-propeller phosphatase (BPP), or histidine acid phosphatases (HAP). The catalytic mechanism of hydrolysis by HAPs is proposed to be as follows: the histidine residue in the conserved motif is used to make a nucleophilic attack on the carbon of interest, and an aspartic acid in the C-terminal of the enzyme stabilises the leaving group by acting as a proton donor. A water molecule is consumed in the hyrolysis of the intermediate, but the phospho-histidine interaction is stable enough for crystallisation, as it has been done e.g., with a heat stable HAP phytase from *Aspergillus fumigatus* Xiang *et al.* (2004).

The number of phytases in wheat is still under investigation. Two enzymes, Phy1 and Phy2, have previously been purified from wheat bran (Lim and Tate 1973) and two isozymes with the N-terminal amino acid sequence EPAXTLTGPSRPV have also been purified, Nakano *et al.* (1999). Based on amino acid sequence and masses of tryptic peptides, a third enzyme with homology to PAP phytases was cDNA cloned from wheat and its homologue from barley Rasmussen *et al.*, (2007). Recently four cDNAs encoding for MINPPs were also cloned. At least two of these show in vitro phytase activity and they are probably expressed during late seed development and germination Dionisio *et al.* (2007). The sequential dephosphorylation of PA by wheat phytase has been studied by several groups Bohn *et al.* (2007), and it has been determined that it acts through both C6 and C3 to yield *myo*-inositol. This is also the case when isolated phytate globoids are used as the substrate, although the activity of the wheat phytase decreases approximately 30% on globoids as compared to when acting on pure PA Bohn *et al.* (2007).

Phytases are present in outer layers of the grain and show diminishing trend towards inner parts. Two-fold variations have been reported in phytase levels in wheat. However, wheat phytase is heat labile and hence genes encoding heat stable phytases from fungi and bacteria can be used for enhancing stability of phytases in wheat. Gregersen *et al* (2005) developed transgenic wheat lines using phytase gene from *Aspergillus fumigatus* (Fig. 4). There was four-fold difference in phytase activity in some of the transgenic seeds as compared to wild-type. At the early and mid-stages of grain filling, the heterologous phytase was primarily synthesized in one or more tissues of the pericarp, seed coat and aleurone, whereas the endosperm was the primary site for phytase synthesis toward the end of grain filling. However, the transgene encoding phytase was not stable in wheat. The reasons for transgene instability are yet to be identified. This demonstrated the feasibility of developing transgenic lines with enhanced phytase levels and hence useful in improving nutritional quality.

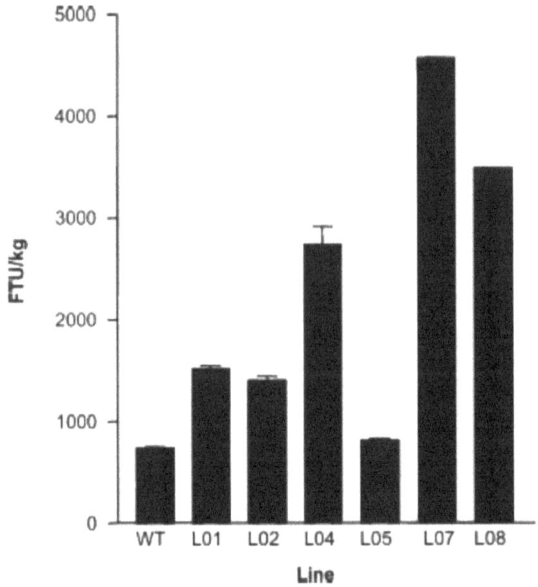

Fig. 4. Phytase activity in mature seeds of wild type and transgenic lines (From Gregersen *et al.* 2005)

Summary

The understanding of the genetic basis of grain quality traits has opened new ways to improve processing and nutritional quality of wheat. The genetic components related to grain texture, gluten strength and extensibility, starch composition, pentosan content yellow pigments, lysine content have been well characterized at molecular level. Many of the genes for these components have been identified and molecular markers developed for them. This has led to the development of molecular marker assisted selection (MAS) strategy. MAS has been successful in selecting desirable recombinants for grain texture, gluten strength, starch quality and others. The understanding of the pathway of biosynthesis of starch, lysine and phytic acid, has opened the possibility of using transgenic approach for improving nutritional quality. Many experiments demonstrated the use of transgenic approach in manipulating gluten proteins and hence processing quality. Recent developments in *Agrobacterium* mediated transformation of wheat have made greater possibility of using transgenic route for improving processing and nutritional quality of wheat.

References

Ahmad, M. (2000). Molecular marker assisted selection of HMW-glutenin alleles related to bread wheat quality by PCR-generated DNA markers. *Theor. Appl. Genet.* 101: 892-896.

Altpeter, F., Vasil, V., Srivastava, V. and Vasil, I.L. (1996). Integration and expression of the high molecular weight glutenin subunit 1A x 1 gene into wheat. *Nature Biotech.* 14: 1155-1159.

Alvarez, M. L., Guelman, S., Halford, M.G., Lustig, S., Reggiardo, M. I., Ryabushkina, N., Shewry, P. R., Stein, J. and Vallejos, R.H. (2000). Silencing of HMW glutenins in transgenic wheat expressing extra HMW subunits. *Theor. Appl. Genet.* 100: 319-327.

Anderson, O.D., F.C. Greene, R.E. Yip, N.G. Halford, P.R. Shewry, and J.M. Malpica-Romero. (1989). Nucleotide sequences of the two high-molecular-weight glutenin genes from the D genome of hexaploid wheat, (Triticum aestivum L. cv.) Cheyenne. *Nucleic Acid Res.* 17: 461–462.

Arruda, P., Kemper, E.L., Papes, F., and Leite, A. (2000). Regulation of lysine catabolism in higher plants. *Trends Plant Sci.* 5: 324–330.

Azevedo, R.A. and Lea, P.J. (2001). Lysine metabolism in higher plants. *Amino Acids* 20: 261-279.

Azevedo. R.A., Arruda, P. Turner, W.L., Lea, P.J. (1997). The biosynthesis and metabolism of the aspartate derived amino acids in higher plants. *Phytochem.* 46: 395-419.

Barrow, F. Rooke, I., Bekes, F., Gras, P., Tatham, A.S., Fido, R., Lazzeri, P.A. Shewry, P.R. and Barcelo, P. (1997). Transformation of wheat with high-molecular-weight glutenin subunit gene results in improved functional properties. *Nature Biotech.* 15: 1295-1299.

Becker, D., Brettchneider, R. and Lorz, H. (1994). Fertile transgenic wheat from microprojectile bombardment of scutellar tissue. *Plant J.* 5: 299-307.

Blauth, S.L., Yao, Y., Klucinec, J.D., Shannon, J.C., Thompson, D.B. and Guittinan, M.J. (2001). Identification of Mutator Insertional Mutants of Starch-Branching Enzyme 2a in Corn' *Plant Physiol.* 125: 1396-1405.

Blechl, A.E. and Anderson O.D. (1996). Expression of a novel high molecular weight glutenin subunit gene in transgenic wheat. *Nature Biotech.* 14: 875-879.

Bohn, L., Josefsen, L., Meyer, A.S. and Rasmussen, S.K. (2007). Quantitative analysis of phytate globoids isolated from wheat bran and characterization of their sequential dephosphorylation by wheat phytase. *J. Agri. and Food Chem.* 55(18): 7547-7552.

Bohn, L., Meyer, A.S. and Rasmussen, S.K. (2008). Phytate: Impact on environment and human nutrition. A challenge for molecular breeding. *J. Zhejiang Univ. Sci.* B. 9(3): 165-191.

Bonnett, D.G., Rebetzke, G.J. and Spielmeyer, W. (2005). Strategies for efficient implementation of molecular markers in wheat breeding. *Molecular Breeding* 15: 75-85.

Briney, A., Wilson, R., Potter, R.H., Barclay, I., Crosbie, G., Appels R. (1998). A PCR-based marker for selection of starch and potential noodle quality in wheat. *Molecular Breeding* 4: 427–433.

Cheng, M., Fry, J.E., Pang, S.Z., Zhou, H.P., Hironaka, C.M., Duncan, D.R., Conner, T.W., and Wan, Y.C. Genetic transformation of wheat mediated by Agrobacterium tumefaciens. *Plant Physiol.* 115: 971-980.

Cheng, M., Hu, T.C., Layton, J., Liu, C.N. and Fry, J.E. (2003). Desiccation of plant tissues post-Agrobacterium infection enhances t-DNA delivery and increases stable transformation efficiency in wheat. *In Vitro Cell Dev Biol-Plant.* 39: 595-604.

Dionisio, G., Holm, P.B. and Brinch-Pedersen, H. (2007). Wheat (*Triticum aestivum* L.) and barley (*Hordeum vulgare* L.) multiple inositol polyphosphate phosphatases (MINPPs) are phytases expressed during grain filling and germination. *Plant Biotech.* J. 5(2): 325-338.

Dubcovsky, J. (2004). Marker-Assisted Selection in Public Breeding Programs. The Wheat Experience *Crop Sci.* 44: 1895-1898.

Falco, S.C., Guida, T., Locke, M., Mauvais, J., Sandres, C., Ward, R.T. and Webber. P. (1995). Transgenic canola and soybean seeds with increased lysine. *Biotech.* 13: 577–582.

Faris, J.P., Fellers, J.P., Brooks, S.A. and Gill, B.S. (2003). A BAC contig spanning the major domestication locus Q in wheat and identification of a candidate gene. *Genetics* 164: 311–321.

Fernazier, R.F., Azevedo, R.A., Ferreira, R.R. and Varisi, V.A. (2003). Lysine catabolism: flow, metabolic role and regulation. *Braz. J. Plant Physiol.* 15 (1)

Feuillet, C., Travella, S., Stein, N., Allar, L., Nublat, A. and Keller, B. (2003). Map-based isolation of the leaf rust disease resistance gene Lr10 from the hexaploid wheat (*Triticum aestivum* L.) genome. Proc. Natl. Acad. Sci. USA 100: 15253–15258.

Forsyth, J.L., Beaudoin, F., Halford, N.G., Sessions, R.B., Clarke, A.R. and Shewry, P.R. (2005). Design, expression and characterization of lysine-rich forms of the barley seed protein CI-2. *Biochem. Biophys.* Acta. 1747: 221–227.

Frisch, M. and Melchinger, A.E. (2001). Marker-assisted backcrossing for simultaneous introgression of two genes. *Crop Sci.* 4: 1716-1725.

Gaziola, S.A., Alessi, E.S., Guimarães, P.E.O., Damerval, C. and Azevedo, R.A. (1999). Quality protein maize: a biochemical study of enzymes involved in lysine metabolism. *J. Agric. Food Chem.* 47: 1268-1275.

Giroux, M.J. and Morris, C.F. (1997). A glycine to serine change in puroindoline b is associated with wheat grain hardness and low level of starch-surface friabilin. *Theor Appl. Genet.* 95: 857-864.

Gregersen, PL., Brinch-Pedersen, H. and Holm, P.B. (2005). A microarray-based comparative analysis of gene expression profiles during grain development in transgenic and wild type wheat. *Transgenic* Res. 14(6): 887-905.

Grusak, M. A. and DellaPenne, D. (1999) Improving the nutrient composition of plants to enhance human nutrition and health. *Annu. Rev. Plant Physiol.* 50: 133-161.

Grusak, M. A., Pearson, J. N. and Marentes, E. (1999) The physiology of micronutrient homeostasis in field crops. *Field Crops Res.* 60: 41-56.

Haefner, S., Knietsch, A., Scholten, E., Braun, J., Lohscheidt, M., Zelder, O. (2005). Biotechnological production and applications of phytases. *Appl. Microbio. and Biotech.* 68(5): 588-597.

Hu, T., Metz, S., Chay, C., Zhou, H.P., Biest, N., Chen, G., Cheng, M., Feng, X., Radionenko, M., Lu, F. and Fry, J. (2003). Agrobacterium-mediated large-scale transformation of wheat (*Triticum aestivum* L.) using glyphosate selection. *Plant Cell Reports.* 21: 1010-1019.

Huang, L., Brooks, S.A., Li. W., Fellers, J.P., Trick, H.N. and Gill, B.S. (2003). Map-Based Cloning of Leaf Rust Resistance Gene Lr21 From the Large and Polyploid Genome of Bread *Wheat Genetics* 164: 655–664.

Jane, G., Muir, E., Yeow, G.W., Keogh, J., Pizzey, C., Bird, A.R., Sharpe, K., O'Dea, K. and Macrae, F.A. (2004). Combining wheat with resistant starch has more beneficial effects on fecal indexes than does wheat bran alone. *Amer. J. Cli. Nutr.* 79(6): 1020-1028.

Jones, H.D., Doherty, A. and Wu, H. (2005). Review of methodologies and a protocol for the Agrobacterium-mediated transformation of wheat. *Plant Methods.* 1:5.

Jung, R. and Falco, S.C. (2000). Transgenic corn with an improved amino acid composition. *In: 8th International Symposium on Plant Seeds,* Gatersleben, Germany.

Kemper, E.L., Cord-Neto, G., Papes, F., Martinez-Moraes, K.C., Leite, A. and Arruda, P. (1999). The role of opaque-2 on the control of lysine degrading activities in developing maize endosperm. *Plant Cell.* 11: 1981-1994.

Kusaba, M. (2004). RNA interference in crop plants. *Current Opinion in Biotech.* 15: 139–143.

Lefèvre, A., Consoli, L., Gaziola, S.A., Pellegrino, A.P., Azevedo, R.A. and Damerval, C. (2002). Dissecting the opaque-2 regulatory network using transcriptome and proteome approaches along with enzyme activity measurements. *Scientia Agricola* 59: 407-414.

Lim, P.E. and Tate, M.E., (1973). Phytases. 2. Properties of phytase fractions F1 and F2 from wheat bran and myoinositol phosphates produced by fraction F2. *Biochimica et Biophysica Acta.* 302(2): 316-328.

Mazur, B., Krebbers, E. and Tingey, S. (1999) Gene discovery and product development for grain quality traits. *Sci.* 285: 372–375.

Morell, M.K., Konik-Rose, C., Ahmed, R., Li, Z. and Rahman, S. (2004). Synthesis of Resistant Starches in Plants. *J. AOAC Inter.* 87 (3): 740-748.

Nakano, T., Joh, T., Tokumoto, E. and Hayakawa, T. (1999). Purification and characterization of phytase from bran of *Triticum aestivum* L. cv. Nourin #61. *Food Sci. and Tech. Res.* 5(1): 18-23.

Raboy, V. (2007). The ABCs of low-phytate crops. *Nature Biotech.* 25(8): 874-875.

Rasmussen, S.K., Johansen, K.S. and Sorensen, M.B. (2007). Polynucleotides Encoding Phytase Polypeptides. US Patent 10/275,311(7,186,817).

Regina, A., Bird, A., Topping, D., Bowden, S., Freeman, J., Barsby, T., Kosar-Hashemi, B., Li, Z., Rahman, S. and Morell, M. (2006). High-amylose wheat generated by RNA interference improves indices of large-bowel health in rats. *PNAS* 103 (10): 3546-3551.

Regina, A., Bird, A.R., Li, Z., Rahman, S., Mann, G., Chanliaud, E., Berbezy, P., Topping, D. and Morell, M.K. (2007). Bioengineering cereal carbohydrates to improve human health. *Cereal Foods World.* 182-187.

Richards, D.E., Hartley, N.M., Murphy, G.P., Devos, K.M. and Jlintham, J.E. (1999). 'Green revolution' genes encode mutant gibberellin response modulators. *Nature*. 400: 256–261.

Roesler, K.R. and Rao, A.G. (1999). Conformation and stability of barley chymotrypsin inhibitor-2 (CI-2) mutants containing multiple lysine substitutions. *Protein Eng.* 12: 967–973.

Rooke, L., bekes, F, Fido, R., Barro, F. Gras, P., Tatham, A.S., Parcelo, P., Lazzeri, P. and Shewry, P.R. (1999). Over expression of a gluten protein in transgenic wheat results in greatly increased dough strength. *J. Cereal Sci.* 30: 115-120.

Shi, J., Wang, H., Schellin, K., Li, B., Faller, M., Stoop, J.M., Meeley, R.B., Ertl, D.S., Ranch, J.P. and Glassman, K., (2007). Embryo-specific silencing of a transporter reduces phytic acid content of maize and soybean seeds. *Nature Biotech.* 25(8): 930-937.

Tosi, P., Masci, S., Giovangrossi, A., D'Ovidio, R., Bekes, F., Larroque, O., Napier, J. and Shewry, P. Modification of the Low Molecular Weight (LMW) Glutenin Composition of Transgenic Durum Wheat: Effects on Glutenin Polymer Size and Gluten Functionality. *Molecular Breeding* 16 (2): 113-126.

Vasal, S.K. (1994). High quality protein corn. In: Specialty Corns, Hallawer AR (ed), pp. 79-120, CRC Press, Boca Raton, FL.

Vasil, V., Srivastava, V. Castillo, A.M., Fromm, M.E. and Vasil, I.K. (1993). Rapid production of transgenic wheat plants by direct bombardment of cultured immature embryos. *Biotechnology.* 11: 1553-1558.

Weeks, J.T., Anderson, O.D. and Blechl, A.E. (1993). Rapid production of multiple independent lines of fertile transgenic wheat (Tritucum aestivum) *Plant Physiology.* 102: 1077-1084.

Wu, H., Sparks, C., Amoah, B. and Jones, H.D. (2003). Factors influencing successful Agrobacterium-mediated genetic transformation of wheat. *Plant Cell Reports,* 21: 659-668.

Xia, G., Fle, C., Chen, H. and Li, Z. (1999). Transgenic plant regeneration from wheat (*Triticum aestivum* L.) mediated by Agrobacterium tumefaciens. *Acta Phytophysiologica Sinica.* 25: 22-28.

Xiang, T., Liu, Q., Deacon, A.M., Koshy, M., Kriksunov, I.A., Lei, X.G., Hao, Q. and Thiel D.J. (2004). Crystal structure of a heat-resilient phytase from Aspergillus fumigatus, carrying a phosphorylated histidine. *J. Mol. Bio.* 339(2): 437-445.

Xue, Z.Y., Zhi, D.Y., Xueb, G.P., Zhang, H., Zhao, Y.X. and Xia, G.M. (2004). Enhanced salt tolerance of transgenic wheat (*Tritivum aestivum* L.) expressing a vacuolar Na+/H+ antiporter gene with improved grain yields in saline soils in the field and a reduced level of leaf Na+. *Plant Sci.* 167: 849–859.

Yahiaoui, N., Sridhumpa, P., Dudler, R. and Keller, B. (2004). Genome analysis at different ploidy levels allows cloning of the powdery mildew resistance gene Pm3b from hexaploid wheat. *Plant J.* 37: 528–538.

Yan, L., Loukoianov, A., Tranquilli, G., Blechl, A., Khan, I.A., Ramakrishna, W. (2004). The wheat VRN2 gene is a flowering repressor down-regulated by vernalization. *Science* 303:1640–1644.

Yan, L., Loukoianov, A., Tranquilli, G., Helguera, M., Fahima, T. and Dubcovsky, J. (2003). Positional cloning of wheat vernalization gene VRN1. *Proc. Natl. Acad. Sci.* USA 100: 6263–6268.

Zhu, X. and Galili, G. (2004). Lysine Metabolism Is Concurrently Regulated by Synthesis and Catabolism in Both Reproductive and Vegetative Tissues. *Plant Physio.* 135: 129–136.

❑❑❑

Chapter 11

Barley Malt, Feed & Food Quality

Introduction

Barley is an important cereal crop grown all over the world and is used mainly for feed and malt purposes and little for human food. Barley is the world's oldest grain, as evidenced by discoveries in ancient cities in the Middle East and North Africa. It has been cultivated for about 8,000 years, and today is the world's fourth largest cereal crop. Considerable evidence is available that points to the role of barley as sustaining food source in the evolution of humans. Additionally, alcoholic beverages of various types and fermented foods prepared from barley are commonly referred to in the ancient literature. As other food crops (wheat, rice, corn and rye) became more abundant, barley was relegated to the status of "poor man's bread". In the United states, barley used as food is 1.5%, as animal feed 65% and in malt and alcohol production 30% (Newman and Newman, 2006). In 1991, barley used as food in European countries was even lower (0.3%) than the United States. In contrast, during the same time period food was the largest for barley in Morocco (61%), China (62%), India (73%), and Ethiopia (79%). However, current consumer interest in nutrition and human health may help restore barley's status in the human diet. Nutritional qualities of barley have been known since ancient times. For example, Ajgaonkar 1972 described how ancient Indian physicians effectively stabilized type II diabetics some 2400 years ago. Treatment was remarkable simple and not really different from recommendations that are given to diabetic patients today, i.e., lose weight, change diet, and increase exercise.

Barley as a food is most commonly identified as pearl barley, traditionally used in soups. Pearl barley has many other food uses, such as casseroles, pilafs or salads. Barley flour and milling fractions such as bran, middlings, shorts and "red dog," can also be used for baked products and breakfast cereals. Barley malt is used for production of beer, whisky, food drinks and medicinal purposes as shown in the chart below. Malt with low nitrogen and low DP is preferred for brewing and distilleries while high nitrogen and high DP content malt extract is used in the manufacture of nutritious baby foods, milk based health drinks, pharmaceuticals, beverages and tonics and confectionary and bakery products. Barley malt is used to impart distinctive colour and flavour to a variety of food stuffs. However, the most extensive use of barley malt worldwide is a source of fermentable sugars for alcoholic fermentation, principally to produce beer and whisky. Twenty five percent of total barley produced in India is utilized for malt purposes of which around 75% is used for brewing and distillation industries and 20% for food drinks as Maltova etc and 5% is used in medicines as diuretics syrup. In this chapter, detailed account of malting, feed and food quality will be discussed (Fig.1).

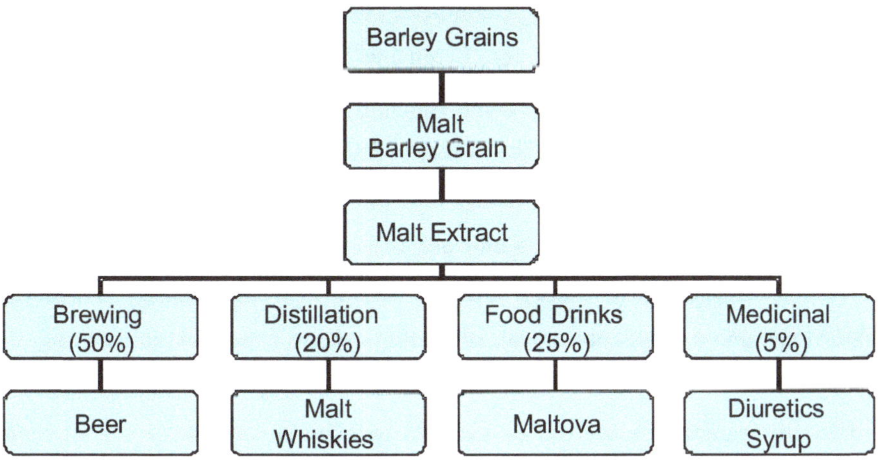

Fig. 1: Utilization of malt for different products in India.

Malting

Simply stating malting is a brief initiation of the normal botanical growth of the grain followed by abrupt interruption with hot air (kilning). The main aim of the malting is to dissolve and degrade the components of barley grain for fermentation. The main steps involved in malting are steeping, germination and kilning. The physical and chemical components in the endosperm determine the ease of modification and thus ultimate malt quality. A loosely structured endosperm with thin cell walls improves the water uptake during steeping and the penetration of enzymes and hormones during germination. The chemical components include protein content and quality, beta glucan content in cell wall, starch hydrolyzing enzymes and lipids. The quality of malt in turn determines its potential to use for different purposes. 60-65% of the weight of malt is undegraded starch. Barley is deficient in certain key enzymes (as α-amylase) and malting increases these levels. During mashing the malt enzymes are mixed with starch to produce maltose and other fermentable sugars. Malt also provides various nutrients for yeast growth, including amino acids, vitamins and minerals. The husk of barley malt provides the filter bed, crucial for forming clear worts during lautering.

Barley is suitable for malting because of the following reasons: (i) Presence of cemented lemma and palea (hulls) which protect embryo during malting and also serve as an aid in filtration process, (ii) The steeped barley has firmer consistency. Various grain and malt parameters have been defined to select varieties with better malt quality. The grain parameters include plumpness (Thin and bold grains), protein content, beta-glucan content, husk content and germinative energy. Important malt parameters are hot water extract, diastatic power, kolbach index, friability, filtration rate and wort viscosity. The characters which determine the quality of malting barley are inherited quantitatively. Genetically, they are complex traits and are considerably affected by environmental conditions. The usability of barley cultivars for the malting industry depends on a complex of morphological, anatomical, physiological and chemical features of grain and technological parameters, which allow us to assess the quality of malt produced from grain. Quality parameters for barley used for production of malt, animal feed and human consumption are presented in table below (Table 1).

Table 1: Quality parameters of barley grain for malting, animal feed and human consumption (according to Gasiorowski 1997)

Quality parameter	Malting barley	Feed barley	Consumption barley
Hull (husk)	desirable, thin	undesirable	undesirable
Content:			
total protein	low (9-12%)	high	high
lysine	-	high	high
starch	high	high	high
fats	-	high	high
ß-glucans	low	low	high
phenolic compounds	low	-	-
Enzymatic activity of malt	high	-	-
Germination capacity	very high	high	high

Based on the quality parameters identified for better malt quality, breeding guidelines have been developed for better malt barley varieties as shown in the (Table 2).

Table 2: Malting Barley Breeding Guidelines Ideal Commercial Malt Criteria

	Two-Row Barley	Six-Row Barley
Barley Factors		
Plump Kernels (on 6/64)	> 80%	> 70%
Thin Kernels (thru 5/64)	< 5%	< 5%
Germination (4ml 72 hr. GE)	> 96%	> 96%
Protein	11.5-13.0%	11.5-13.5%
Skinned & Broken Kernels	< 5%	< 5%
Malt Factors		
Total Protein	11.3-12.8%	11.3-13.3%
on 7/64 screen	> 60%	> 50%
on 7/64 + 6/64 screen	> 90%	> 85%
Measures of Malt Modification		
Beta-Glucan (ppm)	<115	<140
F/C Difference	<1.5	< 1.5
Soluble/Total Protein	40-45%	40-45%
Turbidity (NTU)	< 15	< 15
Viscosity (absolute cp)	< 1.50	< 1.50
Congress Wort		
Soluble Protein	4.9-5.5%	5.2-5.6%
Extract (FG db)	>81.0%	>79.0%
Color (°ASBC)	1.6-2.0	1.7-2.1
Malt Enzymes		
Diastatic Power (°ASBC)	120-140	140-160
Alpha Amylase (DU)	45-60	45-60

The Malting Process

Malting is the limited germination of the barley grain under very carefully controlled conditions. There are three phases: Steeping, germination and kilning. Steeping encourages germination to start, germination prepares the conversion of the starch to sugars, and kilning stops the germination to ensure that very little of the starch is hydrolysed. Conversion of starch to maltose occurs in the mashing process. A good maltster ensures that all the kernels germinate and at the same rate. Good levels of enzymes are produced. The cell walls of the starchy endosperm are degraded and the protein matrix is disrupted. All the three steps in malting are discussed below.

Steeping

Steeping is the first stage of the true malting process and takes 2-3 days in total. The moisture content of the barley is raised from typically 12% to 43-46%. This is achieved by a series of immersions or "wet stands" followed by dry stands. The grain becomes hydrated and starts to germinate producing heat and carbon dioxide. During the wet stands, air is blown through the wet grain and during the dry stands, carbon dioxide is removed. Steeping is the crucial phase of malting. By the end of steeping all the grains should be fully hydrated and should be just beginning to show signs of germination - the formation of a tiny rootlet or "chit". Several factors influence the rate of water uptake. The so called steeliness/mealiness character of the starchy endosperm appears to be particular importance. Mealy endosperm characteristics of good malting barley have a relatively open structure containing many cracks, and the starch granules are relatively loosely packed in the protein matrix. Water defuses more readily through such an open structure than it does through a steely endosperm, which has tight starch-protein packing. Access of the water is restricted by the non permeable covering layers of the kernel, notably the pericarp. In the intact kernel, water enters principally in the embryo region, although some reports indicate that water can enter through distal end of the kernel and also in the ventral region.

Germination

During germination the embryo and the aleurone layer produce enzymes that pass into the starchy endosperm breaking down cell walls and the protein matrix. The endosperm loses its hard structure and becomes "modified". The modification depends upon the rate at which moisture distributes through the starchy endosperm, the rate of synthesis of hydrolytic enzymes, the extent of release of these enzymes ant the structural features of the starchy endosperm that determine its resistance to digestion. The differences in water uptake capabilities may be because of the structure of the starchy endosperm cell walls in terms of the degree of the cross linking of beta-glucans, pentosans and proteins. Starch is then hydrolysed to simple sugars (eg maltose). Nutrients pass from the starchy endosperm into the embryo and support the growth of a new plant. Germination continues after the steeping for a further 4-5 days. The grain is kept cool and carbon dioxide is removed by blowing streams of cool, humidified air through the bed. The grain is turned regularly to prevent rootlets matting and maintain a loosely packed grain bed. The maltster manipulates the germination conditions to vary the type of malt being manufactured. The details of the enzymes involved in the modification of endosperm will be discussed in greater details in subsequent sections.

Kilning

Kilning is the third phase of malting, dries the grain down to 3-6% moisture and arrests germination. Large volumes of hot air are blown through the grain. By varying air flows and kiln temperatures malts of different colour can be produced with varying flavour profiles. At the end of kilning the malt is cooled and the tiny rootlets removed. Through the controlled drying of green malt, the malster is able to arrest modification and render malt stable for storage, ensure survival of enzymes for subsequent employment in processing and introduce desirable flavour and colour characteristics. If stored correctly, malt at 3-% moisture may be stored for several months with no loss of quality.

The basic principal of kilning are that drying should commence at a relatively low temperature to ensure survival of most sensitive enzymes followed by progressive increase of temperature to effect flavour and colour changes and complete drying within the limited turnover time available (typically less than 24 hours) in most reasonably sized kilns. Principal among the changes occurring during the kilning is the browning or Maillard reaction. The interaction of reducing sugars and amino acids produces reductones, which in turn can be converted by polymerisation to the colourful melanoidins or, by alternate routes to the heterocyclic pyrazines, thiophenes, pyrrolles and furans. The oxygen heterocyclins are responsible for toffee or caramel flavours. The pyrazines afford the roasted, coffee like flavours typical of more intensely kilned malts. Kilning also has a profound role to play in the regulation of levels of heat labile S-methylmethionine, which decomposes to dimethyl sulphide, a major flavour attribute of larger-type beers.

Physical Characteristics of Kernels and Malting Quality

Uniformity of Kernels

One of the yield structure components, i.e. 1000 grain weight, is also a grain quality character. 1000 kernel weight belongs to the group of characters which determine plumpness of kernels. Kernels uniform in shape, size, and weight produce uniform, high quality malt. Such kernels, when soaked, absorb water uniformly, begin germinating evenly, and malt identically. The so-called filled or technical fraction of kernels (over 2.5 mm in diameter) is thought to be most suitable for malting purposes. Breeding of cultivars with filled kernels is therefore very important. Brewers prefer highly uniform cultivars with kernels well filled with starch, but not a very large1000 kernels weight. It is important to determine both 1000 kernels weight and percentage of filled grain while breeding new cultivars or testing the existing ones.

Husk Content

The husk consists of two outer layers, which surround and adhere to the entire grain. The palea is the husk layer on the ventral side of the grain and the lemma is at the dorsal or back side of the grain. Recent research has shown a possible association between high levels of malt extract and thin husk. However, there are a number of potential problems associated with thin husk. The husk provides physical protection for the grain and the growing acrospire. Thin husked varieties may therefore have an increased tendency for weather damage and preharvest sprouting, and an increased likelihood of embryo damage and skinning during harvesting and subsequent grain handling. The husk may also help in part to physically restrain the acrospire and the swelling of the grain during germination. Thin husk may therefore cause problems in controlling the germination process during malting, leading to over modification. Husk is also needed to form the filter bed during lautering, and low levels may therefore impact on the brewing process.

In addition, there are related issues of skinning and hull adherence. The high extract Canadian variety, Harrington, although of excellent malting quality has a loosely adhering husk and is highly susceptible to skinning. As with thin husk, skinning of the grain has a number of adverse effects on malting. These include reduced uniformity and germination during the malting process, embryo damage, mould growth and numerous grain handling problems. Hull adherence has in fact been identified by the Brewing and

Malting Barley Research Institute in Canada as the most important quality problem to address (Edney, 1999).

Grain Texture

Observation of the endosperm texture on a transversal section of the kernel provides some information on the content of protein and starch in grain. Kernels with a high content of protein have a translucent or partly translucent transversal section. Kernels rich in starch, with a lower level of protein substances, are powdery. Translucent grain, with the compact structure of the endosperm, absorbs water slowly and not uniformly during steeping, and the resultant malt is low in extractability. Drought could be one of the causes producing apparent translucency, which disappears when grain is steeped and then dried slowly at room temperature. Barley used to produce grits is expected to have translucent grain. Barley cultivars differ in traits related to grain texture. A survey of barley cultivars found that milling energy, another measure of grain hardness, correlates negatively with malting quality. Generally grain hardness is negatively correlated with percentage dry-matter digestibility by ruminants. However, larger particles and slower dry-matter digestibility is associated with higher feed quality. Therefore, the development of softer cultivars may benefit malting-quality traits and the development of harder barleys may benefit feed-quality traits.

Physiological Characters of Kernels Affecting Malting Quality

Viability

Viability is the most important physiological character of kernels of cereal plants. In the case of malting barley, kernels viability determines its technological value. Germination energy and germination capacity are two practical indices of grain viability. Germination energy means uniformity and rate with which kernels can germinate, and it indicates that kernels are ready for malting. Germination energy is determined 72 or 96 hr (depending on the method applied) after kernels are soaked. Malting barley kernels are of high quality if their germination energy is in excess of 95%.

Dormancy and After-Ripening

Dormancy and after-ripening of barley kernels are two physiological processes associated with malting. Morphologically ripe and viable kernels enter the dormancy stage, during which their germination capacity is halted. If kernels fail to enter dormancy at the time they are ripe, sprouting may occur – germination of kernels in spikes. Duration of dormancy is specific for each plant species. Grains of cereal plants have low levels of dormancy for several weeks. Barley kernels have the longest real dormancy phase, lasting from 3 to 9 weeks. Dormancy may become deeper if the weather during the ripening and harvest is unfavourable. After-ripening comprises all biochemical changes taking place in grain dried naturally and harvested. Once this is over, kernels achieve full physiological ripeness, expressed by their germination capacity. Recession of dormancy means that kernels are physiologically ready to germinate. It is typical of good malting barley to produce kernels which reach physiological ripeness early and germinate promptly, uniformly and at a nearly 100% success rate.

Sprouting

Sprouting of grains, pre-mature germination of kernels in spikes during harvest, incurs considerable economic loss due to the lower yield and inferior technological value of grains. Sprouted grain cannot be used as seed material and it becomes poor quality produce for breweries. Resistance to sprouting is a cultivar specific character, genetically determined, but under a strong influence of environmental factors. Resistance to sprouting is a resultant of a complex set of morphological, anatomical, physiological and biochemical factors as well as variable environmental conditions. Morphological and anatomical characters linked to sprouting comprise spike capacity for water retention, structure and chemical composition of hulls, structure and chemical composition of the seed coat. The depth and length of the post-harvest dormancy are among the most important physiological factors related to the development and growth of kernels. Biochemical factors comprise all processes involved in synthesis or changes in activities of hormones and enzymes participating the process of development and maturation of kernels for germination.

One of the early symptoms of germination is α-amylase synthesis, induced by gibberellines, first occurring in the scutellum, and later in the aleurone layer. Enzymes travel from the aleurone layer to the endosperm, where they activate the degradation of starch. The efficiency of starch degradation can be elevated thanks to activity of several other enzymes synthesised during germination, e.g. β-glucanase, β-xylanase, which, by loosening or partly breaking down cell walls, make it easier for α-amylase to penetrate the endosperm. Activity of α-amylase in kernels of cereals is generally considered to be a marker in the assessment of sprouting. Direct or indirect assessment of α-amylase activity is one of the methods for evaluating resistance of grains to pre-harvest sprouting. Because of the very complex nature of the resistance of plants to sprouting, breeders aim to supply breeding materials with many different sources of genetic resistance, so as to produce a combination of several genetic mechanisms which may guarantee post-harvest dormancy.

Chemical Constituents of the Grain and Malting Quality

Protein Content and Composition

Protein content in barley kernels plays important role in imparting malt quality. Assessment of proteins in barley kernels is based on the total protein content and the effect of particular protein fractions on the malting process. The cultivars bred for malting industry should have a low protein content. Albumins and globulins are important as they participate in the creation of several enzymes. These proteins have a relatively high content of exogenous amino acids and, even more importantly, they contain all the other amino acids in well-balanced proportions. Albumins and globulins are completely transferred to aqueous solution during malt extraction, producing a good medium for yeast, which ferments the wort. As regards to storage proteins in the endosperm, glutelins do not pass to the malt extract, whereas about 50% of horedins do. It has been demonstrated that there is a positive correlation between the malting value of barley and the content of albumins and globulins. The ratio between the content of albumins and globulins to the content of horeidns (A+G/H) can serve as the malting value index of barley, as it has been proved that there is a very strong positive correlation between this index and the malting value index according to the EBC (European Brewing Convention) method. There is also a correlation between the content of protein and starch; low protein content coincides with high starch content and vice versa. Barley rich in proteins produces poor quality malt due to low efficiency of malt. Correlations between contents of various chemical components present in kernels and the usability of kernels for technological processes have found practical application in selection programmes. The content of protein is largely dominated by the effect of environment. Changes in climate during vegetation as well as agricultural techniques applied have a considerable influence on this character.

Barley Hordeins

Barley protein comprises a mixture of polypeptides of which hordein accounts for about 35-55% of total proteins (Brennan *et al.* 1998, Shewry *et al.* 1985). Hordeins are a heterogeneous group of polypeptides namely D, C and B. (Fig. 2) D hordeins are high molecular weight proteins accounting for 1-2% of total proteins and consist of only 1 or 2 polypeptides. B & C are the major group of hordeins and show extensive genetic variability. Due to their high polymorphism, they can be used in the identification of variety. The ability to distinguish variety is important because of differences among varieties in their quality and other agronomic properties and protection of proprietary rights of plant breeders. Various methods as electrophoresis, isoelectric focussing, chromatography and molecular techniques have been used for cultivar identification. Genetic analysis show that all hordeins are encoded by structural genes on chromosome 5 (1H) of barley. The genes for D, C, B and γ are present at loci Hor3 (long arm), Hor1, Hor2 and Hor5 (short arm) respectively. In addition a locus HrDG (Hor4), encodes B hordein like proteins. This is located between Hor1 and the centromere. Amino acid composition of hordeins indicate that they are rich in glutamine, glutamate and proline. The combined proportions of these range from 40 to 70% in hordeins. C hordeins are rich in phenylalanine and deficient in sulfur containing amino acids (notably no cysteine) whereas D horsein contains 16 mol% glycine.

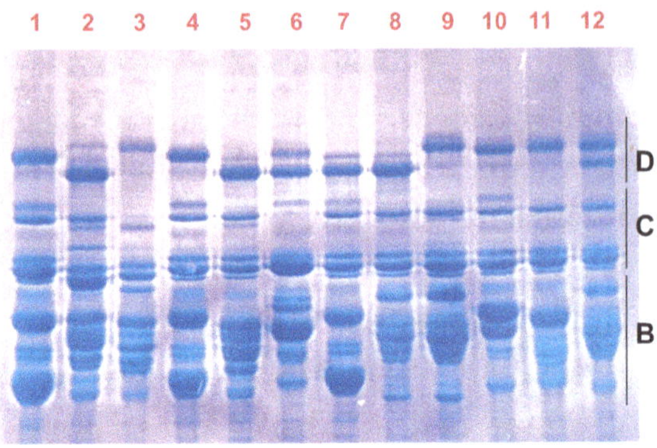

Fig. 2: Hordein profile of barley varieties developed in different centres in India

Hordeins affect malting and brewing quality during steeping and wort filtration and provide amino acids during fermentation. Hordein profile of germplasm lines as well as released varieties in India was developed as shown in the figure. Presence of lower band of D hordein displayed negative correlation (though less significant) with hot water extract. For understanding the role of hordeins in malting quality there is a need to develop specific genetic stocks. However the hordein profile can be used to identify relatedness among different genotypes. The hordein profile study has also indicated that genotypes developed by different breeding enters in India had specific band combinations prevalent. This information can be helpful in designing breeding strategy for developing high yielding malt barley varieties using specific cross combinations.

Starch Content and Quality

Starch is a dominant carbohydrate in grains of cereal plants. The content of starch in barley kernels ranges in between 48 and 70%, with an average value of 65%. High starch content corresponds to high efficiency of malt. Differences in the starch content result from genetic and environmental conditions. Barley starch contains on average 27% amylose and 73% amylopectin. Through selection it is possible to modify percentages of particular fractions. Barley starch, same as rye or wheat starch, is composed of two types of grains: spherical and lenticular. Chemical composition and characters of both small and large grains are identical. The shape of starch grains can alter through mutation. Gelatinization temperature is a starch specific feature. For barley, it varies between 51 and 60°C. Gelatinization temperature of barley starch is similar in range to gelatinization temperatures of other cereals.

Non-Starch Polysaccharides Content

Non-starch polysaccharides, including β-glucans and pentosans (arabinoxylans), are second in quantity among polysaccharides. Their total content is over 10%. These compounds are main components of cell walls, aleurone layer cells and endosperm of mature barley grain. Both groups of compounds play an important role in barley grain processing and animal feed production. β-glucan content is a well-recognized, important index for selection of breeding material for malting barley. Barley usually contains between 3.0 to 7.0 % β-glucan content on dry weight basis. For malting purposes, β-glucan content exceeding 4.5% in is not desirable. In addition, low contents of β-glucans and pentosans mean better quality of fodder barley. On the other hand, barley cultivars used for human consumption should have the highest possible amounts of these compounds, as β-glucans and pentosans constitute an essential part of dietary fibre. The detailed information on β-glucans is provided other sections of this chapter.

Activity of Hydrolytic Enzymes in Malting

Since main aim of the malting is to dissolve and degrade the components of barley grain for fermentation, the combined action of various enzymes as alpha amylases, beta amylases, limit dextrinases and alpha glucosidases determines quality of malt and the amount of reducing sugars released. The ability to mobilize starch is referred to as DP (diastatic power) in the brewing industry and the levels of DP affects both malt yield and brewing quality. These enzymes are known collectively as "diastase" and their performance is known as "diastatic activity". Malt extract is a major quality parameter representing the proportion of the grain solubilized in the wort. Because starch constitutes about 75% of the grain dry weight, this process makes a major contribution to the total amount of the extract. Increasing either the amount of these enzymes and/or their thermostability may beneficially speed up starch granule hydrolysis. β-amylase is relatively thermo labile enzyme compared to α-amylase and limit dextrinase. During brewing extraction of soluble malt components is performed isothermally at ~65°C, or using a ramped temperature profile from ~48°C to 70°C. β-amylase retains maximum activity up to 55°C and thereafter decreases. Therefore, potential for β-amylase with enhanced thermo stability has been recognized. This will lead to the identification of potential source of thermostable β-amylases and their use in breeding for improving malt quality.

Among these enzymes, β-amylase is synthesized during the development of the barley grain, in contrast to major hydrolysing enzymes which are synthesized *de novo* during germination. β-amylase (1,4-α-glucan maltohydrolase; EC3.2.1.2) catalyses the liberation of β-maltose from the non-reducing ends of 1,4-α-glucans. Hydrolytic enzymes are induced to synthesis or become activated by gibberellins produced by the embryo. Vulnerability of the aleurone layer to gibberellins is controlled genetically and may play a regulatory role in breeding works. Because of the great importance of enzymatic transformations in malting processes, it is essential that new malting barley cultivars be potentially capable of producing large amounts of hydrolytic enzymes (especially β-glucanase and α-amylase). Activation of hydrolytic enzymes occurs in following sequence: enzymes breaking down cell walls are activated first, followed by enzymes that break down starch and proteins.

Cell walls of the endosperm in barley grains are composed mainly of β-glucans, amounting to 75% of their weight. The first enzyme to be activated is β-glucanase, in dormant seeds present in small amounts, but as soon as seeds begin to germinate its quantity increases rapidly. β-glucanase causes degradation of the endosperm cell walls (cell wall looseness) and allows active amylolytic and proteolytic enzymes to permeate inside the endosperm cells. On the basis of a degree of cell wall loosening, it is possible to suggest if cultivars or breeding strains are valuable for brewing industry. The extent of cell wall degradation depends on the initial β-glucan content in kernels as well as the activity of β-glucanases. In breeding works, low content of glucans in barley kernels is just as important as their capability of producing larger amounts of β-glucanases in early phases of malting.

Amylases are the enzymes which take part in the breaking down of starch while kernels are germinating. During the germination of kernels, α-amylases are mainly synthesised in the aleurone layer cells, and an increase in the activity of α-amylase in cereal grains is related to *de novo* synthesis. β-amylase has a supportive catalytic function at the hydrolysis of starch. It is present in all tissues of the kernel, with the content equal about 1% of total protein. During the development of the kernel, β-amylase is stored mostly in the aleurone layer cells and in the endosperm. Mature barley kernels contain two forms of α-amylase: free and bound. In the course of malting the level of free $\alpha 1$-amylase increases. Mutual activities of amylases and enzymes which accompany them cause complete hydrolysis of starch and products of starch break-down to free glucose.

The degradation of storage proteins during germination is caused by proteases, which are synthesised in cells of the aleurone layer and the endosperm. Hydrolysis of proteins is initiated by endopeptidases (proteases), which break proteins down to peptides. Next, peptides are broken down by exopeptidases (carboxypeptidases) to amino acids. Endopeptidases are synthesised *de novo* under the influence of gibberellic acid produced by the embryo. Dormant barley kernels have a high level of exopeptidases, which goes yet higher during the germination. Typically, about 40-50% of total protein in kernels is degraded.

The ability to mobilize starch is referred to as DP (diastatic power) in the brewing industry and the levels of DP affect both malt yield and brewing quality. The alternate strategy for barley breeding with high levels of fermentability includes the selection for specific beta amylases. Since only a small sample is needed and there is no requirement of micro-malting, it is amenable to early generation screening which is a distinct advantage over current methods of selecting for fermentability. Malt extract is measure of the percentage of dry matter solubilisation of a malt grist during hot water extraction or mash while extract may be a good indicator of the degree of modification of a malt (i.e., Degradation of protein matrix and cell walls in the endosperm during the germination phase of malting), caution should be exercised when using extract as an indicator of malt potential for fermentable sugar production during mashing.

Brewing

Though brewing industry is already developed in Europe America, recently consumption of beer has increased very fast in Asia Pacific Region (Fig.3). In the brewery, the malted grain is milled to produce relatively fine particles, which are for the most part starch. The particles are then intimately mixed with hot water in a process called mashing. Mashing is geared principally to the controlled hydrolysis of starch to the required degree of sugar formation. The water must possess the right mix of salts. For example fine ales are produced from waters with high levels of calcium. Famous pilsners are from waters with low levels of calcium. Typically mashes have a thickness of three parts water to one part malt and contain a stand at around 65° C, at which temperature the granules of starch are converted by gelatinization from an indigestible granular state into a "melted" form which is much more susceptible to enzymatic digestion. The enzymes that break down the starch are called the amylases. They are developed during the malting process, but only start to act once the gelatinization of the starch has occurred in the mash tub. Some brewers will have added starch from other sources, such as maize or rice, to supplement that from malt. These other sources are called adjuncts.

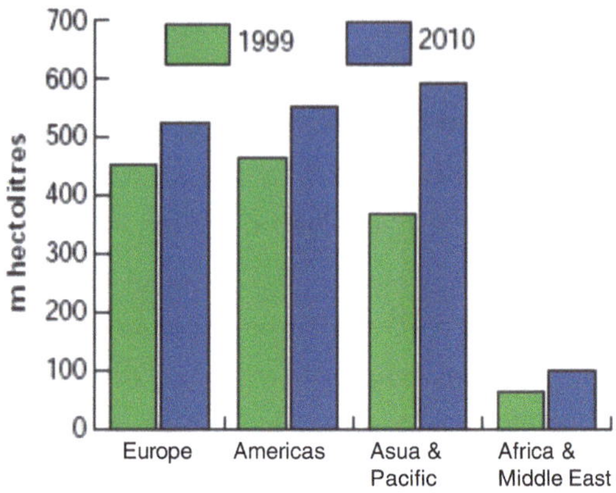

Fig. 3: World beer consumption
Source : World Beer Report 2000, Plato Logic Ltd.

After around an hour of mashing, the liquid portion of the mash, known as wort, is recovered, either by straining through the residual spent grains (lautering) or by filtering through plates. The wort is run to the kettle (sometimes known as the copper, even though they are nowadays fabricated from stainless steel) where it is boiled, usually for 1 hr. Boiling serves various functions, including sterilization of wort, precipitation of proteins (which would otherwise come out of solution in the finished beer and cause cloudiness), and the driving away of unpleasant grainy characters originating in the barley. Many brewers also add some adjunct sugars and hops at this stage. The hops have two principal components: resins and essential oils. The resins (so-called alpha-acids) are changed ("isomerized") during boiling to yield iso-alpha-acids, which provide the bitterness to beer. This process is rather inefficient. Nowadays, hops are often extracted with liquefied carbon dioxide and the extract is either added to the kettle or extensively isomerized outside the brewery for addition to the finished beer (thereby avoiding losses due to the bitter substances' tendency to stick on to yeast).

The oils are responsible for the "hoppy nose" on beer. They are very volatile and if the hops are all added at the start of the boil then all of the aroma will be blown up off the chimney. In traditional lager brewing a proportion of the hops are held back and only added towards the end of boiling, which allows the oils to remain in the wort. For obvious reasons, this process is called late hopping. In traditional ale production, a handful of hops is added to the cask at the end of the process, enabling a complex mixture of oils to give a distinctive character to such products. This is called dry hopping. Liquid carbon dioxide can be used to extract oils as well as resins and these extracts can also be added late in the process to make modifications to beer flavor.

After the precipitate produced during boiling has been removed, the hopped wort is cooled and pitched with yeast. There are many strains of brewing yeast (Saccharomyces cerevisiae), and Brewers guard and look after their own strains because of their importance in determining brand identity. Fundamentally brewing yeast can be divided into ale and lager strains, the former type collecting at the surface of the fermenting wort and the latter settling to the bottom of a fermentation (although this differentiation is becoming blurred with modern fermenters). Both types need a little oxygen to trigger off their metabolism, but otherwise the alcoholic fermentation is anaerobic. Ale fermentations are usually complete within a few days at temperatures as high as 20 °C, whereas lager fermentations at as low as 6 °C can take several weeks. Fermentation is complete when the desired alcohol content has been reached and when an unpleasant butterscotch flavor which develops during all fermentations has been mopped up by yeast. The yeast is harvested for use in the next fermentation.

In traditional ale brewing the beer is now mixed with hops, some priming sugars and with isinglass finings from the swim bladders of certain fish, which settle out the solids in the cask. In traditional lager brewing the "green beer" is matured by several weeks of cold storage, prior to filtering. Nowadays, the majority of beers, both ales and lagers, receive a relatively short conditioning period after fermentation and before filtration. This conditioning is ideally performed at –1 °C for a minimum of three days, under which conditions more proteins drop out of solution, making the beer less likely to go cloudy in the package or glass. The filtered beer is adjusted to the required carbonation before packaging into cans, kegs or glass or plastic bottles.

Proanthocyanidins in Beer

During storage of beer, proanthocyanidins (PA) originating from barley and hops form precipitates with proteins. The formation of colloidal precipitates between the polyhydroxilated proanthocyanidins (PA) and protein leads to haze (clowdiness) formation. This haze development is normally avoided by using different methods for stabilising the finished beer to ensure a brilliant look and long shelf-life of the product. A similar or even better haze stability of the beer can be obtained by the use of PA-free malt. Proanthocyanidins belong to the flavonoid group of polyphenols and occur in the testa layer of the grains of all traditional barley varieties. Barley PAs are polymers of (+) catechin and (+) gallocatechin, products of the flavonoid pathway and accumulates in the maternally derived seed coat of the developing grains. These PA polymers remain in the testa after completion of the malting process and are extracted into the wort during mashing. Therefore, haze formation can be prevented either by addition of hydrolytic enzymes to break down the protein or cause protein precipitation and /or reduction of PA in wort by selective absorption. However these processes are undesirable and expensive. Because most of the PA in beer comes from barley, one approach to alleviating the problem is to develop PA free barley cultivars. PA-free barley was unknown until 1974, when the first PA-free barley mutant was detected at the Carlsberg Laboratory. Beer brewed with malt from this mutant has dramatically improved haze stability. PA-free barley mutants have been used to breed PA-free malting barley varieties. The breeding program has been based on chemically induced mutations in the pathway of proanthocyanidin biosynthesis. Yield performance and disease resistance of the PA-free barley lines have improved a lot in recent years and PA-free varieties are now comparable with the traditional malting barley varieties in the market. First PA free cultivar, Caminant, has been recommended for commercial use in Denmark. In the future, it may be possible to control PA production in transgenic barley using antisense or co-suppression technology.

Beer off-flavours

Off-flaours in beer can be caused by excessive amounts like dimethyl sulphide (DMS) and trans-2-nonenal which are produced during brewing. DMS is formed primarily from S-Methylmethionine (SMM) at high temperature during kilning and wort boiling. SMM is a very common non protein amino acid among flowering plants. Though the physiological role of SMM in plants is not yet fully understood, it can act as a methyl donor in the reaction with homocysteine to form methionine (Met). However, other methyl-transfer reactions using SMM as the methyl donor are still speculative. SMM is believed to function as a storage form for methyl groups. When the capacity to form methyl groups de novo diminishes relative to the capacity to form homocysteine, the methyl groups stored could be returned to the Met pool. This function is important for ethylene formation at the time of senescence in flower tissues.

In the salt-tolerant plant *Wollastonia biflora* (L.) DC, SMM is the first intermediate in the synthesis of DMSP, an important osmoprotectant compound. In plants that do not accumulate DMSP, such as barley (*Hordeum vulgare* L.), SMM presumably has a different metabolic role. It may be converted into Met or, by enzymatic hydrolysis, into DMS and homoserine. The physiological importance of DMS is still speculative but it may contribute to the aroma of flowers. During food processing SMM gives rise to homoserine and DMS by thermal degradation. DMS has an important role in flavoring many cooked vegetables and other foodstuffs, including tea and beer, in which it can be considered undesirable. SMM is synthesized from AdoMet and Met in a reaction catalyzed by MMT (EC 2.1.1.12). This enzyme has been purified to homogeneity from leaves of the DMSP-accumulating plant *W. biflora.* A better knowledge of MMT from a germinating seed such as barley is also of utmost importance from both a physiological and a technological standpoint. It would be possible to regulate the amount of SMM production in germinating grain using anti-sense or co-suppression technology. Since SMM is an essential component of metabolism, large reductions may be harmful to barley growth and develoemnt.

Trans-2-nonenal is another metabolite causing beer haze problem. Lipoxigenase (LOX) pathway and enzyme responsible for trans-2-nonenal have been explored in barley. Hydroperoxidation of linolenic acid and linoleic acuds by lipoxigenase gives rise to 9-hydroperoxide, which is further metabolized to give aldehydes and alcohols that confer off-flaour to beer. Lipoxigenase is present in the embryo in two forms. LOX 1 is stable in the dry grain and survives kilning to cause off flavours during brewing. LOX 2 is not stable. It would be worthwhile to use antisense or co-suppression to reduce LOX1 activity to lessen off-flavour problem.

Malting and Brewing Biotechnology

The details provided about the steps involved in malting and brewing processes demonstrate that there are large numbers of traits which can be manipulated by genetic engineering. Among them the traits related to endosperm texture, protein content, malting enzymes, β-glucan content and β-glucanase activity are potential targets for manipulation by genetic engineering. Many of the genes determining these traits have been characterized in detail and can be used in genetic engineering. A major process during malting and mashing is the conversion of starch to fermentable sugars as glucose and maltose. The ability to mobilize starch is referred to as DP (diastatic power) in the brewing industry and the levels of DP, affects both malt yield and brewing quality. Good quality malt will provide brewer with high levels of extract and produce a wort that is easily fermentable by brewing yeast. The effectiveness of starch breakdown during mashing depends on both the gelatinization temperature and the level of starch degrading enzymes. This process is catalyzed by a group of enzymes such as α-amylase, β-amylase, α-glycosidase and limit dextrinase. α-amylase is relatively stable while β-amylase and limit dextrinase are heat labile at gelatinization temperature of starch. Therefore heat stability of β-amylase is more beneficial than selecting for barleys with higher beta-amylase levels, if one has to increase levels of fermentable sugars. β-amylase is synthesized during grain development and is stored in mature grains. This contrast with α-amylase, limit dextrinase and α-glucosidase which are synthesized de novo during germination. Therefore increasing either the amount and/or thermostability of these enzymes may speedup the starch hydrolysis. Either natural variation in the thermostability of diastatic activity or introducing foreign diastatic genes with enhanced heat stability to compensate for their loss during kilning can be employed in genetic engineering.

Since β-glucans are the major constituents of the cell wall in the endosperm and adversely affect the malt quality, either rduction of and β-glucans and/or their increased break down during malting can be used in improving malt quality. (1-3,1-4)-β-glucanase synthesized in the aeurone cells are major enzymes responsible for the break down of β-glucans during malting. Therefore β-glucans hydrolysis is the function of the levels as well as the thermostability of β-glucanases. Doubling the amount of β-glucanase levels and enhanced heat stability can ensure sufficient break down of β-glucans during malting process. β-glucanase levels can be increased by enhancing the expression levels of genes. The later goal can be accomplished by using native barley genes engineered for greater heat stability or by introducing heat stable β-glucanase genes from fungi, or bacteria. However, selecting foreign genes requires other studies such as pH optima, substrate stability and hydrolysis pattern of the enzyme synthesized from foreign genes.

As discussed in the previous sections, the problem of haze formation and off-flavours in beer can be minimized by reducing proanthocyanidins (PA) and S-methylmethionine (SMM) respectively. Reduction in PA can be accomplished by suppression of genes involved in PA synthesis by RNAi technology or other co-suppression techniques. Similarly the amount of SMM production in germinating grains can be regulated using anti-sese or co-suppression technology. In addition to SMM, lipoxigenase level has also been found responsible for off-flavours during brewing. Similar technique can be used to suppress the activity of LOX genes and reduce the off-flavours in beer (McElroy and Jacobsen 1995).

Feed Quality of Barley

Non-malting uses of barley include feed, food and industrial applications. Barley grains can be used as major source of energy, protein and fibre for ruminants; a major source of energy and protein for swine and to support egg production in laying hens. Use of barley in livestock feeds in different countries varies from a low of 47% to high of about 90%. Digestible energy or metabolizable energy is used as the most important quality criteria in feed barley, particularly in monogastric animals. Other desirable quality traits of feed barley include absence of husk (hullless), high protein and lysine and starch content and low β-glucan content. Low content of β-glucan which is favourable in malting is also good for the feed quality, because β-glucans can hardly be utilized by monogastric animals and high amounts cause wet faeces droppings in poultry feeding.

Higher protein content along with high lysine content and other attributes such as high starch and digestible energy are feed quality attributes. Generally, negative correlation between protein content and lysine content in barley has hampered the development of barley varieties with both high protein and high lysine. To alleviate this problem, high lysine mutants were developed in barley and used in breeding to enhance lysine content in high yielding backgrounds. The feeding experiments of high lysine varieties indicated that the daily growth of pigs, fed with the high-lysine barley without protein supplement at 11.5% protein in the feed, approaches the growth rate on a full feed consisting of normal barley and soybean additives at 15.0% protein. This indicated that the optimal protein content in pig feeding with high-lysine barley would be even less than 15%. This would mean that lesser nitrogen pollution would be excreted from pig manure fed with high lysine barley. Discovery of high lysine mutants in barley such as Hiproly (CI3947) with 4.1% lysine, Notch1 and Notch-2 mutants of N113 with 30% higher lysine discovered by Bansal 1970 and Riso 1508, a mutant of two-rowed Danish barley cultivar Bomi treated with ethylene diamine (Ingversen 1973) with 42% enhanced lysine content led to the development of barley cultivars with higher lysine content.

Twin problems of acidosis and bloat are associated with barley, which can reduce feed intake and weight gain. These problems can be avoided by "slow DMD" barley. The term refers to barley varieties that contain a special trait for a slower rate of dry matter disappearance (DMD). Cattle fed with slow DMD barley, digest the grain more slowly. This reduces the risk of a buildup of volatile fatty acids, which can lead to acidosis, and of gases, which can lead to bloat. Acidosis and bloats cause reduction in the potential productivity of animals and in some cases, these problems can lead to death. This is why feeders refer to barley rations as being too hot. The starch in all cereal grain is held in the kernel in a protein matrix, which is broken down in the digestive process. Slow DMD barley has a different starch-protein matrix structure compared to traditional barley such as Harrington. With slow DMD barley, as with corn, starch digestion is slow because the starch is held more tightly in its protein matrix. Slow DMD barley will not completely eliminate acidosis and bloat problems, but it should greatly reduce them. Slow DMD shifts more of the starch digestion from the rumen to the small intestine which lowers the risk of acidosis and bloat.

The composition of barley and hulless barley relative to corn, wheat, rye, and sorghum is presented in the (Table 3). Relative to corn, barley has significantly higher crude protein (CP), acid detergent fiber (ADF), and neutral detergent fiber (NDF) content. The higher protein content of barley is offset by a somewhat lower energy content than observed for corn. For lactating animals the net energy value reported for barley is 96% of that for corn. The higher CP content of barley can result in savings in costs of supplemental protein, however the relatively lower ruminal escape of barley protein may offset some of this advantage. Amino acids such as lysine, methionine, cysteine, and tryptophan are present in substantially greater concentrations in barley than in corn.

Hull constitutes 10-14 % of the dry weight of the barley grains and contributes largely to the crude fibre content. It primarily consists of cellulose, hemicellulose, lignin, pectins and small quantity of proteins. Husk is highly undesirable because main components of husk as cellulose and hemicellulose are indigestible. Passing of this excess of indigestible bulk through the digestive tract of an animal reduces fed efficiency considerably. Moreover, it has a major detrimental effect on DE of barley particularly for swine. This effect is brought about by the dilution of the available nutrients or by physical or chemical inhibition of nutrient digestion and utilization. Hullless trait is caused by a recessive gene. This results into the hull not adhering tightly to the grain.

The hull then threshes free of the grain to leave a naked grain much like wheat in appearance. Details of hulless barley are given in the next section.

Table 3: Comparative nutrient composition of barley and other grains.

Nutrient	Barley	Hulless barley	Corn	Wheat	Rye	Sorghum
Dry matter, %	88	88	88	88	88	88
Crude protein, %	11.5	13.2	8.8	13.5	12.1	11.0
ADF, %	6.0	2.0	3.0	4.0		9.0
Crude fibre, %	5.0	1.4	2.2	2.6	2.2	2.3
Test Weight (kg/hl)	62		74	76		54
Fat, %	1.9	2.0	3.8	2.2	1.5	2.9
Ash, %	2.3		1.4	1.7	1.9	1.8
TDN, %	74	76	78	78	74	73
NE maint Mcal/kg	1.76		1.97	1.94	1.81	1.79
NE gain, Mcal/kg	1.19		1.36	1.34	1.23	1.21
NE lact, Mcal/kg	1.71		1.78	1.82	1.71	1.62
Bypass protein, %	28		50	25	20	55
Lysine, %	.42	.56	.30	51	.36	.27
Swine DE, kcal/kg	3140	3350	3450	3430	3280	3415
Meth + cys, %	.42	.56	.30	.51	.36	.30
Tryptophan, %	.18	.15	.09	.15	.14	.09
Phosphorus, %	.34	.35	.26	.37	.32	.29

Source: D. Hickling, Canadian International Grains Institute

Barley for Poultry Feeding

Barley is less suited for poultry, particularly broiler chicks, than to swine. It has low metabolizable energy (ME or TME) as compared to wheat and corn. In addition, β-glucans present in both hulled and hull-less barley increase gastric viscosity, interfere with nutrient availability, and cause sticky droppings, leading to litter management problems. However, recently hull-less barley cultivars have been developed with crude fibre content similar to wheat and corn. This combined with the use of commercial enzyme preparations has largely eliminated the deleterious effect of barley β-glucans in poultry. Barley is extensively used as feed for pigs and chickens. However, since it has a suboptimal content of the essential amino acids lysine, methionine and threonine, massive supplementation with other protein sources or amino acids are required to meet the nutritional demands. The second problem is that of prolamins, the major barley storage proteins, which contain large amounts of glutamine and proline. These amino acids cannot be utilized completely by the animals. The excess is degraded and the nitrogen excreted in the urine contributing significantly to the environmental nitrogen load.

Barley as Food (For Human Consumption)

Non-malting utilization of barley includes food, fed and industrial applications other than malting. At present very little barley is used in human foods, however recently greater emphasis is being put on food barley research because of it's nutritional value. Barley furnishes a multitude of dietary nutrients including amino acids, minerals, vitamins, insoluble and soluble dietary fiber and physiologically active phytochemicals. Many genotypes of barley are available for processing into foods and some are more concentrated in various nutrients than others and differ in processing requirements. Concentration of nutrients in barley can be altered by standard breeding practices. Barley can be used in numerous ways to prepare delicious, nutritious and health promoting foods, depending upon the innovation of a food processor. Traditional use of barley as a food has been pearl barley in soups, stews and as a replacement or extender for white rice. Processing techniques, such as milling, flaking and extrusion increase the variety of barley food products. In many countries, barley is used in traditional dishes, such as miso and finely pearled barley as a rice extender, for the production of soy paste and soy sauce in Korea or for the fermented drink Shochu in Japan. Barley is also roasted and used as a tea or coffee substitute, and wheat-barley flour mixtures used for making breads, biscuits, cakes and noodles. In the West Asia-North Africa region, barley is consumed as pearled barley in soups, flour in flat breads and ground grain in porridge. Because of the nutritional and health benefits of barley, its consumption in human foods should be actively promoted.

Barley is the main cereal grain used for the development of functional foods, as it contains several classes of compounds of strong nutritional interest: tocols, -glucans and proanthocyanidins. The positive nutritional properties of barley are numerous. Tocols are well recognised for their biological effects, including antioxidant and hypocholesterolemic activity. Barley is considered a good source of this class of compounds, because it contains both a high concentration of total tocols and a favourable distribution of the most biologically active isomers. Barley has certain health benefits such as attenuation of postprandial blood glucose and insulin, lowering blood lipids and low density lipoproteins (LDL cholesterol). Barley also has a very low glycemic index which helps to prevent high peaks in blood sugar and insulin after meals. This effect is beneficial in the prevention of type II diabetes and controlling blood sugar in both type I and type II diabetes. Most of these benefits are related to soluble fibre content such as β-glucans. With the production of new cultivars and processing methods, cultivars with high dietary fibre components such as β-glucans and pentosans are desirable. β-glucan content of barley ranges from 3 to 10% and thus barley grains are rich source of β-glucans. In addition hulless barley with waxy starch, high β-glucan, good cooking and pearling qualities, large plump grain and lacking pro-anthocyanidin are preferred requirements for human consumption. Screening for germplasm with waxy starch and high β-glucan is commonly performed in the early stages of breeding selection. The potential of β-glucans as functional food ingredients has been reviewed recently by Brennan and Cleary 2005 and Wood 2007. An account of the structure and function of β-glucans is given below.

β-glucans (1-3, 1-4)- β-D-glucans are the predomiant component of cell walls of cereals grains such as barley and oats. β-glucans are different from cellulose polymers with respect to β,1-3 and β,1-4 linkages. In cellulose only β,1-4 linkages are present while in β-glucans both the linkages are present. The endosperm cell wall β-glucan components are linear molecules with approximately 30% (β,1-3)- and 70% (β,1-4)-linkages. The distribution is not random; the β,1-3 linkages always occur singly, and most of the β,1-4 linkages occur in groups of three or four. The intestine does not contain enzyme that can digest β-glucans, so they are by definition a fibre. The composition of the walls of the starchy endosperm and the aleurone are qualitatively similar but quantitatively different. The walls of the starchy endosperm of barley consist of about 70% β-glucan and 20% arabinoxylan, whereas, the aleurone walls contain 26% β-glucan and 67% arabinoxylan and both have similar amounts of glucomannan and cellulose. The rheological properties of β-glucans appear to depend on a number of factors including the ability of β-glucan chains to associate, determined by the proportion of cellotriosyl/cellotetraosyl units and their arrangements and by the degree of polymerization and hence the molecular weight of β-glucan. Factors which affect the rheological characteristics of β-glucan have obvious links to their viscosity behavior and their potential effect on food structure, texture and nutritional properties. For example, reduced solubility of β-glucans has been attributed to higher ratios of cellotriosyl/cellotetraosyl units.

β–glucan

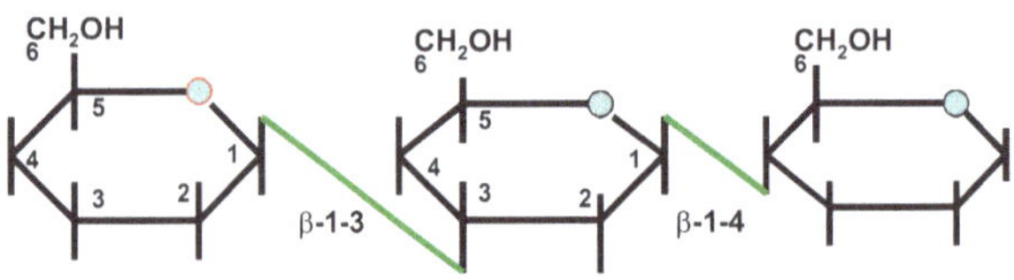

Potential Benefits of β-Glucans as Soluble Fibre

Carbohydrates in foods consist of three forms: fibers, starches, and sugars. Starches are complex carbohydrates having slower digestion than sugars. Grains, vegetables, and legumes are sources of complex carbohydrates, and these foods are also good sources of important nutrients as well as energy. Legumes are also good sources of protein, particularly when combined with grains. Sugars are simple carbohydrates, rapidly digested, and generally considered sources of "quick energy." Dietary fiber is a type of complex carbohydrate that is almost totally indigestible. Although it does not supply energy, its physical nature contributes to health and well-being. Dietary fiber is not a single substance but a complex mixture of substances which varies in different foods. The best sources of fiber are grains, legumes, fruits, vegetables, and seaweed. Dietary fibers can be classified according to water solubility in an extraction process. Most foods of plant origin contain both insoluble and soluble dietary fiber, but the proportions vary with the type of food. For example, wheat bran contains mostly insoluble fiber, whereas barley and oat bran are about half insoluble and half soluble fiber. Insoluble fibers are not soluble in water and are sometimes termed structured fibers. Cellulose, lignin (a wood material), and some hemicellulose are examples of insoluble fibers. Benefits of insoluble fiber are related to gastrointestinal function while health benefits of soluble fiber are due to metabolism of nutrients through the bloodstream. Potential health benefits related to dietary fibres are:

- Increased bulk in the large intestine to prevent constipation. The fiber holds water and improves the movement of the intestinal contents and thus reducing bowel transit time. Whole grains and wheat bran are particularly effective.
- May help prevent diverticular disease and risk of colon cancer.
- May aid in weight control, if high fiber foods are substituted for high-fat, high-calorie foods. High fiber foods provide a feeling of fullness because of their bulk.

- Soluble fibers form a gel in water and consist of pectins, β-glucan and gums. Good sources of soluble fiber are barley, oats, and some fruits and vegetables, especially legumes.
- Lowers blood cholesterol in some people, particularly when eaten as part of a low-fat diet. This decreases risk of heart disease.
- Slows digestion of all carbohydrates, thus helping to control blood sugar. This may be particularly helpful for persons with diabetes or insulin resistance.
- May aid in weight control because soluble fiber may interfere with fat absorption.
- Promotion of the growth of beneficial gut microflora, the prebiotic effect.
- Reduction of the risk factors for degenerative diseases such as obesity, hyperlipidaemia, cardiovascular disease and hypertension.

Many attempts have been made to clarify the mechanism by which dietary fibre and β-glucans have these effects. The potential reduction of glycaemic response following ingestion of dietary fibre depends on the amount and quality of fibre, increased intrinsic viscosity of food in combination with fluids, and hence the gastrointestinal environment; maintenance of the physical integrity of the food material and incomplete starch gelatinization. Other hypothesis is that in the intestine food is incorporated in the viscous β-glucan solution making it more difficult for enzymes in the intestine to degrade food components and causing digestion to take longer. Another hypothesis is that β-glucans form a protective layer along the intestinal wall that acts as a viscous barrier, slowing food uptake from intestine. Substantial clinical evidence from the last 30 years have documented that β-glucans have an effect on blood cholesterol levels and control of lipoprotein metabolism. There are different theories concerning the mechanism of blood cholesterol lowering effect of β-glucans. The cholesterol lowering effect of cereal fibre is considered to result from effects manifest in the upper gastrointestinal tract. These in turn may be related to the ability of cereal fibre to form a gel-like network and alter gastrointestinal viscosity.

One theory is that β-glucans encapsulate bile acids, resulting in their excretion in the feces. Through the excretion in the feces, the body loses bile acids and has to synthesize new ones, which is done in liver. The building block for bile acids is blood cholesterol, which the liver extracts from the blood, decreasing blood cholesterol levels.

Another hypothesis is that fermentation of soluble fibre by bacteria in the large intestine produces propionate. Propionate is then absorbed by the colon cells and goes to the liver, where it is thought to have an effect on cholesterol synthesis. The third theory is that β-glucans interfere with the absorption of lipids, probably by reducing or delaying the emulsification and lipid hydrolysis process. Barley has high concentrations of tocotrienols, compounds which act as antioxidants, and are capable of reducing serum LDL-cholesterol (one of the risk factors in cardiovascular disease). Tocotrienols act by repressing the activity of the first rate-limiting enzyme (HMG-CoA Reductase) in the liver, thus reducing cholesterol synthesis.

β-glucans act as prebiotics by selectively stimulating growth of bacteria in the colon. The growth and metabolism of individual bacteria in colon depend primarily on the substrates available to them, most of which come from the diet. β-glucans are indigestible in the small intestine but are fermented by the bacteria in the colon. Growth of bacteria in the colon beneficially affects a series of intestinal functions by modulating the structure, composition and metabolic activity of mucosa and microflora in the colon. The end product created from prebiotic fermentation in the colon are short chain fatty acids, e.g., butyric acid, that serve as nutrients for mucosal cells. The beneficial effect of β-glucans for weight management is considered by stimulating sensation of satiety. Satiety is complex bodily sensation that signals that the stomach is full and it is time to stop eating. By reducing the desire for food intake, the effect can help in weight control when combined with a healthy balanced diet and adequate exercise. As a result of extended period of digestion nutrients are utilized by the body over a longer period and, thus, may contribute to a longer period of satiety in weight management programmes. Despite the beneficial effects of fibres in diet, following cautions should be considered while taking fibres.

- Increase fiber slowly in the diet. A sudden increase of fiber may cause bloating and gas.
- Use moderation. Excessive fiber may be irritating to some individuals.
- Use a variety of foods as fiber sources rather than fiber supplements.
- Fiber absorbs water in the intestines, so it is important to drink sufficient liquids to enhance movement of the intestinal contents.

Hull-less and Pearled Barley

There are two types of barley: (i) covered, having a hull, and (ii) hull-less, sometimes called nude. Most barleys that are grown for malting in beer making are covered. The covered barley must be pearled before using it for foods. Hull-less barley can be milled into whole meal without pearling and is, therefore, very adaptable for food products. The hull-less barley varieties are also higher in fiber than covered barley. A major factor affecting total dietary fiber in barley is the nude gene (nn) which controls the adherence or non-adherence of the hull to the kernel. Thus, this separates all barleys into one of the following types:

1. Covered, or having a hull that adheres tightly to the kernel;
2. Hull-less, sometimes called "nude," because the hull falls off the kernel, as with wheat, during harvesting.

Most common barleys are of the covered type because of increased yield due to the presence of the hull and demand for this type in malting. However, there is growing interest in the hull-less type, which is very adaptable for human food products because the entire grain can be used without pearling. There is an additional genetic property in some cultivars, waxy starch, which means the starch is almost 100% amylopectin. The combination of hull-less and waxy characteristics produces barleys with very high in soluble dietary fiber especially β-glucans.

Covered barley must be processed to remove the hull. Pearling is part of the milling process that produces blocked (dehulled) barley, pot and pearled barley, barley flakes, and finally barley flour by roller milling of blocked or pearled barley. Blocking is the first stage of processing, where only the major part of the hull is removed. Blocked barley can then be steam rolled, cut into grits, pearled, ground into a meal, or roller milled similarly to wheat. Nutritionally, hull-less or pearl barley is not too different from wheat in its caloric, protein, vitamin and mineral content. As in wheat, there are high protein varieties. Barley differs from wheat, however, in that there are some varieties high in lysine, the essential amino acid that is most limiting in cereals. Barley is very high in dietary fiber, particularly the soluble fiber portion. The soluble portion contains β-glucans, arabinoxylan or pentosans. Barley also contains tocotrienol, an oil component which has cholesterol-lowering activity.

Pot and pearled barley are prepared by gradual removal of hull, bran and germ by abrasive action in a stone mill. Production of pot barley is the first stage of pearling, which may remove7-14% of the weight of the grain. Further abrasion results in the removal of seed coat, aleurone, subaleurone layers and the germ leaving behind central endosperm rich in carbohydrates and proteins. The final product of fine pearl is around 60-70% of the grain. In brief the sequence of pearling process are : cleaning, conditioning or tempering to about 15.5% moisture, bleaching (blue aleurone barley), blocking, aspirations, size grading, groat cutting, grading and sifting and polishing. Barley can be milled into flour and other milling fractions, some of which have a higher concentration of dietary fiber than the wheat grain. These high-fiber milling fractions have many uses, with important healthful benefits. The barley shorts fraction contains about 18% total dietary fiber, which is approximately 50% soluble fiber. Compare this with white wheat flour, which contains only 3% total dietary fiber, including 1% soluble fiber. Flour in any recipe can be supplemented with whole barley flour or a milling fraction such as shorts to increase fiber.

Barley Starch

The food industry, too, has the potential to become an important user of barley, but it needs more information on the functionality of barley and it's components to ensure their effective utilization. In addition to β-glucans, barley can also be an excellent source of starches. Several genotypes of barley containing large and small starch granules, and various amylose-amylopectin ratios (waxy, normal, high-amylose barley) have been developed and are awaiting greater uses in the food and other industries. Incorporation of barley starch into blends with other starches and/or non-starch polysaccharides may provide the food industry with attractive ingredients having unique, desirable functional properties. Barley components might, therefore, be excellent value-added products and would increase significantly the value of barley as a crop and diversify its utilization.

Waxy cultivars tend to have high total dietary fibre and produce flour with excellent food thickening properties due to their high water absorption capacity. Both high and low amylose barley starches exhibit extremes in swelling power and viscosity compared to normal starch. These characteristics may be useful for specific products within the food and non-food industry. Waxy barley has a bright white endosperm, which is desirable for pearled barley particularly when used as a rice extender. The amylose content of barley starch can be determined by iodine staining or enzyme assay using Megazyme kit.

High Lysine Mutants

High-lysine barley is characterized by an increased content of the amino acid lysine of its seed protein compared to that of normal barley. The basic gene symbol *lys* and *sex* have each been proposed for the genes of the barley high-lysine mutants. The first high-lysine barley discovered was Hiproly (Munck *et al.* 1970). The increased lysine content of Hiproly is due to a single recessive gene designated as *lys.* Most of the high-lysine mutants developed have shrunken endosperms, and the gene controlling the shrunken endosperm trait is designated as sex. However, the close association between the high-lysine trait and the shrunken endosperm has been reported. The two traits may be the pleiotropic effects of one gene. There are different ways to suppress the shrunken endosperm trait while retaining the high-lysine content of the seeds. If the high-lysine character is present in materials with plump seeds, which is a goal of many barley breeding efforts, the *sex* symbol will no longer be meaningful. It was proposed in another study that the high-lysine gene of Risø 1508 be designated *lys3a* and that its allelic genes, in Risø 18 and 19, be designate *lys3b* and *lys3c,* respectively Jensen (1979a). The designation *lys* was advanced because the mutants were selected and best known for their increased lysine content, and also because the symbol *lys* was considered to have priority in that the similar gene in Hiproly was so designated (Munck 1972). A series of papers were published by Ulrich and Eslick to classify high lysine genes in different mutants.

Generally these mutant genes vary in their mechanisms and most result in decreased synthesis of lysine poor prolamins fraction and compensatory increases in other more lysine rich fractions, which may include elevated levels of free lysine. A considerable amount of time and effort and money has been spent to incorporate genetic resources of lysine into commercial cultivars with acceptable yield potential, with largely disappointing results. In barley two lines as Hiproly and Riso 1508 were used in breeding as source of high lysine. Though, these sources failed to result in development of commercially successful varieties. However, High lysine proteins from these sources were exploited in genetic engineering studies (Shewry 2007).

Low Phytic Acid

Phytate is a salt of phytic acid (myo-inositol-1,2,3,4,5,6-hexakiphosphate) and it is the most abundant form of phosporus (P) in seeds where it is used as a phosphorus reservoir during seed germination. Phytate is indigestible by humans and non-ruminant livestock. Phytic acid in feed grain seeds can bind with micronutrients (e.g., iron, manganese and zinc) and macronutrients (e.g., calcium and magnesium). Excessive non-digested phytate from feed could end up in animal waste (manure). This could pollute the environment and cause eutrophication of waterways. In a study, samples of 423 barley, 114 wheat, 38 triticale, 24 oat, and 8 rye genotypes samples were analyzed for phytate-P and total P on a wet and dry basis. The phytate-P percent of total P, on a dry matter basis, ranged between 2.6 and 70.2 for barley, 39.3 and 84.1 for wheat, 44.6 and 84.8 for triticale, 43.5 and 64.3 for oats, 52.1 and 63.5 for rye (Veum *et al.* 2007). The widest variability of phytate-P was determined in barley compared to other cereal species. Results of the current study show that there is variability in phytate-P content in the genotypes of cereals analyzed. This variability could be exploited for the benefit of livestock feed and human food. In addition, low phytate barley mutants have been developed through mutation breeding. The *lpa* mutations are recessive and translocation of gene products or metabolites does not appear to complement the loss of gene function in homozygous mutant (*lpa*/*lpa*) seeds from heterozygous (Lpa/lpa) maternal plants.

Of the three low phytate mutants investigated, the *lpa1-1* locus shows a 50% decrease in phytic acid and has been mapped to the long arm of chromosome 2H. Grain produced by plants homozygous for the barley low phytic acid 2-1 (lpa2-1) allele of the barley lpa2 locus exhibits a 70% decrease in phytic acid, and has been associated with an increase in both Pi and inositol pentakisphosphate, an intermediate from the phytate biochemical pathway. Barley lpa2 has been mapped to chromosome 7HL. Two further mutants, M635 and M955, show 75% and 95% decreases in seed phytic acid respectively, again with concomitant increases in Pi. The inheritance of the mutations in M635 and M955 fits a single-gene model (Dorsch *et al.* 2003), but neither mutation has been located to a specific chromosome.

Unlike normal genetic and environmental effects (which predominantly affect total seed P), these *lpa* mutations show dramatic reductions in the ratio of phytic acid P to inorganic P. Two types of biochemical phenotypes have been detected when screening for *lpa* mutations in barley and maize. The causal mutations for these two biochemical phenotypes circumscribe two complementation groups that are similar in both species. The first, and most common, type of mutation (designated *lpa1*) shows a decrease in phytic acid P and a molar-equivalent increase in inorganic P. The second type of mutation (designated *lpa2*) also shows a decrease in phytic acid P, but does not show molar-equivalent increases in inorganic P (relative to loss of phytic acid P). Instead, a significant portion of the total seed P remains bound in lower inositol polyphosphates (e.g., *myo*-inositol penta*kis*phosphate or IP5) in the *lpa2* mutants. Nevertheless, all *lpa* phenotypes appear to be the result of single-gene, recessive mutations that result in significant reductions of phytic acid P (i.e., 50% or more) and at least several fold increases in inorganic P (i.e., 5-10 fold for *lpa1* and 3-4 fold for *lpa2*) relative to normal kernels. Therefore, each of these two complementation groups show distinct phenotypes which may be useful separately, or perhaps even combined. These *lpa* mutations have the potential to significantly reduce the environmental and nutritional problems associated with cereal phytic acid.

To understand how *lpa* mutations affect metabolic and developmental processes during seed growth, gene expression experiments were performed using oligonucleotide microarrays. Differential gene expression was assayed at 7 d after anthesis in developing seeds homozygous either for the *lpa* mutation M955, or for its corresponding nonmutant, wild-type allele. Homozygosity for M955 blocks phytate accumulation throughout seed development, resulting in a ~90% reduction in mature seed phytate. Consistent and substantial differential expression was observed for 38 genes (probesets) representing various cellular processes and unknown functions. The majority of differentially expressed genes showed decreased expression with a much smaller proportion of upregulated genes. No changes in expression of genes thought to be directly involved in the synthesis of phytate were observed. Reduced expression was observed for functions important to carbohydrate and cell wall metabolism, cytokinin and ethylene signaling, and transport functions. These results reflect the interconnection of phosphorus, inositol phosphate, and sugar metabolism, and signaling networks regulating these metabolic and developmental pathways. The M955 *lpa* mutation appears to affect seed development and function through carbon transport and starch synthesis and may account for yield reductions previously reported for this mutation.

Recently molecular markers have been identified linked with lpa locus. A RFLP marker (ABC153) located in this region has been converted to a dominant SCAR marker. Segregation analysis in the doubled haploid population from the cross CDC McGwire by low phytate mutant Hvlpa1-1 has confirmed linkage between the SCAR marker and *lpa1-1* with 15% recombination. The mutation in a second low phytate mutant, M 635 (75% phytate decrease), has been mapped to the long arm of chromosome 1H via linkage with ISSR marker LP75. Analysis with ABC153 confirms this locus is not allelic to *lpa1-1*, and the locus has been designated *lpa3-1*. Both markers are polymorphic in a cross involving a third mutant M 955 (95% phytate decrease). Initial results indicate that the mutation in M 955 and locus *lpa3-1* are in the same region on chromosome 1H, and screening of a larger population is in progress to determine if the mutations are allelic.

Summary

Though barley is used mainly for feed and malt purposes, recently greater emphasis has been put on it's use as human food because of positive health inducing factors in barley. Genetic basis of malting quality traits has been well established and malt quality parameters have been standardized internationally. Malt quality improvement has been a success through conventional plant breeding efforts. However, recently molecular tools have been used in improving malting quality. Since starch hydrolysis is the main chemical reaction during malting and brewing process, the enzymes of starch hydrolysis have been characterized. The knowledge will be useful in controlling the levels of enzymes of starch hydrolysis to produce desired quality of malt and beer. Barley has certain health benefits such as attenuation of postprandial blood glucose and insulin, lowering blood lipids and low density lipoproteins (LDL cholesterol). Barley also has a very low glycemic index which helps to prevent high peaks in blood sugar and insulin after meals. Most of these benefits are related to soluble fibre content such as â-glucans. In addition low phytic acid mutants and high lysine mutants have been identified in barley. The mutants are useful source for improving nutritional quality.

References

Ajgoanker, S.S. (1972). Diabetic Mellitus as seen in the ancient Ayurvedic medicine. *In: Insulin and Metabolism.* Ed. J. S. Bajaj. Association of India, Bombay.

Bowen, D.E., Souza, E.J., Guttieri, M.J., Raboy, V. and Fu, J. (2007) A low phytic acid barley mutation alters seed gene expression. *Crop Sci.* 47: 149-159

Brennan C.S. and Cleary L.J. (2005). The potential use of cereal (1-3, 1-4)- β-glucans as functional food ingredients. *J. Cereal Sci.* 42: 1-13.

Brennan, C.S., Smith,D.B., Harris,N. and Shewry, P.R. (1998). The production and characterization of Hor3 null lines of barley provides new information of the relationship of D hordein to malting performance. *J. Cereal Sci.* 28: 291-299.

Dorsch, J.A., Cook, A., Young, K.A., Anderson, J.M., Bauman, A.T., Volkmann, C.J., Murthy, P.P.N. and Raboy, V. (2003). Seed phosphorus and inositol phosphate phenotype of barley low phytic acid genotypes. *Phytochemistry.* 62: 691–706.

Edney, M.J. (1999). *Proceedings of the Canadian Barley Symposium'99,* p.117-123. Winnipeg, Manitoba, Canada.

Ertl, D.S., Young, K.A., Raboy, V. (1998). Plant genetic approaches to phosphorus management in agricultural production. *J. Environ Qual.* 27: 299–304

Eslick, R.F. and Hockett, E.A. (1976a). A second locus for high lysine barley. *Barley Genetics III:* 634.

Eslick, R.F. and Hockett, E.A. (1976b). In: Description of genetic stocks. *Barley Genetic Newsletter* 6: 140.

Jarvi, A.J. and Eslick, R.F. (1971). In: Description of genetic stocks in the barley genetic stock center at Fort Collins, Colorado. *Barley Genetics Newsletter.* 1:193.

Jensen, J. (1979b). Chromosomal location of one dominant and four recessive high-lysine genes in barley mutants. *In proc.: Int. Symp. of Seed Protein Improvement in Cereals and Grain Legumes.* Neuherberg, Federal Republic of Germany, 4-8 Sept. 1978, IAEA (in press).

Kennedy, J., Okine, E. and Khorasami, R. Barley as a grain and forage source for ruminants. web site: *D:|BARLEY|barley ruminants.htm.*

McElroy, D. and Jacobsen, J. (1995). What is brewing in barley biotechnology? *Biotechnology.* 13: 245-248.

Munck, L. (1972). High lysine barley - a summary of the present research development in Sweden. *Barley Genetics Newsletter* 2: 54-59.

Munck, L., Karlsson, K.E., Hagberg, A. and Eggum, B.O. (1970). Gene for improved nutritional value in barley seed protein. *Science* 168:985-987.

Newman, C.W. and Newman R. K. (2006). A brief history of barley foods. *Cereal Foods World.* 51(1): 4-7.

Roslinsky, V., Eckstein, P.E., Raboy, V., Rossnagel, B.G. and Scoles, G.J. (2007). Molecular marker development and linkage analysis in three low phytic acid barley (*Hordeum vulgare*) mutant lines. *Molecular Breeding.* 20: 323–330.

Shewry, P.R., Bunce, N.A.C., Kreis, M. and Forde, B.G. (1985). Polymorphism at the Hor1 locus of barley (*Hordeum vulgare* L.). *Biochem. Genet.* 23: 391-404.

Ullrich, S.E. and R.F. Eslick. (1977). Inheritance of the shrunken endosperm character, *sex3c* of Bomi Risø mutant 1508 and its association with lysine content. *Barley Genetics Newsletter* 7:66-73.

Ullrich, S.E. and Eslick, R.F. (1978c). Allelism evidence for barley high lysine, shrunken endosperm xenia (*sex*) mutants. *Barley Genetics Newsletter* 8:109-112.

Ullrich, S.E. and Eslick, R.F. (1978d). Chromosome location evidence for Risø induced high lysine shrunken endosperm mutants of barley. *Barley Genetics Newsletter* 8:114-125.

Veum, T.L., Ledoux, D.R., Bollinger, D.W., Raboy, V. and Cook, A. (2002). Low-phytic acid barley improves calcium and phosphorus utilization and growth performance in growing pigs. *J. Anim Sci.* 80:2663–2670

Veum, T.L., Ledoux, D.R. and Raboy, V. (2007). Low-phytate barley cultivars improve the utilization of phosphorus, calcium, nitrogen, energy, and dry matter in diets fed to young swine. *J. Animal Sci.* 85:961-971.

Wood, P.J. (2007). Cereal β-glucans in diet and health. *J Cereal Science.* 46: 230-38.

❏❏❏

Chapter 12

Processing & Nutritional Quality of Rice

Introduction

Rice is grown in more than 100 countries in the world and is an important crop for nutritional security of large part of the population. World rice production in 2007 was approximately 645 million tones. Around 50 countries have an annual production of more than 1 million tones. Ninety percent of the total rice is produced in Asian countries, with two countries, China and India, growing more than half the total crop. Top ten rice producing countries are China, India, Indonesia, Bangladesh, Vietnam, Thailand, Myanmar, Japan, Philippines, and Brazil. In Bangladesh, Cambodia, Indonesia, Lao PDR, Myanmar, Thailand, and Vietnam, rice provides 50-80% of the total calories consumed. Major part of the rice is consumed locally where produced and only 6-7% of world rice production is traded internationally.

Quality of rice is not always easy to define as it depends on the consumer and the intended end use for the grain. Grain quality is not just dependent on the variety of rice, but also on the crop production environment, harvesting, processing and milling systems. The many diverse uses of rice both domestically and for export, require that quality be evaluated according to its suitability for specific end uses. Quality is based on a combination of subjective and objective factors. Whether rice is acceptable for an intended use is determined by quality testing based on a fixed set of criteria. The importance of quality criteria largely depends on the choice of consumers. Rice grain quality is complex trait involving nutritional quality, appearance quality, cooking and eating quality. Each one of these components consists of many attributes whose values are determined by not only physico-chemical properties but also by the history and cultural traditions of people in different communities consuming rice. Rice, unlike most other cereals, is consumed as a whole grain. Therefore, physical properties such as size, shape, uniformity, and general appearance are of utmost importance. Cooking quality is judged by the amylose content, alkali-spreading score and gel consistency. Furthermore, because most rice is milled, the important physical properties are determined primarily by the milled endosperm. A number of interrelated features determine the quality of paddy like environmental conditions, crop production practices, soil conditions, the harvest and milling practices. All of these quality traits including nutritional quality will be discussed in detail.

Paddy Quality and Grading

Paddy quality is determined by the appearance and purity of grains. Purity refers to material other than paddy and includes chaff, stones, weed seeds, soil particles, rice straw, stalks, etc. These impurities come from the field or from the drying floor. Unclean paddy increases cleaning and processing time. Foreign matter reduces milling recoveries and quality by increasing the wear and tear on the machinery. In addition to purity of the grains, other characteristics such as moisture content, number of immature and cracked grains and discolouring of the grains are used in checking the quality of paddy. The amount of immature paddy grains in a sample has a major affect on head rice yield and quality. The immature rice kernels are very slender and chalky and this results in excessive production of bran, broken grains and brewer's rice. The optimal stage to harvest grain is at about 20-24% grain moisture. Overexposure of mature paddy to fluctuating temperature and moisture conditions leads to development of fissures and cracks in individual kernel. Cracks in the kernel are the most important factor contributing to rice breakage during milling. Cracks reduce milled rice recovery and head rice yields. Yellowing is caused by over-exposure of paddy to wet environmental conditions before it is dried. This results in a combination of microbiological and chemical activity that overheats the grain. Based on the above characteristics, quality standards can be defined as a quantitative way by which we measure and compare certain quality characteristics and grade accordingly. The objectives of establishing standards and grades are:

- To ensure only edible rice reaches the consumer.
- To improve post harvest practices so as to eliminate or reduce waste.
- To improve agronomic practices to increase farm yields.
- To improve processing practices for better milling recoveries and for market expansion.
- To protect consumer from price/quality manipulation.

In relation to the first objective, characteristics such as the moisture content, foreign material, seeds and the discoloured (damaged) grains are important considerations in assuring that only edible rice reaches the consumer. By setting the standards for degree of milling, broken rice content, moisture and damaged grains, the second objective is addressed. Better threshing, drying and improved storage facilities can improve rice quality. The third objective provides incentives to the farmer/agricultural scientist to optimize production by considering standards for chalkiness, varietals purity, foreign seeds, immature grains and red rice. The fourth objective provides a measure of the miller's success in delivering higher recovery and allowing market expansion. The characteristics considered are the standards for degree of milling, broken rice, paddy kernels and foreign matters. Finally, the standards which clearly identify to the consumer the true value of their purchases will provide the protection required against the possibility of unfair trading practices.

Kernel Shape, Size and Form

Grain characteristics such as shape, size, translucency and colour have direct bearing on the marketability, and hence commercial success of modern rice cultivars. Rice that has been harvested from the plant with its hull (husk) intact is known as rough or paddy rice. The hull is not eaten by humans and sometimes burned for use as an energy source. When the hull is removed from rough rice it is called brown (unmilled) rice. However, not all dehulled rice is brown in color. The outer bran layer of the grain and embryo (germ) is what gives rice its color and can vary from light yellow to red to dark purplish black. Rice bran and germ contains greater amounts of dietary fiber, vitamins, minerals and other health-related components than the white center portion of the kernel (endosperm). But those outer portions of the kernel also contain more lipid (fats) material, making brown rice more susceptible to becoming rancid (spoiling). Brown rice, therefore, has a shorter shelf life compared to milled white rice. Storage under cool conditions will lengthen its shelf life. Cooked brown rice has higher fiber content and is chewier in texture than its white rice counterpart and is described as having a slightly nutty flavor. Rice that has had its bran and hull layers removed by milling is called white, table, polished, or milled rice. White rice cooks faster than brown rice and has a longer shelf life. In the U.S., most white rice is coated with iron, niacin, thiamin and folic acid to enhance its nutritional quality.

Rice is primarily classified according to its grain shape. However, within grain shape categories there are differences in cooking qualities that are determined by the chemical make up of the grain and affect cooked grain texture. In most Asian countries, commercial cultivars belong to the medium or long grain class. Long and slender grained Basmati cultivars of India and Pakistan command premium process in the international market while short and bold grained cultivars are preferred in Japan. In temperate areas short grain varieties of Japonica type are prevalent. Extra long grain types are preferred in Thailand. Grain length and shape are independently inherited. Some of the grain tissues are maternally inherited and others are derived after fertilization. The lemma and palea of rice hull are maternal tissues. The seed size and shape are determined by the shape and size of hulls.

While grain size and shape can be visually classified, more exact measurements are needed to classify them for critical comparison of varieties. Different standards are followed for evaluation of grain shape and size in different countries. Though, broadly rice kernels are classified into fine, medium and coarse. In India a committee was appointed to provide guidelines for classification of kernel shape and size. The committee suggested 5 groups based on length and length/breadth ratio of kernels. Similar international classification has also been developed by IRRI in 1996. Based on L/B ratio, Food Corporation of India gave a scheme of classifying rice varieties into three groups as common ((L/B ratio <2.5), fine (L/B ratio 2.5-3.0) and super fine (L/B ratio >3.0). The various grain shape and specialty rice categories are described below.

Long Grain. Long grain rice that is approximately three times longer than it is wide. A conventional U.S. long grain rice has 19 to 23% grain amylose content. After cooking, it is firm and fluffy (not sticky). Consumers in areas of the world such as North and South America, Southern China, Europe, and the Middle East often prefer this type of rice.

Medium Grain. The medium grain rice category describes milled rice that is from 2.1 to 2.9 times longer than it is wide. Medium grain rice is generally has an amylase content of 16-18% and after cooking is soft, moist and sticky in texture. This type of rice is in generally preferred by people from Japan, Northern China and North and South Korea.

Short Grain. Rice that is less than two times longer than it is wide is classified as short grain. In general short grain rice has cooking quality and amylose content similar to that of rice in the medium grain category. Because this type of rice is used for making sushi some call it sushi rice.

Specialty Rice

Rice that has cooking or processing quality different from the standard market classes described above is known as specialty rice. These are used for special styles of cooking as specific products and are consumed in different parts of the world.

Arborio

This rice type originated in Italy where it has traditionally been used for making risotto. Arborio rice is classified as a medium grain, but it has fairly firm internal texture and a unique creamy exterior. It often has a very large white chalky center that is thought to be responsible for its ability to take up the flavor of the stock or sauce it is cooked in.

Basmati Rice

It is a long grain rice with moderately firm cooked texture. It is dry and not sticky after cooking and has an aroma often described as being popcorn like. This category is unique in that its grains become very long and thin (extreme elongation) after cooking. Basmati rice is originated in India and Pakistan. Today, however, rice that has these same unique quality traits is also grown in the U.S.

Jasmine-Type

This style of rice is originally from Thailand. Jasmine style rice has long grains that when cooked are soft and cling to each other. It is considered aromatic rice because it possesses a distinctive aroma often reported to be popcorn-like.

Aromatics

This type of rice is a long grain aromatic rice whose aroma is said to be popcorn like. Although the aroma is similar to Jasmine and Basmati-style rice, its texture mimics that of conventional U.S. long grain firm cooking rice.

Japanese Premium Quality

Japanese Premium Quality rice is similar to conventional U.S. medium grain rice in terms of grain length and amylose content. However, it differs from standard medium grain rice because of their glossiness, lack of flavor, sticky but smooth texture, and softness after cooling. These properties have been traditionally desired by some people of Japanese and Korean descent.

Toro-Type

This is long grain rice that after cooking is soft in texture and the kernels cling to each other. The apparent amylose content is similar to that of conventional U.S. medium grain rice. Toro- Type rice is primarily used in certain ethnic (i.e. Cajun) style cooking.

Waxy (Glutinous, Sweet or Sticky)

This style of rice can be in the long, medium or short grain form, and is eaten both milled and unmilled. Milled waxy rice appears opaque (solid white), as opposed to non-waxy rice, which is translucent. Waxy rice has very little amylose and cooked milled waxy rice is extremely soft and sticky. When the bran is left on, waxy rice is slightly chewy and flavoury. Flour made from waxy rice is also used in products such as candy, salad dressings, baked crackers, and snack foods. It is a ceremonial rice used in areas of Asia.

Wild Rice

Wild rice is not rice nor is it wild. It is a grass, which is native to North America. It used to be just a natural grass found in shallow lakes and waterways, but it is now grown commercially in the U.S. Its nutty, chewy texture and dark brown to black color provide its appeal.

Superior Processing Quality

Some long-grain rice has been developed to remain intact, firm and fluffy after parboiling or canning and thus produces a better appearing rice for use in canned soups and frozen dinners. This type of rice also has less solids loss and kernel splitting after processing compared to conventional U.S. long grain rice and has about 26% grain amylose content. Parboiled rice is produced by steeping and cooking the rice while the hull is still on the kernel. As a result, the color from the hull and the nutrients from the bran layer penetrate the grain during cooking. After parboiling the rice is dried and then milled, producing a slightly amber colored grain.

Parboiled Rice

Rough rice that has been exposed to some combination of soaking in water and exposure to steam, dried, and then milled, is said to have been parboiled or converted. This process results in the natural vitamins and minerals being transferred from the rice bran layer into the starchy endosperm. Parboiling is thought to have originated in India and Pakistan more than 2,000 years ago. It came into use in the U.S. during World War II because it fit the military's need for nutritious food that had a long shelf life. Long grain is the type of rice that is generally parboiled for consumption as table rice. Medium grain rice is also parboiled and ground into flour for use as an ingredient in food products.

Rice Flour and Baby Foods

Rice flour is widely used in the production of baby food. It is ideal as an initial food for babies because it is hypoallergenic. Being hypoallergenic, plus having the ability to prevent and correct dehydration has resulted in beverage mixes being developed which are rice-based and used in the treatment of diarrheal diseases such as cholera. Many of the problems associated with the production of infant cereal can be traced to the starch fraction of the rice flour. Enzymes are used to control the reconstitution properties of instant cereals. The enzymes are deactivated at 158° F. Long grain starch requires temperatures greater than this to begin thickening. Thus the temperature required to thicken with long grain starch would deactivate the needed enzymes. Whereas, short and medium grain flour thickens at 150° F. Additionally, flour with higher amounts of fatty acids is preferred. Fatty acid content changes with storage time and conditions. Thus, flour produced from aged short or medium grain varieties stored under specific conditions is preferred.

White rice that has been ground into a flour or meal is used in many different types of food products around the world. The relatively bland flavor of rice makes it well suited for use in products with mild flavors. It has advantages over other grains in that it will not obscure what natural flavors are present and less added flavor is needed. Waxy rice is often used to make baked crackers, which are light and crispy. When a firmer less delicate baked cracker is desired medium grain rice is often used. Fried snack foods made using a blend of waxy rice flour and other grains will tend to be crisper and take up less fat than if made without the rice flour.

Rice as Breakfast Cereals

Low amylose rice with low gelatinization temperature is preferred. Thus, short and medium grains are typically used. Short grain rice with moisture content of 13% and a fat content of less that 1% is required for puffing. Medium and long grain varieties are unsuitable because of irregular shapes once puffed. For Kellogg's Special K cereal, translucent, highly milled grains with a fat content of less than 0.5% are used. Aged rice is preferred due to different processing characteristics.

Milling Quality

The quality characteristics of milled rice are classified both physically, and chemically. One of the most important aspects of rice grain quality is its milling yield. During the process of milling, first the hull is removed from rough rice using a huller to produce brown rice. About 20% of paddy rice is hull. The hull is removed using either *rubber-roll sheller* or disk sheller. R*ubber-roll sheller* is most commonly used nowadays. In *rubber-roll sheller,* rough rice is passed between two rubber-coated rols that turn in opposite directions and having different speeds. The pressure and the sheer exerted by the rolls remove the hull. The pressure varies depending upon the cultivar as different cultivars require different pressures for adequate milling. In disk sheller, a horizontal abrasive-surfaced wheel rotates on a vertical axis just below a stationery abrasive-surfaced wheel. The abrasive action of the wheels on the rice hull sands of part of the hull and separates the brown rice. After separation, the hull is removed by aspiration, and the remaining rice is separated from the brown rice.

Hulls are tough, fibrous and have low nutritive value. They are made up of interlocking of lemma and palea. Proximate analysis shows that hulls are high in ash content (~20%), cellulose (~30%), pentosans (~20%), and lignin (~20%). In addition, smaller amounts of protein and fats are present. The predominant component of ash in rice is silica. The most common use of hulls is to mix them with rice bran in the ration of 2:1. Other uses of hulls are as fertilizer, as a fuel, as filter aids or abrasives and also the production of furfural.

Dehulling is followed by milling of brown rice by pearling to remove bran parts. In the pearling process, dry calcium carbonate is used to help in removing the bran. During pearling, mixing roller causes the grains to rub against each other and abrading of the bran. This results into broken and whole kernels. The broken kernels are removed from the whole kernels using screens sized for use on long, medium or short grain varieties to produce whole grain rice. Head rice (unbroken milled kernels) milling yield is the percentage of whole kernels recovered after milling and removal of the broken kernels. Producers are paid less for broken kernels than for whole. Milling recovery of rough rice is an estimation of the quantity of head rice and total milled rice that can be produced from a unit of rough rice. After milling loose bran is removed by an aspirator. Milled rice is then polished.

Milling yield is determined by the moisture contents in the grain. It is essential to mill the paddy with proper moisture for obtaining high head rice yield. Paddy is at its optimum milling potential at 14 per cent moisture content (wet weight basis). The high moisture contents make the grains too soft to withstand the hulling pressure thus resulting into breakage and pulverization, while too dry grains become brittle and break easily. Moisture contents and temperature are also critical as these determine small fissures or full cracks in grain structure. In addition, degree of milling is influenced by grain hardness, size and shape, depth of surface ridges, bran thickness and mill efficiency. Normally the hull content is 20-22 % of the rough rice, it can vary from 18% to 26%. Bran and embryo account for 8-10%. Head rice accounts for around 70% of rough rice.

Over exposure of mature paddy to fluctuating temperatures and moistures can lead to the development of fissures and cracks in kernel. Cracks are the most important factor contributing to rice breakage during milling. This result in reduced milled rice recovery and head rice yields. In addition, the amount of immature paddy grains in a sample has a major affect on head rice yield and quality. The immature rice kernels are slender and chalky, resulting in excessive production of bran and broken grains. The optimal stage to harvest grain is at about 20-24 per cent grain moisture. If the harvest is too late, many grains are lost through shattering and are cracked during threshing, which causes grain breakage during milling.

Grains of different sizes and shapes make it difficult to adjust the hullers and polishers to produce whole grains resulting in low husking efficiencies, higher percentage of re-circulated paddy, non-uniform whitening, and lower grade of milled rice. Difficulties at milling result in reduced capacity, excessive breakage, lower milled rice recovery and reduced head rice. The size and shape (length-width ratio) is a varietal property. Long slender grains normally have greater breakage than short, bold grains, and have a lower milled rice recovery. The grain dimensions also dictate to some extent the type of milling equipment needed. For instance, the Japanese designed milling equipments may be better suited to short-bold, japonica grains, whereas Pakistani equipments are suitable for longer slender and Indica type. Varieties possessing medium slender and translucent grains give high head rice yield. Milling percentage can be enhanced by parboiling. Varieties with high grain protein content suffer less damage during milling. Sun cracking caused by alternate drying and wetting of grains due to delayed harvest also cause more breakage. High gelatinization temperature types are less prone to cracking.

Milling of rice increases its shelf life and provides consumers with a physical property they have come to desire, whiteness. Therefore the goal of milling is to remove as much of the colored bran and germ as possible. The quantity of milled rice with respect to brown rice is defined as milling degree. Milling degree affects recovery and influences consumer acceptance. Apart from the amount of white rice recovered, milling degree influences the colour and also the cooking behaviour of rice. Un-milled brown rice absorbs water poorly and does not cook as quickly as milled rice. The water absorption rate improves progressively up to about 25 per cent milling degree after which there is very little effect. A high milling degree means that the milled rice is very white with relatively light milling.

However, milling results in the loss of vitamins, minerals and dietary fiber. Most of the vitamins and minerals are concentrated in the bran parts which are removed during milling, and hence lower nutritional value of milled rice. Rice bran and polish are the by-products from rice milling. The bran is the outer layers of the pericarp; and the polish is the inner layers containing aleurone cells and small amounts of starchy endosperm. In less developed countries, where rice is a major component of the people's diet, such nutritional losses may significantly impact human health. As a result, in some areas, the government has encouraged the production of under-milled rice to improve nutritional well being of its population. Because of excellent source of vitamins and minerals, rice bran can be used in animal feeds. In addition, oil can be extracted from rice bran.

Translucency and Chalkiness of Rice Kernels

Essentially all segments of the rice industry require clear, vitreous, translucent endosperms. Although the preferences for rice grain characteristics vary with different consumer groups, long and slender rice is generally preferred by many consumers in China, USA and most of the Asian countries. Consumers also have a preference for rice that is transparent and not chalky. If part of the milled rice kernel is opaque rather than translucent, it is often characterized as "chalky". Chalkiness disappears upon cooking and has no effect on taste or aroma, however it downgrades milled rice. Excessive chalkiness is caused by interruption during the final stages of grain filling. Chalky areas of the grain are a result of air spaces in between the starch granules that make up the endosperm. Variation in kernel whiteness and transparency can be due to differences in rice varieties, cultural management methods, weather conditions during the crop year, and storage conditions of the harvested rice. Consequently, breeding programs routinely incorporate intensive selection for these traits.

Appearance quality is mostly determined by grain shape as specified by grain length, grain width and length-width ratio and the translucency of the endosperm.

Chalkiness is undesirable in virtually all instances. It detracts from overall appearance, uniformity, and generally results in lower mill yields because chalky grains tend to be weak and break easily. An exception to this is parboiled rice where the process tends to strengthen grains. However, non-uniformity in chalkiness may result in over processing some of the grain and under processing of others. Chalkiness in rice is caused by many factors, included are: (i) harvesting at too high moisture; (ii) uneven ripening of the grains in a panicle, leading to excessive numbers of immature grains; (iii) high temperatures during ripening; and (iv) genetic factors.

Soft textured and whit spots occurring in the middle part on the ventral side are called white belly (Abdominal white). A white chalky region extending to the edge of the ventral side and towards the centre of the endosperm is called a white core. A long white streak on the dorsal side is called the white back. In some varieties the grain tends to break more frequently at these chalky portions or pit left by the embryo when it is milled. Rice samples with damaged eyes have poor appearance and low market value. Since starch granules are loosely packed in chalky areas and presence of air spaces between them, chalky areas are more prone to damage during milling. The heritability of chalkiness is low and hence various agronomic practices and preharvest handling can be used to influence chalkiness.

Procedures of Measuring Milling Quality

The above discussion of milling quality shows that there are various measurements used in defining quality. The brief description of these terms rae given below for better understanding.

Milling Degree

Milling degree is computed based on the amount of bran removed from the brown rice. To obtain the weight of brown rice, dehull the paddy samples using the Laboratory Huller. It is the percentage of milled rice with respect to brown rice.

Milling Recovery

Dehulled samples are milled using Abrasive Whitener. The milling recovery is computed by dividing the weight of milled rice recovered by the weight of the paddy sample.

Broken Grain

Using the Grain Grader, separate the broken grain from the whole grains. Compute the percentage of the head rice and brokens using the following equations:

$$\% \text{ Brokens} = \frac{\text{Wt of broken grains}}{\text{Wt of paddy samples}} \times 100$$

$$\% \text{ Head rice} = \frac{\text{Wt of whole grains}}{\text{Wt of paddy samples}} \times 100$$

Chalkiness

A visual rating of the chalky proportion of the grain is used to measure chalkiness based on the Standard Evaluation System (SES) scale. The chalky grains are segregated and weighed and the % chalky grains are calculated and the scale is determined as per the table below.

Scale	% area of chalkiness
1	less than 10
5	10-20
9	more than 20

Molecular Markers Associated with Milling Quality

Genetically milling quality is controlled by both triploid endosperm genotypes and maternal diploid genotype, as well as by the cytoplasmic genomes Li *et al.* (2004). Environmental factors and G X E interactions are also known to influence rice milling quality. Rice milling quality is controlled by both main-effect and epistatic loci, and results from different mapping populations can be quite different. On the contrary, grain shape includes several grain morphological traits under direct selection that are highly correlated with milling quality. Typically, increased grain length, grain length to width ratio or length to thickness ratio are negatively associated with grain milling quality, while increased grain width and thickness tend to result in increased milling quality (Xu *et al.* 2004b, Wang *et al.* 2005). Zheng *et al.* (2007) reported one major QTL on chromosome 7 that had large effects on rice grain shape and milling quality. They dissected the components of rice milling quality, i.e. brown rice rate, milled rice rate and head rice rate by the QTL mapping approach using a set of 231 random rice introgression lines and 160 SSR markers. A total of 10 genomic regions were found to be associated with rice grain shape and milling quality traits. This offers an opportunity for marker aided improvement of rice milling quality and QTL cloning.

Cooking and Eating Quality

Though the grain appearance of milled rice is important to the consumer, which makes it also important to the producer and the miller, chemical constituents in the grain determine rice quality. The consumer prefers rice with a clear endosperm and pays premium rice for it, even though capacity disappears during cooking and does not affect eating quality. Grain with opaque areas in the endosperm, caused by the loose packing of the starch and protein particles, breaks more readily during milling than clear grain, greatly reducing its market value. Cooking and eating quality characteristics of rice are determined by the chemical constituents of the grain especially starch that makes 90% of the milled rice. Chemical tests are generally conducted after storage of few months because aging increases water absorption and volume expansion during cooking, resulting in a flakier cooked rice than freshly harvested grains. There are five different tests for assessing rice quality namely alkali test, amylose test, gel consistency test, kernel elongation test and aroma test. In addition protein content, parboiling stability and amylographic paste viscosity are used as quality criteria. These physical and chemical characteristics are given in the (Table 1).

Table 1: Chemical and physical quality characteristics of long, medium and short grain rice types in US as reported by B. D. Webb 1985. In: Rice Chemistry and Technology.

	Characteristics	Long grain	Medium grain	Short grain
1	Amylose	23-26	15-20	18-20
2	Alkali Spreading Value	3-5	6-7	6-7
3	Gelatinization temp. (°C)	71-74	65-68	65-67
4	Gelatinization Class	Intermediate	Low	Low
5	Water uptake (ml/100g)	121-136	300-340	310-360
6	Protein (%)	6-7.5	6-7	6-6.5
7	Parboiling Stability, Solid Loss (%)	18-21	31-36	30-33
	Amylographic Paste Viscosity			
8	Peak	765-940	890-820	820-870
9	Cooked 10 min at 95° C	400-500	370-420	370-400
10	Cooled at 50° C	770-880	680-760	680-690
11	Brewing Cookability, sec	120	5-15	5-10

Gelatinization Temperature

Cooking and eating quality in rice is mainly determined by the starch composition, especially by three physicochemical characteristics of the starch i.e., amylose content (AC), gel consistency (GC) and gelatinization temperature (GT). The time required for cooking rice is determined by its gelatinization temperature. Gelatinization temperature is the temperature at which the rice absorbs water and starch granules swell irreversibly. Gelatinization temperature is measured by alkali spreading value (ASV). The degree of spreading value of individual milled rice kernels in a weak alkali solution (1.7% KOH) is closely related with gelatinization temperature. Rice with low gelatinization temperature disintegrates completely, while rice with medium GT shows partial disintegration. Rice with high GT remains largely unaffected in the alkali solution. Based on gelatinization temperatures, milled rice is classified as low (below 70°C), intermediate (between 70°-74°C), and high (greater than 74°C).

Starch granule gelatinization involves the heat driven transformation of granules in excess of water from an ordered state to a disordered state. The initial stages of gelatinization involve the uptake of water lowering the glass transition temperature in the amorphous regions and causing a crystalline region to begin to melt. Further heating and water uptake results in the mobilization of amylose and the initiation of its leaching from the granule and the complete melting of the crystalline regions, marked by the loss of birefringence. Further swelling of the granules occurs, typically producing greatly enhanced viscosity in the starch water paste. The temperature at which pasting occurs is called pasting temperature. Pasting temperature provides an indication of the minimum temperature required to cook a rice sample. Environmental conditions such as temperature during grain development influence gelatinization temperature. A high ambient temperature during grain ripening results into starch with high GT. The overall process of gelatinization is similar in both normal and waxy starch granules, although by definition, amylose is not present in a waxy starch and cannot be lost by leaching. These observations suggest that the properties of the amylopectin fraction of starch are most critical in controlling gelatinisation and granule swelling in cereal starches. Amylose content has a major influence on starch granule swelling, mediated through the reduced mobility of amylose/ lipid complexes in the granule that restrict water movement and swelling.

Most Japonica varieties have a low a gelatinization temperature. Most tropical Indica varieties have intermediate or low gelatinization temperatures. Gelatinization temperature directly affects the physical properties of the starch granule, which in turn influences the quality ratings of cooked rice. Rice with a high gelatinization temperature becomes excessively soft and tends to disintegrate when overcooked. Under standard cooking procedures, this type of rice tends to remain undercooked. It requires more water and time to cook than those with low or intermediate gelatinization temperature. Thus, gelatinization temperature correlates positively with the time required to cook rice. Soaked milled rice of high gelatinization temperature elongates less during cooking than rice with low or intermediate gelatinization temperature.

Gel Consistency

Gel consistency measures the tendency of cooked rice to harden on cooling. The gel consistency test is based on the consistency of the rice paste and differentiates rice with high amylose content. The cooked rice tends to be less sticky if it has hard gel consistency, harder gel consistency is associated with harder cooked rice and this feature is particularly evident in high-amylose rice. When the gel consistency is soft, the cooked rice has a higher degree of tenderness. This is a preferred characteristic. Gel consistency test is based on the consistency of a cold 4.4% milled rice paste in 0.2N KOH. The length in culture tube of the cold gel held horizontally for 0.5 to 1 hour measures consistency. Gel consistency is classified as follows: hard with the length of the gel in the range of 27-35 mm, medium hard in the range of 36-40 mm, medium 41-61 and soft 61-100 mm. Varietal differences in gel consistency exist among varieties of similar high amylose content (more than 25%). The gel consistency test separates the high-amylose rice into three categories: very flaky rice with hard gel consistency (length of gel = 40 mm or less), flaky rice with medium gel consistency (length of gel = 41 to 60 mm) and soft rice with medium gel consistency (length of gel > 61 mm). Gel consistency of rice is normally soft, when the amylose content is less than 25%. Gel consistency is controlled by single major gene.

Amylose Content

Starch makes up about 90 per cent of the dry matter content of milled rice. Starch is a polymer of amylose and amylopectin. Amylose is a linear polymer of glucose. The amylose content of starches usually ranges from 15 to 35 per cent. Amylose content is considered the single most important character for predicting rice cooking and processing quality. High amylose content rice shows high volume expansion (not necessarily elongation) and high degree of flakiness. High amylose grains cook dry, are less tender, and become hard upon cooling. In contrast, low-amylose rice cooks moist and sticky. Intermediate amylose rice is preferred in most rice-growing areas of the world, except where low-amylose japonicas are grown. High amylose content in the endosperm is usually associated with dry, fluffy, and separated cooked rice grains and represents the key determinant of poor cooking and eating quality. 1-2 % amylose is classified as waxy and 3-9 % as very low amylose content. 10-20% amylose content is classified as low, 20-25% amylose content is classified as Intermediate. 25-30% amylose content is classified as high. Amylose content of milled rice is determined by using the calorimetric iodine assay index method.

Amylose content is genetically controlled by a major gene, the Waxy (*Wx*) gene. In addition, amylose content is also affected by several modifying genes and environmental factors such as temperature. The *Wx* gene encodes the granule-bound starch synthase that is responsible for amylose synthesis in the endosperm. Rice strains having the *Wxb* gene contain 15–20% amylose, whereas rice strains having the *Wxa* gene contain 20–25% amylose. The *du* loci modify amylose content by controlling expression of the *Wxb* gene and mutations at the *du* loci cause a reduction in amylose content.

Waxy starches (amylose free) are available with a range of viscosity and stability characteristics. Waxy rice starches are known for their freeze/thaw stability, creamy fat-like mouthfeel and high water binding capacity. Some varieties that exhibit unique process stability without the use of chemical modifying agents are available. Waxy rices do not expand in volume, are glossy and sticky and remain firm when cooked. These rices are the staple food of people in Northern Ireland and Laos.

Although the *amylose extender* (*ae*) mutant shows increasing levels of apparent amylose content, this increase is caused by the altered amylopectin structure. The *ae* mutation is found in the gene for starch branching enzyme (SBEIIb) (Nishi *et al.* 2001). Though, other mutation has been characterized in rice with enhanced amylose content in the endosperm that does not affect the amylopectin structure. The mutants were generated in a variety Snow Pearl (Higashi *et al.* 1999) whose amylose content is very low (about 5%) and the endosperm is chalky when seeds matured at a normal temperature (26 °C). On the other hand, amylose content was about 18% and the endosperm appeared semi-translucent, when seeds matured at a cool temperature

(20 °C). Because amylose content in rice seed influences the degree of transparency (Suzuki *et al.* 2002), a mutant with enhanced amylose content in (semi-translucent) seeds matured at a normal temperature could be easily distinguished from the original variety with low-amylose (chalky) seeds.

Studies indicated that a single nucleotide polymorphism (G or T) located at the first nucleotide of the splice donor site of *Wx* intron 1 is responsible for mutant allele. A CAPS marker was subsequently developed, and with this molecular marker-assisted selection (MAS) can be employed. The introgression of the *Wx*-TT locus of rice cultivars with good quality intermediate AC into elite backgrounds becomes more precise. The elite lines can be further used in improving quality of rice hybrids by transferring desired genes into male sterile lines. The MAS was used in introducing *Wx*-TT locus into their relevant male-sterile lines (LTF-A and ZS-A) to generate improved indica hybrids. In the selected lines LTF(tt)-B and ZS(tt)-B, the AC was reduced to a relatively low level (15%). Consequently, the hybrids crossed from the selected lines had dramatically reduced amylose levels. In the field trials, the agronomic performance in the improved lines and their hybrids was examined and compared to the originals. The other key factors involved in rice cooking and eating quality were also improved in the selected lines and their hybrids (Liu *et al.* 2006).

Aroma (Fragrance)

Scented rice is an important commodity worldwide and command premium prices over non-scented varieties. Middle East countries prefer strong aroma. They feel that rice with out aroma is like food without salt. For consumers in Europe, a trace of aroma is objectionable trait, because for them any scent signals spoilage and contamination. The Indian subcontinent has the 'natural gift' of basmati rice that has been accepted as the best scented, long and slender grain rice in the world markets and fetches high prices. High milling returns and good cooking quality are often associated with aromatic or scented rice. However, aromatic cultivars have often undesirable agronomic characters such as low yield and susceptibility to pest and diseases. In addition to basmati, many indigenous scented rice varieties are also grown that excel in aroma and cooking qualities. Most of these varieties are medium or short-grain varieties and hence are not traded at international level. However, in domestic markets, their demand is far greater than the long-grain ones.

Fragrance can be detected by testing the associated flavour in individual seeds or assessing the aroma of leaf tissue or grains after either heating in water or reacting with solutions of KOH or I2-KI. More than 100 volatile compounds have been identified in basmati rice. Among these compounds, 2-acetyl-1-pyrroline (2AP) is the principal aroma compound in both basmati- and jasmine-style fragrant rice. AP is found in all parts of the rice plant, except for the roots, and is also found, at concentrations up to 100-times lower, in non-fragrant varieties.

Many studies of the genetic control of aroma trait in rice have been reported. Some of the authors concluded that the trait undergoes monogenic inheritance while others favoured the idea of two or three dominant or recessive genes controlling the trait. Lorieux *et al.* (1996) reviewed the genetics of aromatic fragrance and concluded that a single recessive gene was responsible for the production of fragrant rice plants. This single recessive fragrance gene (*fgr*) was linked to the RFLP clone RG28 on chromosome 8, at a genetic distance of 4.5 cM. Lorieux *et al* (1996) confirmed the close linkage between RG28 and *fgr* (5.8 cM) and also identified two quantitative trait loci for fragrance, one on chromosome 4 and the other on chromosome 12. Subsequently PCR based marker for fragrance was developed. A small mono-nucleotide repeat, that was polymorphic between a pair of fragrant and non-fragrant cultivars, was identified and developed into a co-dominant PCR-based marker. These PCR-based markers will be highly useful in distinguishing fragrance-producing alleles from non-fragrance-producing alleles at the *fgr* locus.

SSR marker analysis confirmed the previous mapping result, and placed the *fgr* locus between RM8264 and RM3459 with the physical distance of ~800 kb. The *fgr* region was then saturated with high-density markers developed by exploiting sequence diversities between *indica* and *japonica* rice subspecies. After mapping with segregant populations consisting of totally 2891 individuals, the *fgr* locus was restricted to an interval of 69 kb flanked by the left marker L02 and the right marker L06. Furthermore, the *fgr* locus was confirmed by simultaneous investigation of both genotypes and 2AP levels for the key recombinants and their offspring. Sequence analysis of the *fgr* region revealed three candidate genes encoding respective eukaryotic-type carbonic anhydrase, 3-methylcrotonyl-CoA carboxylase beta chain, and betaine aldehyde dehydrogenase (BADH). Bradbury *et al.* (2005) showed that a single recessive gene was responsible for aroma. This gene is a defective allele of a gene encoding BADH. The deletion observed in exon of this gene generates a premature stop codon and presumably results in loss of activity causing 2ACP accumulation.

Nutritional Quality

Protein Quality: Lysine and Tryptophan Content

Rice serves as an important source of dietary protein for humans in developing countries. The protein content is lower in rice (on an average 8%) as compared to wheat (on an average 12%). The demand of rice for dietary proteins is expected to increase dramatically in future with the rise in world population. Though, rice is superior to many other cereals in terms of protein quality with higher lysine content (3.8 to 4.5%), the nutritional quality of rice endosperm storage proteins is not optimal because the proteins are deficient in certain amino acids. The major food supply for humans and livestock comes from plants, which contain very low proportions of some of the 'essential' amino acids, mostly Lys, Met, Thr and Trp. Large efforts are therefore being devoted to increasing the content of these essential amino acids in crop plants that serve as human foods and animal feeds. Many of the pathways of the metabolism of essential amino acids in plants have been elucidated, and genes encoding most of the enzymes in these pathways are now available. Modification of amino acid metabolic pathways by metabolic engineering, coupled with expression of recombinant genes encoding proteins enriched in essential amino acids, has contributed significantly to increased accumulation of essential amino acids in plants. In addition mutants have also been identified in cereals including rice with enhanced lysine content. However, these approaches have also exposed difficulties associated with achieving optimal levels of essential amino acids, as well as impaired plant growth and productivity, implying that deeper understanding of regulatory networks associated with amino acid metabolism is needed to improve the nutritional quality of plant foods and feeds.

Improving Lysine Levels by Metabolic Engineering

Most efforts to improve Lys production in plants by metabolic engineering have utilized bacterial enzymes that are much less sensitive to Lys inhibition than their counterpart in plants. Lysine accumulation is controlled both by synthetic and catabolic processes (Fig.1). The main enzymes of lysine biosynthesis as aspartate kinase (AK) and dihydrodipicolinate synthase (DHPS) are inhibited by lysine itself. In addition, lysine is catabolized by Lys-ketoglutarate reductase (LKR) and saccharopine dehydrogenase (SDH). Therefore, by reducing feed back inhibition of AK and DHPS and simultaneously reducing LKR and SDH activity, lysine accumulation can be enhanced. To inhibit feed back inhibition of DHPS, recombinant constructs encoding a plastid-targeted bacterial

DHPS enzyme have been expressed in a number of model and crop plants. These have been done using either constitutive cauliflower 35S promoter, or in a seed-specific manner using a variety of promoters derived from genes encoding seed storage reserves. (Galili, 2002, and references therein). Constitutive expression of the bacterial DHPS caused significant over-accumulation of free Lys in vegetative tissues and also associated with increased levels of various Lys catabolites (Mazur *et al.* 1999), implying that Lys catabolism is also an efficient metabolic pathway in seeds of soybean and rapeseed.

Unexpected results were obtained when a bacterial DHPS was expressed in a seed-specific manner in transgenic maize seeds. Unlike soybean and rapeseed, in which the major storage organ is the embryo, the major storage tissue in cereal seeds is the endosperm and Lys catabolism is considered to operate in several specific peripheral cell layers of the endosperm (Arruda *et al.* 2000). However, expression of the bacterial DHPS in the endosperm did not cause any major increase in seed Lys levels, while its expression in the developing embryo caused significant increases in seed Lys levels (Mazur *et al.* 1999). Moreover, Lys overproduction in the maize embryos was associated with an increase in Lys catabolic products (Mazur *et al.* 1999), despite the fact that Lys catabolism was suggested to be minimal in this tissue (Arruda *et al.* 2000). This illustrates again the complexity of amino acid metabolism in plant seeds.

The significant activity of Lys catabolism in plant seeds suggests that metabolic engineering of high-Lys crops would require its reduction or elimination. However, the exact procedure to manipulate Lys catabolism depends on whether this pathway is essential for plant growth and development. To study the significance of Lys catabolism, an *Arabidopsis* mutant with knockout of the *LKR/ SDH* gene, encoding the first two linked enzymes of the Lys catabolism pathway, was generated (Zhu *et al.* 2001). This knockout exhibits a morphologically normal phenotype under favourable growth conditions and also possesses enhanced Lys levels in the seeds (Zhu *et al.* 2001). Analysis of natural high-Lys maize mutants has also confirmed a regulatory role of Lys catabolism in balancing Lys concentration in seeds (Kemper *et al.* 1998, Azevedo *et al.* 2003, 2004 a,b).

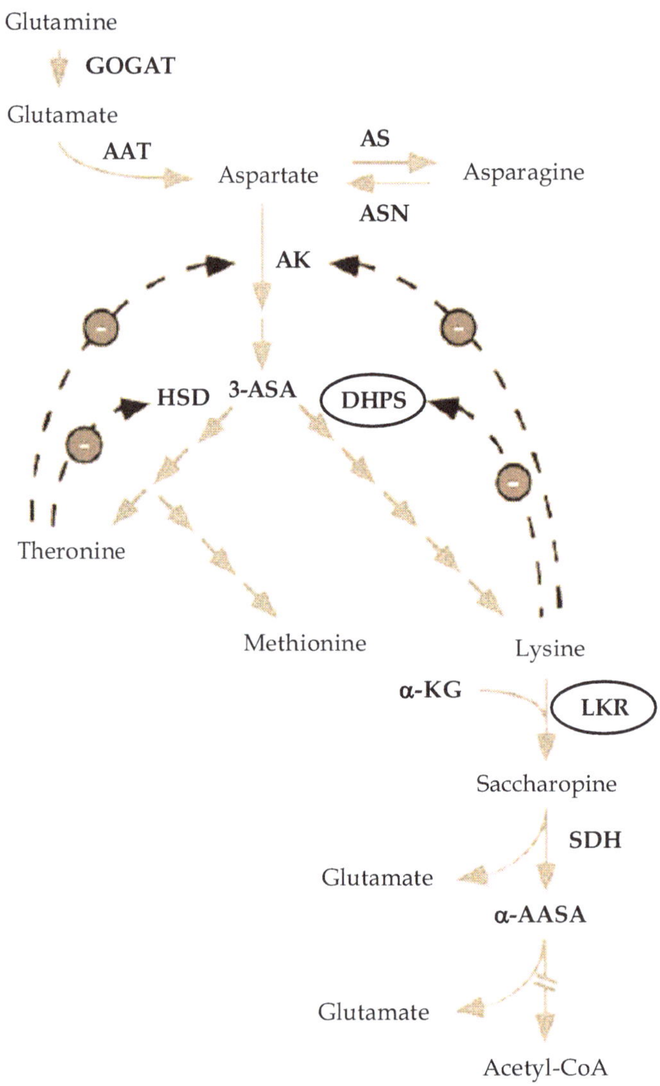

Fig. 1. Schematic diagram of amide amino acid metabolism (top), the Asp-family pathway (center), and the pathway of Lys catabolism (bottom). Dashed arrows with a "-" sign represent end-product feedback inhibition loops. GOGAT, Glu synthase; AAT, Asp aminotransferase; AS, Asn synthase; ASN, asparaginase; AK, Asp kinase; DHPS, dihydrodipicolinate synthase; HSD, homoserine dehydrogenase; LKR, Lys-ketoglutarate reductase; SDH, saccharopine dehydrogenase; 3-ASA, 3-aspartic semialdehyde; a-KG, a-ketoglutarate; a-AASA, a-amino adipic semialdehyde. Broken arrow at the bottom indicates several enzymatic steps leading to the production of acetyl- CoA. The enzymatic steps manipulated in this report (DHPS and LKR) are circled. (Zhu and Gallili 2004).

In rice, transgenic plants with a lysine-feedback-insensitive maize DHPS gene under the control of CaMV 35S and the rice glutelin GluB-1 promoter for over-expression and seed-specific expression were developed (Lee *et al.* 2001). The transgenic plants were fertile and expressed the DHPS gene abundantly or specifically in rice seeds. The transgenic lines (TC lines) containing mutated DHPS controlled by CaMV 35S promoter possessed higher mutated DHPS transcript levels and in vitro DHPS activities in seeds than those of TS lines containing the mutated DHPS gene driven by a seed-specific promoter, GluB-1. The content of free lysine in immature seeds of both TC and TS lines was higher than that of wild-type plants. The content of free lysine in mature seeds of TC lines was still higher than, but that of TS lines was similar to, that of wild-type plants. From a comparison of DHPS and lysine-ketoglutarate reductase (LKR) expression levels it was concluded that the presence of the foreign DHPS gene leads to an increase of LKR activity, resulting in enhanced lysine catabolism. However, over-expression of the mutant DHPS gene in a constitutive manner overcomes lysine catabolism and sustains a high lysine level in mature rice seeds.

Metabolic Engineering for Tryptophan (TRP) in Rice Seed

Metabolic manipulation of plants to improve their nutritional value is a primary goal of plant biotechnology. Essential amino acids such as Lys, Met, Thr, and Trp contribute substantially to the nutritional quality of plant-based foods for humans and domestic animals, but the amounts of such amino acids are limited in many crops. Dietary supplementation with Trp increases the growth rate of pigs and poultry. An inadequate supply of Trp leads to a marked reduction in food intake in pigs as a result of the reduced synthesis of serotonin in the brain (Henry *et al.* 1992, Seve 1999). Indeed, Trp has been used as a pharmaceutical agent in the treatment of depression (Massey *et al.* 1998). The ability to increase the level of Trp in food crops by metabolic engineering is thus desirable from both nutritional and clinical viewpoints. The achievement of an increase in the amino acid content of seeds by genetic engineering, however, has been limited to Lys (Zhu and Galili 2003) and Met (Lai and Messing 2002). The metabolic pathways of tryptophan synthesis have been identified and the genes for each step characterized. The details of biosynthesis pathway are given in (Fig.2).

Phosphoenolpyruvate (PEP)

+

Erythrose 4-phosphate

↓ DHAP synthase

3-Deoxy-D-arabinoheptulosonate-7-phosphate (DAHP)

⇣

Shikimate

⇣

Chorismate

Chorismate

↓ Anthranilate synthase (α and β)
(Feedback inhibited by tryptophan)

Anthranilate

↓ Phosphoribosyl anthranilate transferase (PAT)

N-(5`-phosphoribosyl) anthranilate

↓ Phphoribosyanthranilate isomerase (PAI)

-(oCarboxyphenyl amino) -1-Deoxyribulose 5 -phosphate

↓ Indole 3-glycerol phosphate synthase (IGPS)

Indole 3-glycerol phosphate (IGP)

↓ Tryptophan synthase (α and β)

L Tryptophan

Fig. 2: Shikimate pathway for tryptophan synthesis in plants. Dashed arrow indicates many steps in the pathway.

Chorismate is the final precursor of the three aromatic amino acids such as phenylalanine, tryptophan and tyrosine. Except for an isozyme of chorismate mutase, all the enzymes of aromatic amino acid synthesis are present in plasmids. The seven step synthesis of chorismate begins with the condensation of two intermediates of carbohydrate metabolism, phosphoenol puruvate from glycolysis and erythrose 4-phosphate from the pentose phosphate pathway. This reaction is catalyzed by 3-Deoxy-D-*arabino*-heptulosonate-7-phosphate (DHAP) synthase. The reaction is followed by three enzymatic steps involving 3-Dehydroquinate synthase, 3-Dehydroquinate dehydratase and Shikimate dehydrogenase to produce shikimate. Shikimate is phosphorylated by the activity of shikimate kinase to produce shikimate 3-phosphate, which is converted into chorismate using 5-enolpyruvylshikimate-3-phosphate (EPSP) synthase and chorismate synthase enzymes in the chloroplast. Chorismate is converted to tryptophan using various steps, which have been studies extensively at genetic levels.

Conversion of chorismate to tryptophan has significance beyond amino acid biosynthesis. Plants use this pathway to produce precursors of numerous secondry meabolites, including the hormone auxin, indole alkaloids, phytoalexins, cyclic hydroxamic acids, indole glucosinolates and acridone alkaloids. Details of these metabolites are given in standard plant biochemistry books. Here, an account of tryptophan biosynthesis and the manipulation of the pathway for improving tryptophan content will be discussed. The first step is catalyzed by anthranilate synthase (AS) which is the committing step in tryptophan biosynthesis. The enzyme is feedback inhibited by tryptophan itself. Feedback insensitive mutants of anthranilate synthase have been identified in plants and used in transgenic experiments.

The accumulation of free Trp in plants has been achieved by the introduction of genes encoding feedback-insensitive subunits of anthranilate synthase (AS), which catalyses the conversion of chorismate to anthranilate. This approach has thus yielded increased levels of Trp in the roots of the forage legume *Astragalus sinicus* (Cho *et al.* 2000) and in the leaves of tobacco (Zhang *et al.* 2001). A mutant *OASA1* gene, *OASA1D* [formerly referred to as *OASA1(D323N)*], that encodes a feedback-insensitive subunit of rice AS has previously been generated. Rice calli and leaves as well as potato plants and tubers that express *OASA1D* accumulate large amounts of free Trp. Manipulation of metabolic pathways in plants by genetic engineering thus has the potential to improve the nutritional value of crops and to allow the production of desired natural products in higher plants. The proteins encoded by the introduced genes might affect not only the abundance of the target metabolite, however, but also, through changes in metabolic networks, that of related compounds. Moreover, such changes may influence the physiology or morphology of plants. Little is known, however, of the consequences of such manipulation for metabolite profiles and agronomic traits in plants.

Metabolism of Trp in plants is associated with the generation of a range of secondary compounds such as indole alkaloids and indole acetic acid (IAA). Manipulation of the Trp biosynthetic pathway might thus be expected to influence the synthesis of such metabolites and thereby elicit a pronounced change in metabolite profile. Indeed, overexpression of Trp decarboxylase, which catalyses the conversion of Trp to tryptamine, resulted in a marked reduction in the amount of indole glucosinolate in canola plants and a decreased abundance of Trp, Phe, and chlorogenic acid in potato. Evaluation of transgenic plants of improved nutritional value for effects of the transgene on agronomic traits and metabolite profiles is thus essential. However, the possible effects of the accumulation of Trp or of any other essential amino acid in seeds on agronomic traits and metabolite profiles have not been determined to date. Demonstration that Trp is the only major metabolite that accumulates in transgenic rice seeds, for example, would be likely to increase their acceptability for consumption by humans and farm animals.

Wakasa *et al.* 2006 developed stable transgenic line of rice with two fold enhanced tryptophan content. Expression in rice (*Oryza sativa* L.) of a transgene (OASA1D) encoding a feedback-insensitive α-subunit of rice anthranilate synthase resulted in the accumulation of tryptophan (Trp) in calli and leaves. It is shown here that the amount of free Trp in the seeds of such plants is increased by about two orders of magnitude compared with that in the seeds of wild-type plants. The total Trp content in the seeds of the transgenic plants was also increased. Two homozygous lines, HW1 and HW5, of OASA1D transgenic rice were generated for characterization of agronomic traits and aromatic metabolite profiling of seeds. The marked overproduction of Trp was stable in these lines under field conditions, although spikelet fertility and yield, as well as seed germination ability, were reduced compared with the wild type.

Micronutrient Density and Their Bioavailability

The prevalence of iron deficiency is estimated to be about 30% of the world population (WHO 1992), making iron by far the most-widespread nutrient deficiency world-wide. The major consequences are poor pregnancy outcome including increased mortality of mother and children, reduced psychomotor and mental development in infants, decreased immune function, tiredness and poor work performance. The amount of bio-available iron is dependent both on iron intake and on its absorption. Dietary iron in developing countries consists primarily of non-haem iron, whose poor absorption is considered as a major factor in the etiology of iron-deficiency anaemia. Cereal and legume staples are rich in phytic acid, which is a potent inhibitor of iron absorption. In addition, the intake of foods that enhance non-haem iron absorption, such as fruits, vegetables or muscle tissue, is often limited. The most-widely recognised strategies for reducing micronutrient malnutrition are supplementation with pharmaceutical preparations, food fortification, dietary diversification and disease reduction. For various reasons, none of these have been very successful in reducing the prevalence of iron-deficiency anaemia in developing countries.

An alternative more sustainable approach would be the enrichment of food staples either by plant breeding or by genetic engineering. Large genetic variability of Fe and Zn concentration has been reported in rice. In 1995 the research conducted at IRRI indicated the range in Fe and Zn concentrations in brown rice within the eight sets of genotypes ($n = 1,138$) tested in one study was 6.3–24.4 mg/g for Fe and 13.5–58.4 mg /g for Zn. Thus, within those genotypes tested, there was approximately a fourfold difference in Fe and Zn concentrations, suggesting some genetic potential to increase the concentrations of these micronutrients in rice grain. The highest grain-Fe concentrations (i.e., ranging from ~18–22 mg/g) were found in several aromatic rice varieties, such as Jalmagna, Zuchem and Xua Bue Nuo. These same aromatic lines also contained the highest grain/Zn concentrations (ranging from ~24–35 mg/g). Additional research using F2-derived populations demonstrated that the aromatic trait was not pleiotropic for grain-Fe or grain-Zn concentrations, and, therefore, this trait may be used to screen for high Fe and Zn levels in rice grain, but the linkage is broken at a low frequency.

Molecular Approach for Enhancing Micronutrient Content

Increasing seed ferritin, the natural iron store, had been suggested as a means to increase the iron content (Theil *et al.* 1997). Ferritin is the iron-storage protein found in animals, plants and bacteria, which can store up to 4500 iron atoms in a central cavity. Recent studies have shown that iron from animal and plant ferritin can be utilized by anaemic rats and man (Beard *et al.* 1996, Skikne *et al.* 1997). However, increasing iron intake will not be successful in eliminating iron-deficiency anemia unless the diet is also low in iron-absorption inhibitors or contains enhancers of iron absorption and utilization. The major inhibitor, phytic acid, can be readily degraded in cereal and legume foods by the addition of exogenous phytases either during food processing (Hurrell *et al.* 1992) or during digestion, increasing iron absorption dramatically. In the same way, muscle tissue, through the action of the cysteine-containing peptides formed on digestion, improves iron absorption from cereal- based meals Cook *et al.* (1997). There are thus three different approaches to increase the amount of iron absorbed from rice by the introduction of the ferritin gene, thermotolerant phytase from *Aspergillus fumigatus* and by over expressing the endogenous cysteine rich metallothionein-like protein. All three approaches were used by Lucca *et al.* (2001). Genes were regulated by an endosperm-specific promoter, to ensure and restrict expression to the endosperm, the tissue constituting the milled rice grains.

Antinutritional Factor (Phytic Acid)

Phytic acid (PA, also known as myo-inositol hexa kisphosphate) is the primary storage form of phosphorus (P) in seeds, accounting for up to 85% of the seed total P. The negatively charged phosphates in phytic acid strongly bind to metallic cations (e.g. K, Mg, Mn, Fe, Ca, and Zn) to form a mixed salt called phytin or phytate that accumulates in seed protein bodies either dispersed throughout the bodies or in dense inclusions called globoids. Phytin is broken down following germination and provides the P, myo-inositol, and mineral cations that are needed for seedling growth and development. PA also plays an important role in animal and human nutrition. Monogastric animals such as pigs, poultry, and fish cannot digest PA. As a result, animal feeds are supplemented with P, and a large fraction of PA is excreted, subsequently accumulating in soil and water which leads to the pollution of waterways. The ability of PA to bind divalent minerals, resulting in the formation of stable complexes that prevent the absorption of these important nutrients from foods or feeds can lead to micronutrient deficiencies (e.g., iron and zinc), particularly in third world countries where diets are primarily seed-based. In developing countries, high-phytate grain-based diets are often feared to exacerbate iron and zinc malnutrition, while in developed countries, where livestock is fed primarily with grain-based feed, the excreted phytate contributes to environmental P pollution by being washed into surface waters (Lott *et al.* 2000). These concerns have collectively provided a strong impetus to developing low phytic acid (*lpa*) crops, in which the PA–P content is significantly reduced in seeds.

Several *lpa* mutant lines have been generated by chemical and physical mutagenesis as well as genetic transformation in several crops, including maize (Raboy *et al.* 2000, Pilu *et al.* 2003, Shi *et al.* 2005), barley (Ramussen and Hatzak1998), wheat (Guttieri *et al.* 2004), soybean (Wilcox *et al.* 2000) and rice (Larson *et al.* 2000). Genetic analysis has shown that the *lpa* phenotypes resulted from single locus mutations, and three types of *lpa* mutation, designated as *lpa1*, *lpa2* and *lpa3*, have been already identified and mapped to different positions of various chromosomes in maize (Raboy *et al.* 2000, Pilu *et al.* 2003, Shi *et al.* 2005), barley (Larson *et al.* 1998) and soybean (Walker *et al.* 2006). Human feeding studies showed that, in comparison with its wild type counterpart, *lpa* maize diets increase the absorption of several important minerals (Adams *et al.* 2001). Similarly, reduction of P content by 10–85% in wastes and increase of P bioavailability by 2–5 times were observed in several animal feeding studies using either *lpa* maize or barley.

In maize, where low PA mutants have been partially characterized, normal levels of seed total P accompany low levels of PA. Two mutations affecting phytic acid levels, *lpa1* and *lpa2*, have been identified. Seeds of homozygous *lpa1* plants have reduced PA in comparison to the wild type and a molar increase in inorganic P (Pi). Seeds of homozygous *lpa2* plants have reduced PA and a molar increase in Pi and inositol phosphates (Raboy *et al.* 2001). The maize *lpa1* mutation maps near one of the seven maize myo-inositol phosphate synthase genes, MIPS, which encodes MIPS, an enzyme involved in the first step of inositol phosphate biosynthesis. In maize, the MIPS gene and protein expression is reduced in lpa1 seeds compared to the wild type (Raboy *et al.* 2000, Pilu *et al.* 2003, Shukla *et al.* 2004). Sequence analysis of the MIPS gene that maps near the *lpa1* mutation did not reveal any lesions within the coding sequence (Shukla *et al.* 2004), and the nature of the *lpa1* mutation in maize still needs to be determined. More progress has been made with respect to the *lpa2* type mutation, which has been shown to affect a gene encoding an inositol phosphate kinase. This gene is expressed in the maize embryo where phytic acid is known to accumulate (Shi *et al.* 2003).

A rice low phytic acid mutant, *lpa1*, has recently been isolated from a gamma-irradiation induced mutant population (Larson *et al* 2000). This mutant is similar to maize and barley *lpa1* mutants in that it has a low phytic acid content and high Pi. Seeds of homozygous *lpa1* plants have a 45% reduction in seed PA with a molar equivalent increase in Pi. This rice *lpa1* mutant has been shown to be the result of a single recessive mutation mapping to rice chromosome 2L (Larson *et al.* 2000). As the rice ortholog of MIPS has been mapped to chromosome 3, they concluded that the rice *lpa1* mutation is not in the rice MIPS gene. Using a recombinant inbred line (RIL) mapping population and published microsatellite markers, a fine genetic map of the rice *Lpa1* locus was constructed and additional markers were developed using publicly available rice DNA sequences, and the locus has been delimited to a 47-kb region containing at least eight putative open reading frames (ORFs) (Andaya and Tai 2005).

Unlike the normal genetic and environmental affects that result in quantitative variation in seed total P, *lpa* mutants show large effects on the partitioning of P into phytic acid P, Pi, and lower Ins polyPs. The *lpa* mutations of maize and barley are characterized by two chromatographic phenotypes that are circumscribed by two respective genetic complementation groups, designated *lpa1* and *lpa2*, in each species. A wide range of phytic acid reductions are observed among different *lpa1* and *lpa2* mutants. However, all *lpa1* mutants are qualitatively identical in that phytic acid P reductions are counterbalanced by molar-equivalent increases in Pi. The *lpa1* loci have been mapped to maize chromosome 1S (centromere-distal; Raboy *et al* 2000) and barley chromosome 2H (Larson *et al.* 1998). In contrast, the phytic acid reductions in *lpa2* mutants are matched by increases of Pi and lower Ins polyPs, in approximately equal parts. These *lpa2* loci have been mapped to maize chromosome 1S (centromere-proximal; Raboy *et al.* 2000 and barley chromosome 7H Larson *et al.* (1998).

Pro-Vitamin A Content (Golden Rice)

Golden rice is a variety of rice engineered to produce β-carotene (pro-vitamin A) to help combat vitamin A deficiency. Vitamin A deficiency causes symptoms ranging from night blindness to xerophthalmia and keratomalacia, leading to total blindness. Beta-carotenes have also been implicated in reducing the risk of cancer in human beings. Vitamin A deficiency also increases susceptibility to malaria and diarrhoeal disease, and reduces the bioavailability of micronutrients, including iron, important to the health of all family members (WHO, 2002). Genetically increasing the levels of vitamin A and its precursors in dietary staples is especially relevant for poor families living in isolated rural areas for whom vegetable sources are often in short supply. In Asia vitamin A deficiency is associated with poverty related predominant consumption of rice, which lacks pro-vitamin A in the edible part of the grain. Because no rice cultivar produces this pro-vitamin in the endosperm, an improvement in the endogenous quantity of β-carotenes will enhance the nutritional value of rice. This can be achieved by recombinant DNA technology by manipulating genes involved in the pathway of β-carotene biosynthesis rather than conventional breeding as shown in the (Fig. 3).

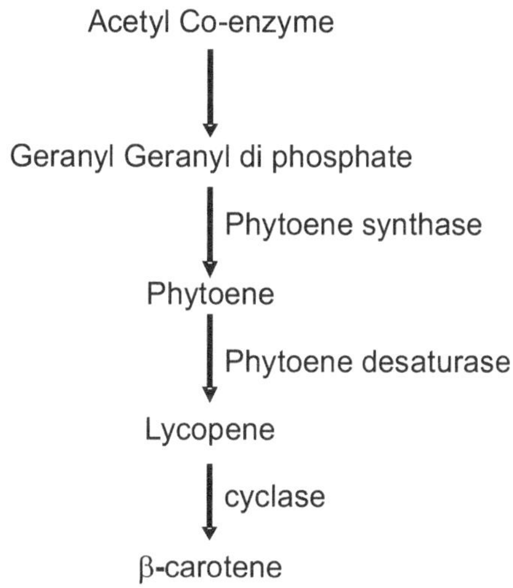

Fig. 3: Two key enzymes of β-carotene synthesis, namely Phytoene synthase and Phytoene desaturase used in improving β-carotene content in rice.

First Golden Rice was developed by Ye *et al.* (2000) using daffodil gene encoding phytoene synthase (*psy*) and bacterial phytoene desaturase (*crtI*) originating from *Erwinia uredovora* under the control of the endosperm specific glutelin and constitutive CaMV (cauliflower mosaic virus) 35S promoter, respectively. However in this pioneering work, a maximum level of 1.6 microgram per gram total carotenoids was achieved. The limited production of pro-vitamin A in Golden Rice was major hurdle to the success of this particular solution for vitamin A deficiency. To enhance β-carotene further, Golden Rice 2 was developed by introducing phytoene synthase isolated from maize along with *Erwinia uredovora* carotene desaturase *(crtI)* (Paine *et al.* 2005). This was based on the hypothesis that daffodil gene encoding phytoene synthase used to develop the golden rice was limiting in beta-carotene accumulation. Introduction of maize *psy* increased carotenoid accumulation substantially to the level of 37 microgram per gram. This was 23 times higher than the original Golden rice and also had a preferential accumulation of β-carotene. This demonstrated that modern molecular tools can improve nutritional quality of rice by incorporating genes from diverse sources.

Summary

Rice grain quality is complex trait involving nutritional quality, appearance quality, cooking and eating quality. Each one of these components consists of many attributes whose values are determined by not only physico-chemical properties but also by the history and cultural traditions of people in different communities consuming rice. Rice, unlike most other cereals, is consumed as a whole grain. Therefore, physical properties such as size, shape, uniformity, and general appearance are of utmost importance. Chemically starch is the most important constituents in the grain influencing milling and cooking quality. Recently molecular markers related to milling and cooking quality have been identified. Currently greater emphasis has been put on understanding the nutritional quality of rice. Nutritional quality is governed by the traits such as micronutrient density and bioavailability, anti-nutritional factors, protein quality with respect to essential amino-acid content and pro-vitamin A content. Current developments on molecular biology demonstrate that nutritional quality can be improved by genetic engineering approach.

References

Adams, C., Raboy, V., Krebs, N., Westcott, J., Lei, S. and Hambidge, M. (2001). The effect of low-phytic acid corn mutants on Zn absorption. *FASEB* J 73:80–85.

Andaya, C.B. and Tai, T.H. (2005). Fine mapping of the rice low phytic acid (*lpa 1*) locus. *Theor Appl Genet* 111: 489–495.

Arruda, P., Kemper, E.L., Papes, F., and Leite, A. (2000). Regulation of lysine catabolism in higher plants. *Trends Plant Sci.* 5: 324–330.

Azevedo, R.A., Damerval, C., Landry, J., Lea, P.J., Bellato, C.M., Meinhardt, L.W., Le Guilloux, M., Delhaye, S., Toro, A.A., Gaziola, S.A. and Berdejo, D.B.A. (2003). Regulation of maize lysine metabolism and endosperm protein synthesis by opaque and floury mutations. *Eur. J. Biochem.* 270: 4898–4908.

Beard, J.L., Burton, J.W. and Theil, E.C. (1996) Purified ferritin and soybean meal can be sources of iron for treating iron deficiency in rats. *J. Nutr.* 126: 154–160.

Bouis, H. (1996) Enrichment of food staples through plant breeding: a new strategy for fighting micronutrient malnutrition. *Nutr Rev* 54: 131–137.

Bourgis, F. Guyot, R. Gherbi, H. Tailliez, E. Amabile, I. Salse, J., Lorieux, M. Delseny, M. and Ghesquière, A. (2008). Characterization of the major fragance gene from an aromatic *japonica* rice and analysis of its diversity in Asian cultivated rice. *Theor. Appl. Genet.* 117: 353–368.

Bradbury, L.M.T., Fitgerald, T.L., Henry, R.J., Jin, Q. and Waters, D.L.E. (2005a). The gene for fragrance in rice. *Plant Biotech J.* 3: 363–370.

Bradbury. L.M.T., Henry, R.J., Jin, Q., Reinke, R.F. and Waters, D.L.E. (2005b). A perfect marker for fragrance genotyping in rice. *Molecular Breed.* 16: 279–283.

Chen, S., Wu, J., Yang, Y., Shi, W. and Xu, M. (2006). The *fgr* gene responsible for rice fragrance was restricted within 69 kb. *Plant Sci.* 171(4): 505-514.

Cho, H.J., Brotherton, J.E., Song, H.S. and Widholm, J.M. (2000). Increasing tryptophan synthesis in a forage legume *Astragalus sinicus* by expressing the tobacco feedback-insensitive anthranilate synthase (ASA2) gene. *Plant Physiol.* 123:1069–1076.

Cook, J.D., Reddy, M.B., Burri, J., Juillerat, M.A., Hurrell, R.F. (1997). The influence of different cereal grains on iron absorption from infant cereal foods. *Am. J. Clin. Nutr.* 65: 964–969.

Galili, G. (2002) New insights into the regulation and functional significance of lysine metabolism in plants. *Annu. Rev. Plant Biol.* 53: 27–43.

Garland, S., L. Lewin, L., Blakeney, A., Reinke, R. and Henry, R. (2000). PCR-based molecular markers for the fragrance gene in rice (*Oryza sativa.* L.) *Theor Appl Genet* 101: 364–371.

Guttieri, M., Bowen, D., Dorsch, J.A., Raboy, V., Souza, E. (2004). Identification and characterization of a low phytic acid wheat. *Crop Sci.* 44: 418–424.

Henry, Y., Sève, B., Colléaux, Y., Ganier, P., Saligaut, C. and Jégo, P. (1992) Interactive effects of dietary levels of tryptophan and protein on voluntary feed intake and growth performance in pigs, in relation to plasma free amino acids and hypothalamic serotonin. *J. Animal Sci.* 70: 1873–1887.

Higashi, T., Saitoh, S., Takita, T., Yamaguchi, M., Sunohara, Y., Yokogami, N., Ikeda, R., Tamura, Y., Oyamada, Z., Kowata, H., Inoue, M., Matsumoto, S. and Kataoka, T. (1999). Breeding of a new rice cultivar with low amylose contents, "Snow pearl". *Bull. Tohoku Natl. Agric. Exp. Stn.* 95: 1-12.

Hurrell, R.F., Juillerat, M.A., Reddy, M.B., Lynch, S.R., Dassenko, S.A., Cook, J.D. (1992). Soy protein, phytate, and iron absorption in humans. *Am J Clin Nutr.* 56: 573–578.

Juliano, B.O. (1985). Rice Chemistry and Technology, 2nd Edition, American Association of Cereal Chemists, Saint Paul, MN, USA.

Kemper, E.L., Cord-Neto, G., Capella, A.N., Goncalves-Butruile, M., Azevedo, R.A. and Arruda, P. (1998). Structure and regulation of the bifunctional enzyme lysine-oxoglutarate reductase-saccharopine dehydrogenase in maize. *Eur. J. Biochem.* 253: 720–729.

Lai, J. and Messing, J. (2002) Increasing maize seed methionine by mRNA stability. *The Plant Journal* 30: 395–402.

Larson, S.R., Rutger, J.N., Young, K.A. and Raboy, V. (2000) Isolation and genetic mapping of a non-lethal rice (*Oryza sativa* L.) low phytic acid 1 mutation. *Crop Sci.* 40: 1397-1405.

Larson, S.R., Young, K.A., Cook, A., Blake, T.K. and Raboy, V. (1998). Linkage mapping of two mutations that reduce phytic acid content of barley grain. *Theor. Appl. Genet.* 97: 141-146.

Lee, S.I., Kim, H.U., Lee, Y.H., Suh, S.C., Lim, Y.P., Lee, H.Y. and Kim, H.I. (2001). Constitutive and seed-specific expression of a maize lysine-feedback-insensitive dihydrodipicolinate synthase gene leads to increased free lysine levels in rice seeds. *Molecular Breeding.* 8(1): 75-84.

Li, Z.F., Wan, J.M., Xia, J.F., Zhai, H.Q. and Ikehashi, H. (2004). Identification of quantitative trait loci underlying milling quality of rice (Oryza sativa) grains. *Plant Breeding* 123: 229-234.

Liu, Q., Li, Q.F., Cai, X.L., Wang, H.M., Tang, S.Z., Yu, H.X., Wang, Z.Y. and Gu, M.H. (2006). Molecular Marker-Assisted Selection for Improved Cooking and Eating Quality of Two Elite Parents of Hybrid Rice. *Crop Sci.* 46: 2354-2360.

Lorieux, M., Petrov, M., Huang, N., Guiderdoni, E. and Ghesquière, A. (1996). Aroma in rice: genetic analysis of a quantitative trait. *Theor. Appl. Genet.* 93: 1145–1151.

Lott, J.N.A., Ockenden, I., Raboy, V. and Batten, G.D. (2000). Phytic acid and phosphorus in crops seeds and fruits: a global estimate. *Seed Sci Res* 10: 11–33.

Lucca, P., Hurrell, P. and Potrykus, I. (2001). Genetic engineering approaches to improve the bioavailability and the level of iron in the rice grains. *Theor. Appl. Genet.* 102: 392-397.

Massey, K.A., Blakeslee, C.H., Pitkow, H.P. (1998). A review of physiological and metabolic effects of essential amino acids. *Amino Acids.* 14: 271–300.

Mazur, B., Krebbers, E., and Tingey, S. (1999). Gene discovery and product development for grain quality traits. *Science.* 285: 372–375.

Nishi, A., Nakamura, Y., Tanaka, N. and Satoh, H. (2001). Biochemical and genetic analysis of the effects of *amylose-extender* mutation in rice endosperm. *Plant Physiol.* 127: 459-472.

Paine, J.A., Shipton, C.A., Chaggar, S., Howells, R.M., Kennedy, M.J., Vernon, G., Wright, S.Y., Hinchliffe, E., Adams, J.L., Silverstone, A.L. and Drake, R. (2005). Improving the nutritional value of golden rice through increased pro-vitamin A content. *Nature Biotech.* 23(4): 482-487.

Pilu, R., Landoni, M., Cassani, E., Doria, E. and Nielsen, E. (2005). The maize *lpa241* mutation causes a remarkable variability of expression and some pleiotropic effects. *Crop Sci.* 45: 2096–2105.

Pilu, R., Panzeri, D., Gavazzi, G., Rasmussen, S.K., Consonni, G. and Nielsen, E. (2003). Phenotypic, genetic and molecular characterization of a maize low phytic acid mutant (*lpa 241*). *Theor. Appl. Genet.* 107: 980–987.

Raboy, V. (1997). Accumulation and storage of phosphate and minerals. In: Larkins BA, Vasil IK (eds) Cellular and molecular biology of plant seed development, pp 441–477. Kluwer, Dordrecht.

Raboy, V. (2001). Seeds for a better future: 'low phytate' grains help to overcome malnutrition and reduce pollution. *Trends Plant Sci.* 6: 458–462.

Raboy, V., Gerbasi, P.F., Young, K.A., Stoneberg, S.D., Pickett, S.G., Bauman, A.T., Murthy, P.P.N., Sheridan, W.F. and Ertl, D.S. (2000). Origin and seed phenotype of maize low phytic acid 1–1 and low phytic acid 2–1. *Plant Physiol* 124: 355–368.

Rasmussen, S.K., Hatzak, F. (1998). Identification of two lowphytate barley (*Hordeum vulgare* L.) grain mutants by TLC and genetic analysis. *Hereditas.* 129: 107–112.

Sève, B. (1999) Physiological roles of tryptophan in pig nutrition. *Adv. Exp. Med. and Bio.* 467: 729–741.

Shi, J.R., Wang, H.Y., Hazebroek, J., Ertl, D.S. and Harp, T. (2005). The maize low-phytic acid 3 encodes a myo-inositol kinase that plays a role in phytic acid biosynthesis in developing seeds. *Plant* J. 42: 408–419.

Shi, J.R., Wang, H.Y., Wu, Y.S., Hazebroek, J., Meeley, R.B. and Ertl, D.S. (2003). The maize low-phytic acid mutant lpa2 is caused by mutation in an inositol phosphate kinase gene. *Plant Physiol.* 131: 507–515.

Shukla, S., Vantoai, T.T. and Pratt, R.C. (2004). Expression and nucleotide sequence of an INS(3)P1 synthase gene associated with low-phytate kernels in maize (*Zea mays* L.). *J. Agric. Food Chem.* 52: 465–4570.

Skikne, B., Fonz, D., Lynch, S.R. and Cook, J.D. (1997). Bovine ferritin iron bioavailability in man. *Eur. J. Clin. Invest.* 27: 228–233.

Suzuki, Y., Sano, Y. and Hirano H-Y. (2002). Isolation and characterization of a rice mutant insensitive to cool temperatures on amylose synthesis. *Euphytica.* 123: 95-100.

The World Health Report (2002). Reducing risks, promoting healthy life.

Theil, E.C., Burton, J.W. and Beard, J.L. (1997). A sustainable solution for dietary iron deficiency through plant biotechnology and breeding to increase seed ferritin control. *Eur. J. Clin. Nutr.* 51: 28-31.

Tozawa, Y., Morino, K., Komatsu, A., Yamada, T., Terakawa T. and Miyagawa, H. (2006). High-level tryptophan accumulation in seeds of transgenic rice and its limited effects on agronomic traits and seed metabolite profile. *J. Exp. Bot.* 57(12): 3069–3078.

Wakasa, K., Hasegawa, H., Nemoto, H., Matsuda, F., Miyazawa, H., Tozawa, Y., Morino, K., Komatsu, A., Yamada, T., Terakawa, T. and Miyagawa, H. (2006). High level tryptophan accumulation in seeds of transgenic rice and its limited effects on agronomic traits and seed metabolite profile. *J. Exp. Bot.* 57: 3069–3078.

Walker, D.R., Scaboo, A.M., Pantalone, V.R., Wilcox, J.R. and Boerma, H.R. (2006). Genetic mapping of loci associated with seed phytic acid content in CX1834-1-2 soybean. *Crop Sci.* 46: 390–397.

Wang, D.Y., Zhang, X.F., Zhu, Z.W., Chen, N., Min, J., Yao, Q., Yan, J.L. and Liao, X.Y. (2005). Correlation analysis of rice grain quality characteristics. *Acta Agron. Sin.* 31: 1086-1091.

Wilcox, J.R., Premachandra, G.S., Young, K.A. and Raboy, V. (2000). Isolation of high inorganic P, low-phytate soybean mutants. *Crop Sci.* 40: 1601-1605.

Xu, Z.J., Chen, W.F., Ma, D.R., Lu, Y.N., Zhou, S.Q. and Liu, L.X. (2004b). Correlations between rice grain shapes and main qualitative characteristics. *Acta Agron. Sin.* 30: 894-900.

Ye, X., Al-Babili, S., Klöti, A., Zhang, J., Lucca, P., Beyer, P. and Potrykus, I. (2000). Engineering the provitamin A (β-carotene) biosynthetic pathway into (carotenoid-free) rice endosperm. *Science.* 287: 303–305.

Zhang, X-H., Brotherton, J.E., Widholm, J.M. and Portis, Jr. A.R. (2001). Targeting a nuclear anthranilate synthase α-subunit gene to the tobacco plastid genome results in enhanced tryptophan biosynthesis. Return of a gene to its pre-endosymbiotic origin. *Plant Physiol.* 127: 131-141.

Zheng, T.Q., Xu, J.L., Li, Z.K., Zhai, H.Q. and Wan, J.M. (2007). Genomic regions associated with milling quality and grain shape identified in a set of random introgression lines of rice (*Oryza sativa* L.) *Plant Breeding.* 126: 158-163.

Zhu, X. and Galili, G. (2003). Increased lysine synthesis coupled with a knockout of its catabolism synergistically boosts lysine content and also transregulates the metabolism of other amino acids in *Arabidopsis* seeds. *The Plant Cell.* 15: 845–853.

Zhu, X. and Galili, G. (2004). Lysine Metabolism Is Concurrently Regulated by Synthesis and Catabolism in Both Reproductive and Vegetative Tissues. *Plant Physiol.* 135: 129-136.

Zhu, X., Tang, G, Granier, F., Bouchez, D. and Galili, G. (2001). A T-DNA insertion knockout of the bifunctional Lysine-Ketoglutarate Reductase/Saccharopine Dehydrogenase gene elevates lysine levels in arabidopsis seeds. *Plant Physiol.* 126: 1539-1545.

❑❑❑

Chapter 13

Processing and Nutritional Quality of Maize

Introduction

Maize (*Zea mays*, or corn as it is known in some countries) is the principal cereal crop in tropical and subtropical regions throughout the world. The increasing use of maize as a staple food reflected the much higher yields per hectare, compared with wheat, rye and barley. Because maize was cheap, it became the dominant food and main source of dietary energy and protein for poor people, particularly those in rural and underprivileged segments of society. Maize is considered to be a native of the Americans. There are different views about the introduction of maize in India. One view is that maize was introduced into India by the Portuguese during the seventh century. Other view is that maize was introduced in Pre-Colombian era through the Atlantic Arabian trade route. This is supported by the more recent variability that has been observed in the maize collections made from the north eastern Himalayan region. However, the historic records regarding the cultivation of maize in India date back only to the Maratha Empire.

Maize in India is an important cereal, and both its area and production have steadily increased during the past two decades. In terms of world acreage, India stands only next to the USA, Brazil, China and Mexico With regards to production India ranks eleventh after USA, China, Brazil, Argentina, Mexico, France, South Africa, the USSR, Romania, Yugoslavia. In terms of area and production, it ranks only next to rice, wheat and jowar. The states of Uttar Pradesh, Bihar, Madhya Pradesh, Rajasthan and Punjab account for over 75 per cent of the area and production of this cereal in the country. Each of the districts of Bahraich, Gonda and Bulandshahr in Uttar Pradesh; Monghyr, Saran and Darbhanga in Bihar; Udaipur and Bhilwara in Rajasthan; and Panchmahal in Gujarat put annually over one lakh of hectares in maize. These nine districts account for a quarter of a national area and production of maize in India. Being an important cereal, over 85 per cent of its production in the country is consumed directly as food in various forms; the chapatis are the commonest preparation, whereas roasted ears, popcorn and porridge are other important forms in which maize is consumed. The use of maize in animal feed, particularly for poultry, and in the starch industry is increasing. Green maize plants also furnish a very succulent fodder during spring and monsoon, particularly in northern India.

Virtually every part of maize has economic value. Grains are used as human feeds, fermented to produce a vide range of food and beverages, fed to livestock and industrial uses in the production of starch, oil, sugar, cellulose and ethanol. Leaves, stalks and tassels can be fed to livestock either as green manure in the form of fodder or silage and dried as stover. Silage is made by fermentation of chopped green cornstalks and leaves. Even the roots can be used for mulching incorporated into soil to improve the physical structure. The grain also has many industrial uses such as corn syrup, ethanol and transformation into plastics and fabrics. Maize ethanol is being used as an additive in gasoline for motor fuels to increase the octane number with lower pollutants, and reduce petroleum use. Utilization of maize as biofuel has generated hot debate across the world. Details will be discussed in the section of biofuels.

Sweetcorn with high content of sugars and low content of starch is served like vegetable. Kernels of certain varieties explode when heated, forming fluffy pieces called popcorn are eaten as a snack. Maize can also be prepared as hominy, in which the kernels are bleached with lye; or grits, which are coarsely ground corn. These are commonly eaten in the Southern parts of Unite States and earlier used by the native Americans. Another common food made from maize is corn flakes. The floury meal of maize (masa) is used to make corn bread and Mexican tortillas. Some forms of the plant are occasionally grown for ornamental use in the garden. For this purpose, variegated and coloured leaf forms as well as those with colourful ears are used. Additionally, size-superlative varieties, having reached 31 ft (9.4m) tall, or with ears 24 inches (60cm) long, have been popular for at least a century. Maize is also used as fish bait called "dough balls" particularly popular in Europe. Stigmas from female corn flowers, known popularly as corn silk, are sold as herbal supplements. An unusual use for maize is to create a *Maize Maze* as a tourist attraction. This is a maze cut into a field of maize. Corncobs can be hollowed out and treated to make inexpensive smoking pipes.

Maize Classes

Numerous cycle of selection by the farmers and breeders have led to the proliferation of major grain types, which differ widely in grain shape, size, colour, hardness, taste, ease of processing and storage quality. International market grain types are characterized in terms of grain hardness and colour. Approximately 85% of global maize crop consists of yellow grained material, 10-12% white grained and the rest consists of red, blue, purple or black grained kernels. Based on kernel characteristics, maize can be further classified into following classes:

Flint Corn: Flint corn has thick, hard and vitreous endosperm, which surrounds a small centre of soft endosperm. Flint kernels are smooth and rounded with no denting.

Flint corn is excellent for breakfast cereals and some snack foods. Flint corn is grown extensively in Argentina and to a lesser extent in other South American countries and Europe.

Hard Endosperm/ Food Grade Corn: Food grade corn hybrids are typically high yielding corn hybrids with special quality traits that are desirable for certain uses such as dry milling or alkaline cooking. Used for snack foods, breakfast cereals, tortillas, etc.

High Amylose Corn: High amylose corn produces a larger volume of amylose starch. It is used primarily for special starch needs, such as quick-setting confectionery gums (jelly beans), and in certain glues.

High Oil Corn: As the name implies, high oil corn contains as much as 100% increased oil concentration. Used for livestock feed because of its high energy and protein content.

High Starch Corn: Although limited hybrids are currently available, high starch corn would be a boon for industrial applications such as ethanol and biodegradable plastics.

Improved High-Lysine Corn: Ideal for livestock because of its excellent nutritional value. Lysine, an essential amino acid, is low in normal yellow corn.

Waxy Corn: Waxy corn is high in amylopectin and low in amylose starch. The starch in this corn is ideal for a variety of special needs. It is used for thickening foods, cherry pie filling, for example, and for bonding in paper and other applications.

White Corn: Typically used for food, examples include tortillas, corn chips and other snack foods. Nebraska is one of the leading white corn producing states in USA.

Popcorn maize: Popcorn maize is an early variation of flint maize which has been selected over the years for the ability of the kernels to pop during rapid cooking. The individual kernels pop because of the build-up of internal pressure during rapid heating.

Sweet Corn: Sweet corn is sweet and tender with a creamy texture, low starch content, and pleasant "sweet corn"- like aroma. Other traits related to the appearance of whole ears seem to influence the likelihood that consumers will buy them for fresh consumption. In general, consumers appear to prefer dark green husks and flag leaves and well-filled ears with straight rows.

Grain Quality

Grain quality means different things to different people. Grain quality largely depends on the grain type and its end use. It includes a range of properties that can be defined in terms of physical (moisture content, test weight, kernel size, total damaged kernels, heat damage, broken kernels, stress cracking, breakage susceptibility), sanitary (fungi and mycotoxin count, insects and insect fragments, rodent excrements, foreign material, toxic seeds, pesticide residue, odor, dust), and intrinsic (milling yield, oil content, protein content, hardness, density, starch content, feed value, viability, storability) quality characteristics. The quality properties of a grain are affected by its genetic traits, the growing period, timing of harvest, grain harvesting and handling equipment, drying system, storage management practices, and transportation procedures. Often premiums are offered by the processors for better quality grains. Thus, it is important to understand how these quality characteristics are affected so that producers can maintain and deliver these grain quality attributes, and receive deserved premiums.

Maize kernels have both translucent and opaque endosperms. The translucent endosperm is tightly compact, with few or no air spaces. The starch granules, polygonal in shape are held together by a matrix protein. In opaque endosperm, the starch granules are spherical and covered with matrix protein that does not contain protein bodies. Many air spaces are to be expected from its opacity. Chemical analysis of the separated opaque and translucent endosperm showed that that the two contained equal amount of protein but that the protein types varied. White and yellow food corn hybrids are dent corn with specific endosperm (starch) traits, often referred to as hard endosperm. White food corn has a white kernel color and whiter starch. Both yellow and white food corns contain high amounts of vitreous (having a "glasslike" texture) endosperm relative to the amount of floury endosperm. Specific quality traits for alkaline-processed foods include ease and completeness of pericarp (seed coat) removal, hardness of endosperm, kernel crown configuration, uniform sized kernels, high specific gravity, kernel red streak resistance, and a white cob. White corn typically has higher test weight than commodity corn. White and yellow food corns are typically contracted and sold to dry-mill processors and used in alkaline cooking processes for making masa, tortilla chips, snack foods, and grits. White food grade has a limited wet milling use for food grade starch and paper uses.

Maize kernels with high test weight, maximum hard endosperm, minimum stress cracks, no mold, and large uniform kernel are desired by the processors. Corn-based snack foods, such as corn and tortilla chips, represent one of the fastest growing segments of salty snack food market. White corn products are in particularly high demand. Before grinding corn into *masa* needed for chip production, each kernel's hull must be removed during the cooling, soaking, and washing process. Some corn hybrids have thick hulls, which are more difficult to remove. Thus, hybrids with thin hulls are preferred by processors. Stress cracked, broken and damaged kernels take up moisture more rapidly and disintegrate during the cooking, soaking, and washing process. Typical snack food corn specifications may include less than 20 parts per billion of aflatoxin, and maximum values of 20% moisture at harvest, 10% defects, 30% stress cracks, and 5% dead germ.

Sweet Corn

The endosperm is the primary storage tissue of the seed. Its physical and chemical makeup effects not only the crop's eating quality but also the kernel's ability to function as a seed. The types of sugars made by the plant and stored in the endosperm and the rate at which they are converted to other compounds, especially starch, continue to be manipulated through genetics and breeding. So also are the factors that control kernel texture and, to a lesser extent, aroma. Breeding efforts are consistent with the fact that flavor and texture drive consumer appeal of fresh and whole-kernel processed corn and that consumers generally prefer corn that is sweet and tender with a creamy texture, low starch content, and pleasant "sweet corn"-like aroma. Over the years, varieties with different types and levels of sugars at harvest maturity, rate of sugar conversion to starch and moisture loss after maturity, and pericarp characteristics (e.g., tenderness, color) have been available. Compared to normal corn, standard sugary or "su" varieties accumulate more sugars and water-soluble polysaccharides (phytoglycogen) and less starch.

However, su-type varieties rapidly convert sugars to starch after harvest maturity, leaving narrow harvest and market windows. As a result, su-type varieties are rarely grown in the U.S. Instead, sugary enhanced (se)- and shrunken2 (sh2)-type varieties dominate since they are more sweet and tender and have longer harvest and market windows than su-type varieties. Se-type varieties tend to bear the creamy texture provided by relatively high levels of phytoglycogen which are lacking in most sh2-type varieties. As a group, se- and sh2-type varieties have overlapping ranges of total endosperm sugar level at harvest maturity although the varieties with the absolutely highest sugar levels tend to be of the sh2 type. Rates of kernel sugar conversion to starch and dry-down also may be similar among se- and sh2-type varieties as a group but, as in kernel sugar levels, varieties with the slowest sugar-to-starch conversion and kernel dry-down rates tend to be of the sh2-type. A variety's genetic makeup strongly but not completely determines its potential eating quality. Although not as clearly defined as various genetic influences, environmental factors and management are also known to influence crop quality. For example, high temperature stress during pollination and kernel development can reduce the number and eating quality of kernels, partly through undesirable effects on texture. Likewise, low and high moisture stress are thought to reduce crop quality. Excessive nitrogen fertilization can also reduce the number and quality of sweet corn kernels.

Corn as Breakfast Cereal

Corn dry milling results in a range of degermed corn products from flaking grits to corn flour. These products may be used in breakfast cereals, snack foods, prepared mixes, batter and breading mixes, and low calories/high fiber foods. For many years, the popular ready to eat cereal has been corn flakes. Large grits produced during corn dry milling are used for making flakes. Large grits retain their identity through the process, each producing a single flake. Corn grits are pressure cooked with a solution containing sugar, malt and salt. After cooking, lumps are broken and partially dried to 20% moisture level. At this stage moisture is not uniform with drier outside and moist inside. Therefore, it gives temper time to allow the moisture to equilibrate. Tempering is followed by flaking. Grits are rolled over with high pressure and subsequently toasted for 50 seconds at 300 °C. Toasting not only dehydrates but also browns and blisters the product. After cooling, flakes may be sprayed with a solution of vitamins and minerals. Corn is also used as break fast cereal that requires cooking. The product is called "grits" or occasionally "hominy grits". Grits are of small pieces of endosperm and made from white corn.

Popcorn is the snack food used in many countries. It has delicate corn flavour and delightful texture. The quality of popped corn is determined by the volume, shape, tenderness and flavour. During popping the volume of the corn generally increases to bout 30 times. Tenderness is generally correlated with popped volume. Popping is caused by the vaporization of moisture in the kernel during heating. Moreover, pericarp of the popcorn acts as a pressure vessel, allowing the water in the kernel superheated.

At a certain temperature, the pressure becomes so large that the pericarp ruptures, allowing the endosperm to expand. In addition the degree of expansion in popcorn has been shown to be correlated to the amount of translucent endosperm. This endosperm is tightly packed and contains no air spaces, whereas the opaque endosperm has many air spaces.

During popping, pericarp appears to undergo no change other than the obvious fracture. At the outer edge of the kernel moisture loss is excessive. Deeper in the endosperm the expanded structure becomes more evident. Opaque endosperm acts much differently than does the translucent endosperm during popping. The starch granules in the opaque endosperm appear to stay intact and do not expand during popping. The moisture vaporizes into void surrounding the granules, creating large voids. In the translucent endosperm, the moisture vaporizes into the hila of the starch granules and expands the granules. In addition to popcorn, corn nuts and corn curls are corn based snacks. Corn nuts are made from corn types with large opaque and white kernels. Corn curls are made from corn flour with a cooker extruder.

Masa - *Masa* is the Mexican word for "dough". It refers to the corn dough used to make tortillas, tamales, as well as other traditional Mexican dishes. For production of *masa,* corn is first heated in water containing calcium hydroxide (~0.5 to 1.0 %). The temperature is brought to around 90 °C for < 1 hr and then cooled to room temperature. This produces *nixtamal* (heat treated alkaline water steeped corn) which is washed to remove hulls and excess calcium hydroxide and then ground. The resulting wet-ground product is called *masa. Masa* is the basic material for the production of large numbers of products. It is pressed either by hand or mechanically into round flat, pancake shape and baked to form tortilla. Masa can also be extruded and fried into corn chips. Other types of corn chips and taco shells are made by frying a tempered tortilla in deep fat.

Corn quality traits for masa

Colour	Bright white or bright yellow
Texture	Harder texture is usually preferred. It provides more flexibility in cooking/ steep times and temperatures
Pericarp removal	Easy pericarp removal is needed. Pericarp interferes with sheeting. None or very little pericarp is used in making chips while some for tortillas.

Corn Milling Quality

Milling is the process of grinding grains to make them palatable. Generally it involves separation of bran, germ and endosperm parts. Germ is removed to reduce oil content which makes the product rancid. Depending upon the requirements, dry and wet millings are used to mill maize grains. Details are given below.

Dry Milling

This process involves physical breaking of maize grains into various fractions. It is the process to separate anatomical parts of the grains as cleanly as possible. Its objective is to make palatable and thus desirable food products. In addition to that, milling involves production of different particle sizes depending upon the crop. For example fine four is produced from wheat and rye, endosperm should be in large piece in case of barley and rice while large grit is desirable from corn.

The main objectives of dry milling are:

- To obtain maximum yield of grits with the least contamination with fat and black specks of the tip cap, to recover as much as possible the endosperm as meal,
- To produce the maximum amount of flour and
- To obtain the maximum amount of oil.

The man steps involved in dry milling are: cleaning, tempering or conditioning and roller milling. Cleaning is the removal of foreign materials such as stones, metals, sticks etc using magnetic separator, aspirator and scourers. Tampering consists of adding water to dry grains and resting them for a period of time. Tampering is done to toughen the bran and thus make it resist to be broken into small pieces and to soften the endosperm to make it easier to grind. Conditioning implies the use of wheat in conjuction with water to mellow the endosperm. In milling hull and germ are removed with out reducing the endosperm into small particles. After the germ and hull are removed, the endosperm is reduced to grits of the desired size by roller milling. The dry corn is basically a break system, as flour is not the desired product. Hulls are sold as animal feed, and the germ is processed to recover the valuable oil. Products from dry milling are maize meal, flour and maize grits. Other by - products are fractions, characterized by various dimensions and sizes which affect composition and utilization. For instance grit fractions ranging from 0.6- 1.2 mm (mesh) are used domestically for many food preparations. In the United

State of America whole hominy (with the whole endosperm) are popular and may be cooked in milk or water and served as a vegetable. Hominy grits (the endosperm broken) are used for hot cereals and for the preparation of ready -to-eat corn cereals. Also the brewing industries use brewers' grits in beer production. Maize meal with fraction of 0.6 -0.2mm is used in meal mixes, making of maize bread and maize muffins. Of recent, it is used for some extruded maize snack products. Maize flour with fraction size of less than 0.2mm is a valuable ingredient in pancake mixtures, baby foods, cookies and biscuits. It is used for ice cream cones, ready-to-eat cereals, battler bread mixtures and for other baking products. Maize flour can be combined with soy-milk to produce high nutrient products such as corn-soy-milk (CSM) and corn-soy-blend (CSB).

Wet Milling

Dry milling is primarily concerned with separation of the anatomical parts of the grain while wet milling goes one step further and separates some of those parts into their chemical constituents. Thus the main objective of the wet milling of maize is to obtain starch, oil and other components, which are useful in other areas such as livestock feeds. Corn starch can be converted to sweeteners such as dextrose and high fructose corn syrup. Wet milling is a maceration process that brings about both physical and chemical changes in the constituents of the endosperm. Basically, it involves the separation of the starch and the protein into separate components from the endosperm cells. In order to facilitate this process, the maize grains must be dried to the desired level to enable the germ to float during the separation process. Thus the grains for wet milling should be dried at a temperature not exceeding 54°C (130°F). Maize dried at higher temperatures brings about changes in the protein, preventing the grains from swelling accordingly during steeping and it also tends to hold to the starch. On the other hand, if the grains are dried at a lower temperature, the germ becomes rubbery and tends to retain high oil content Kent, (1975). Thus the crucial point here is that the grains must be dried at the correct temperature.

Steeping of the maize grains in warm water is necessary because it softens them and thus assists in separating the hull, germ and fiber. Steeping involves submersing the corn in water containing 0.1-0.2% sulfur dioxide at 48-52 °C for 30 to 40 hr. Sulfur dioxide is used for two reasons. First, it aids in stopping the growth of putrefactive organism and second it reacts with disulfide bonds in the matrix proteins of the corn.

This reduces the molecular weight of proteins and makes them more hydrophilic and soluble. As a result release of starch from protein matrix is much easier and with higher yield. Corn that has been dried at excessively higher temperature gives lower amount of soluble protein when steeped with bisulfite than does unheated corn. After steeping the steeped water contains about 60 g of solubles per litre. Steeped water is further concentrated to 55% solids and mixed with the bran and spent grains used for animal feeding. The solids from the steep can also be used in the growth medium for production of desired microorganisms. In addition solids from steep contain higher content of protein, lactic acid, ash content and phytic acid and B vitamins.

After steeping, the softened grain is coarsely ground in water by an attrition mill to free the germ without breaking. Germ is separated from the remainder of the kernel with a liquid cyclone separator, *hydroclone*. The separation is based on density as germ has lower density because of higher oil content. The recovered germ is washed free of adhering starch, dried and used for producing oil. The remaining material is then sieved and the hulls and endosperm chunk are ground. The objective is to free starch, protein and fibre from each other. Fiber is removed and used for animal feed. The remaining stream contains the starch and proteins in water. Starch and proteins are separated from each other by centrifuge or hydrocyclones. Less dense gluten containing 60-70% proteins is then centrifuged to remove water , dried and used for animal feed. Starch is further purified using centrifuges or hydrocyclones. Starch at this stage contains about 0.3% proteins and is ready for chemical modification to syrup or dried. The general quality goals for the four wet milling products are: less than 0.3% protein in the starch, 70% protein in the gluten, minimum bound starch in the fiber, and more than 45% oil in the germ. The wet miller prefers a soft endosperm because it has a shorter steeping time, which increases throughput, and its starch fraction separates more easily from the protein. High stress-cracked corn increases the amount of broken kernels, and reduces oil and starch recovery. Additionally, in order to optimize starch yield, corn kernel temperatures must be kept below 140°F during drying.

Grain Hardness

Grain hardness affects various aspects of the growth and processing of cereal grain from resistance to fungal infection to cooking quality. Harder kernels produce higher percentage of prime products such as large grits suitable for flaking grits. Hardness is defined as the amount of vitreous endosperm in the kernel relative to the amount of floury endosperm present in the maize kernel. It is primarily genotypic trait and has lesser effect of the environment. High soil nitrogen and drought like growing conditions increase kernel harness to some extent. Hard kernel takes longer time to grind and results into coarser particles than soft kernels. Hardness can be measured either by visual inspection or mechanical device by cutting transversely or longitudinally. There are three main classes of maize kernels based on hardness.

Floury Maize

Floury maize is characterized by lack of hard or vitreous kernels and is grown primarily in parts of South America. During drying down, the endosperm uniformly shrinks and results into no denting on the kernel.

Flint Maize

It has a thick, hard vitreous kernel with small amount of soft or floury endosperm. It is grown extensively in Argentina and to a lesser extent in South American countries.

Dent Maize

It is a cross between floury and flint maize and is most widely grown. Dent corn gets its name from a characteristics dent in the crown area of the kernel which occurs during drying and maturation. The dent results from collapse of the protein matrix structure in the floury endosperm region, which is generally in the centre of the kernel.

Genetic studies demonstrate that prolamins play an important role in determining grain hardness with the γ-prolamins being the most important. Maize prolamins have been variously classified as 22 kDa α-, the 19 kDa β-, 27 kDa γ- and 10 kDa δ-zeins based on mass and solubility. Experiments have demonstrated that soft endosperm of maize are richer in γ-zeins compared with vitreous endosperm and that vitreous endosperm contain upto twice the amount of α-zeins than was present in soft endosperms. The role of prolamins in determining grain strength has been further verified by the study of *opaque-2* mutants. *Opaque-2* mutants are soft and the concentrations of prolamins are significantly reduced in the endosperm relative to those present in normal endosperm. The mutation affects primarily the 19 kDa proteins and the 22 kDa protein is synthesized only in trace amount. As a result *Opaque 2* seeds have about 50% less zein than to normal genotypes. The transcription of α-zeins genes is reduced significantly in o2 mutants, with expression of those encoding the 22 kDa proteins being blocked.

It would appear that these prolamins help to shape the protein bodies and form disulphide bonds within themselves or with other proteins. The γ-prolamins form the cement while the α-prolamins are the bricks. Both prolamins are present in greater proportions in hard grains and in the vitreous portion of hard grains. Genetic and environmental effects on the amounts of the different prolamins and on their distributions within the protein body and in different parts of the endosperm also determine grain hardness. Grains that will be hard appear to deposit prolamins and antifungal proteins earlier and in greater amount than do soft grains. The cell wall composition is also different between the two types of grains while there is a higher proportions of amylose in the starch of hard than in that of soft grains. Most of the differences that exist between hard and soft grains also exist between the outer and inner portion of the grain. It is postulated that there might be a master gene controlling the onset of strength in grains by simultaneously altering the levels of various apparently unrelated biochemical events. It is also suggested that solute availability may play an important role in the regulation of expression of genes for hardness-related proteins.

Nutritional Quality

Quality Protein Maize (QPM)

In normal maize genotypes, zeins usually account for 50 to 70% of the endosperm proteins and are characterized by a high content of glutamine, leucine and proline. Since zeins are essentially devoid of lysine and tryptophan, they dilute the contribution of these essential amino acids from the other types of endosperm proteins, which are collectively called 'non-zeins'. The non-zein fraction contains enzymes, structural polypeptides and membrane-associated proteins. In normal maize, proportions of various endosperm storage protein fractions, on an average, are: albumins (3%), globulins (3%), zeins (60%) and glutelins (34%). Significantly, all fractions other than zeins are balanced in amino acid content and are quite rich in lysine and tryptophan. Suppression of lysine-deficient zein fraction without drastically altering the contribution of other fractions could be, thus, seen as a feasible approach to bring about improvements in the amino acid balance in maize grain.

In the early 1960s, special interest was manifested in the search for gene mutants that could provide better quality protein in the maize endosperm. In 1963, researchers at Purdue University, USA, discovered that a mutation, designated *opaque-2* (*o2*), made grain proteins in the endosperm nearly twice as nutritious as those found in normal maize. In fact, *o2* mutation was first described by Jones and Singleton in the early 1920s, but the nutritional significance of the mutation was first discovered by Mertz and coworkers. This was soon followed by the discovery of another mutation, *floury-2* (*fl2*) that also has the ability to alter endosperm nutritional quality. However, the euphoria over the discovery of *o2* and its direct utilization in breeding programmes was soon tempered with the realization that pleiotropic effects of this mutation, namely a soft endosperm that results in damaged kernels, an increased susceptibility to pests and fungal diseases, inferior food processing and generally reduced yields, were not easily overcome. Search for better genes has continued and additional mutants are known that improves protein quality of corn endosperm (*opaque-7, opaque-6, floury-3,* and *opaque-11*) but no other mutant has been found to offer any additional advantages over *opaque-2* gene (Vasal 2001). Although these mutant genes vary in their mechanism, most result in decrease synthesis of lysine poor prolamin fraction and compensatory increase in other more lysine rich fractions, which may include elevated levels of free lysine.

Different approaches have been used to improve the agronomic quality of opaque-2 genotypes. Breeders at the International Maize and Wheat Improvement Center in Mexico have continued working for improvement of protein quality in maize, converting *opaque-2* maize into varieties that have high nutritional quality, high yields, appearance of normal maize, greater hardness than *opaque-2*, equal or superior pest and disease resistance. This enhanced *opaque-2* is called quality protein maize (QPM). QPM also contains nearly twice the lysine and tryptophan, higher amounts of histidine, arginine, aspartic acid, and glycine, and lower amount of glutamic acid, alanine and leucine. Amino acid content of cereals, *opaque-2* and quality protein maize grain is shown in (Table 1). Furthermore, some QPM hybrids contain as much as 13.5% protein (CIMMYT 2001).

Table 1: Comparison of the contents of essential amino acids (g/100g protein or g/16g N) in cereal grains and flours with the FAO recommended levels of children and adults (CIMMYT 2001).

	Wheat Grain	Flour	Barley Grain	Oat Groat	Rye Grain	Rice Milled	Maize Cornflour	FAO Recommendation Children	Adults
Histidine	2.3	2.2	2.3	2.2	2.2	2.4	2.7	2.6	1.6
Isoleucine	3.7	3.6	3.7	3.9	3.5	3.8	3.6	4.6	1.3
Leucine	6.8	6.7	7.0	7.4	6.2	8.2	12.5	9.3	1.9
Lysine	2.8	2.2	3.5	4.2	3.4	3.7	2.7	6.6	1.6
Cysteine	2.3	2.5	2.3	1.6	1.9	1.6	1.6	4.2	1.7
Methionine	1.2	1.3	1.7	2.5	1.4	2.1	1.9		
Phenylalanine	4.7	4.8	5.2	5.3	4.5	4.8	5.0	7.2	1.9
Tyrosine	1.7	1.5	2.9	3.1	1.9	2.5	3.8		
Threonine	2.9	2.6	3.6	3.3	3.3	3.4	3.7	4.3	0.9
Tryptophan	1.1	1.1	1.9	ND	1.1	1.3	0.6	1.7	0.5
Valine	4.4	4.1	4.9	5.3	4.8	5.8	4.8	5.5	1.3

Since the discovery of the *opaque-2* gene, various studies on protein quality were conducted in rats, children and adults. Both, metabolic and growth studies in children have demonstrated the superiority of QPM over normal maize (Graham *et al.* 1989, Bressani 1991). Serna-Saldivar *et al.* (1992b) studied the bone and plasma composition of rats fed raw grain and tortillas from QPM and regular maize. Femurs of the rats fed with tortillas weighed more, were thicker, longer, denser and stronger ($P<0.05$) than those fed with raw grain. Among rats fed with tortillas, QPM produced denser, stronger, longer and thicker bones with more ash and Ca. This correlates to the positive effect of L-lysine on Ca intestinal absorption and decrease in urine excretion reported by Civitelli *et al.* (1992). Other studies exhibited that protein efficiency ratio (PER) of the raw QPM (2.25) was the highest followed by QPM tortilla (1.87), while values for raw food grade maize (1.37) and tortilla (1.37) were significantly lower ($P<0.05$). the above data clearly demonstrated that QPM can improve the nutritional value of maize. However, traditional method of breeding QPM maize is very difficult. Recently molecular studies have demonstrated the utility of molecular markers in QPM breeding. The details are provided in (Table 2).

Table 2: Protein composition and lysine content of selected opaque mutants of maize (From Shewry, 2007).

Line	Total protein (%)	Zein protein (%)	Non-Zein protein (%)	Non-protein N %	Lys %
W64A+	12.1	8.2	2.1	0.6	1.5
W64AO1	12.8	8.5	2.1	0.7	1.7
W64AO2	10.1	2.9	3.6	2.3	3.8
W64AO5	11.5	6.4	2.0	1.5	2.1
W64Ash4O9	12.2	7.3	2.8	1.3	2.0
W64AO11	12.0	6.4	3.4	1.4	2.8
W64AMc	11.7	7.2	2.7	1.4	2.1
W64ADeB30	12.0	4.5	3.8	1.7	2.9
W64Afl2	11.8	5.9	3.7	0.9	2.8
W69Mfl1	12.7	7.5	NA	NA	4.1
W69Mfl3	11.9	4.9	NA	NA	3.9
BA hybrid	13.5	NA	NA	NA	2.0

Molecular Marker Assisted Selection for Lysine Improvement

The recessive character of the *o2* allele raises an additional challenge to the breeding strategy of backcross conversion as discussed above. Forward molecular marker selection for the *o2* allele has been suggested Dreher, (2003) and implemented in attempt to achieve a faster conversion Babu *et al.* (2005), and would appear to be an appropriate use of marker assisted selection (MAS). However, without concurrent selection on aa-modifiers, protein quality can drop considerably even within *o2o2* backgrounds. In the Babu *et al.* (2005) study, tryptophan content as a percentage of total protein decreased from 1.05% in the QPM donor line to 0.78 - 0.85% in the BC2F2 families. The reduction in protein quality when not selecting for aa-modifiers has also been observed within the CIMMYT QPM breeding program, and is not surprising considering the wide variation of lysine levels in *o2o2* genotypes of different genetic backgrounds Moro *et al.* (1996). In order to bring to bear the full effectiveness of QPM genotype MAS (through reduced breeding time, selection out of main environment etc), a suite of effective markers linked to modifying loci of both endosperm hardness and amino acid levels needs to be identified. In addition, it is probable that additional traits will need to be coupled with selection for QPM to be cost competitive with traditional methods. If the goal of multiple trait selection can be met then total MAS costs will be much more inline with traditional screening methods.

Accurate, rapid and cost effective measurement of grain amino acid content is a critical component of QPM breeding. While protein quality assessments via a ninhydrin (Zarkadas *et al.* 2000) and enzyme linked immunosorbent assay (Moro *et al.* 1995) have been used, the reliability of the assays can be low. Colorimetric method of Hernandez and Bates (1969) can be used but the reliability of this procedure is unduly dependent on the type of glacial acetic acid used in the process, which can vary greatly from one source to another. The colorimetric procedure, therefore, has limited use in countries with unreliable sources of this reagent. Research in substitution of glyoxylic acid for acetic acid in tryptophan quantification is ongoing at CIMMYT and any future assay that can lower the sample cost and reduce measurement error will be immediately useful in QPM breeding programs. The recessive character of the *o2* allele raises an additional challenge to field implementation of QPM. If QPM is pollinated by pollen from normal maize, the resulting grain will lose the higher protein quality. Several years of field tests have shown that contamination of QPM open pollinated varieties is less than originally feared (Cordova, 2000). However, to ensure that the higher nutritional benefits of QPM are maintained, additional training of collaborators in partner countries with a focus on seed production techniques to maintain variety and line purity must also accompany QPM dissemination.

Transgenic Approach

A future strategy to mitigate the contamination effects on the recessive character of the *o2* trait may lie with a transgenic approach. In separate evaluations, RNA interference (RNAi) was employed in gene silencing studies to reduce 22-kDa (Segal *et al.*, 2003) and 19-kDa alphazeins (Huang *et al.*, 2004; Huang *et al.*, 2005) with antisense transformation constructs. In these studies, lysine content was increased in the transgenic lines by 15 – 20%, although well below the 100% increase that is often observed in o2 genotypes. In a more recent however, an improved construct using double-strand RNA (dsRNA) was employed to increase the efficiency of gene suppression of both 19-kD and 22-kD alpha-zein gene families (Huang *et al.*, 2006). As a percentage of total protein, up to 5.62% of lysine and 1.22% of tryptophan were achieved in transgenic lines compared with 2.83% of lysine and 0.69% of tryptophan in wild-type, which is above what is typically achieved with QPM lines and near or above FAO guideline requirements of lysine and tryptophan content (5.8 and 1.1% respectively) for children (Table 2). In these studies, the reduction of alpha-zein protein synthesis induced an opaque endosperm phenotype, which (as with the opaque phenotype caused by the *o2* allele) would need to be modified to a hard endosperm phenotype utilizing modifying loci. The dominant nature of the gene silencing transgene would be advantageous over the recessive *o2* allele when deployed in farmers' fields, as the high lysine/tryptophan levels would be essentially maintained even under relatively high amounts of pollen contamination from conventional maize. However, with this transgenic approach to improving protein quality, constraints on deployment of these materials in some countries, due to legislative barriers and lack of consumer acceptance would have to be confronted.

Mineral and Vitamin Content of Maize

Enhancing mineral and vitamin content of cereals has become primary goal in many breeding programmes. For healthy human beings, there is minimal requirement of certain micronutrients and vitamins. The main source of micronutrients for most low income groups are foods of plant origin. A recent UN report indicate that more than half of human population in the world suffers from micronutrient deficiency: most seriously affected are children and women as they have comparatively larger requirements. Micronutrient deficiency generally led to impairment of mental development, physical growth and learning capacity. Four of the micronutrients such as Fe, Zn, iodine and vitamin A have been reported to be the major cause of mortality in developing countries. Iron deficiency ranks among the most widespread nutrient deficiency, estimated to affect over 2 billion people across the world. Iron deficiency anemia (IDA) has been linked to maternal and prenatal mortality, and to impairment of cognitive skills and physical activity. Zn is an essential trace element influencing gene expression as well as cell development and replication. Because it significantly increases risk of diarrhea, pneumonia and malaria, Zn deficiency has been linked to morbidity and mortality of children younger than 5. The global average prevalence of Zn deficiency was estimated at 31% with the most severe in Africa and South Asia. Vitamin A deficiency is related to night blindness, corneal scaring and blindness, weakening of immune system and micronutrient deficiency. Vitamin a deficiency (VAD) is implicated in the death of 1 million people every year (Rice *et al.* 2004, Stein *et al.* 2005). Globally VAD in children is associated with an estimated 24% of diarrhea mortality, 20% of measles-related mortality, 20% of malaria incidence and mortality and 20% maternal mortality. Thus, enhancement in micronutrients content of maize can be useful in improving health of large populations in the world. Bio-fortification is a valuable component of an interim integrated food system approach aimed at reducing micronutrient deficiency until a long term solution as diet diversification is achieved.

In general very little variation exists for iron levels in grains and moderate variation for Zn concentration in grains of maize. Analysis of multiple samples from maize and wheat by CIMMYT across environment and genotypes suggest that the levels of Fe and Zn are higher in wheat than in maize. All yellow genotypes of maize contain carotenoids, although, the fraction of provitamin activity is typically small (10 to 20%) compared to zeaxanthin and lutein (30-50 %). However, there is considerable variation in the ratio of total provitamin A to total carotenoids (Table 3). The variation can be used in breeding maize to enhance provitamin A concentration and hence nutritional quality. Certain assumptions have been made to calculate total requirements of provitamin A. The assumptions are 50% retention during processing and cooking, a 12:1 bioconversion rate of β–carotene to retinol, 200 and 400 g daily maize consumption by children and women, respectively and 250 and 500 ug retinol daily requirements for children and women respectively. These assumptions led to tentative targets of 15 ug/g of provitamin A for maize. Like-wise targets have been identified by HarvestPlus for Fe and Zn based on different assumptions. The tentative targets for Fe and Zn are 40 ug/g and 15 ug/g above the concentrations in the most widely grown variety for individuals consuming 200 g of maize daily.

Table 3: Average, maximum and minimum values found in maize samples at CIMMYT, and HarvestPlus target values for provitamin A, carotenoids, Iron and Zn in Maize (ug/g) (From Ortiz-Monasterio *et al.* 2007)

	Lutein	Zea xznthin	β-crypto xanthin	Cis β-carotene	Trans β carotene	α-carotene	Total Provit A	Iron	Zn
Average	7.08	8.87	1.41	0.95	0.79	0.25	2.58	20	25
Maximum	32.31	34.88	6.13	7.59	5.81	2.31	8.80	39	47
Minimum	1.33	0.38	0.00	0.00	0.00	0.00	0.24	11	15
Harvest plus							>15	>60	>60
Target							(>8.5)	(>40)	(>40)

Niacin Deficiency in Maize (Pellagra)

The nutritional disease pellagra, which is caused by a deficiency in niacin, is associated with maize-based diets in certain parts in the world. In the early twentieth century, hundreds of thousands of people in the southern United States suffered from pellagra, a serious disease that affects people whose diet does not include enough niacin. Pellagra occurs in cultures that depend on corn as a dietary staple, because the niacin in corn is difficult to digest and is often removed during processing. Clinically, the disease is identified by the three Ds-dermatitis, diarrhoea and dementia and neurological disorders. If untreated, pellagra typically leads to death in four or five years. The puzzle started to be solved when it was noted that pellagra was rare in Mexico despite widespread consumption of maize. The reason appeared to lay in the different way in which the grain was prepared in Mexico. Niacin exists in a bound form (niacytin) that is not biologically available to monogastric (single-stomach) animals.

The people of the Aztec and Mayan civilisations softened the corn to make it edible with an alkaline solution-limewater. This process liberated the bound niacin (also known as niacytin) and the important amino acid tryptophan, from which niacin can be formed, making both "bioavailable" for digestion. Niacin is required for adequate cellular function and metabolism as an essential component of nicotinamide adenine dinucleotide (NAD) and nicotinamide adenine dinucleotide phosphate (NADP). These compounds are important coenzymes for glycolysis, protein and amino acid metabolism, pyruvate metabolism, pentose biosynthesis, generation of high-energy phosphate bonds, glycerol metabolism, and fatty acid metabolism. In case of deficiency, all sorts of cell functions become deranged.

Niacin is absorbed from food in the stomach and small intestine. A small quantity of niacin is produced endogenously from tryptophan, an essential amino acid. Food that is rich in tryptophan and deficient in niacin will not give rise to clinically manifest deficiency. Alcoholics and people with hyperthyroidism are at higher risk of pellagra. The conversion from tryptophan to niacin is more difficult in people with vitamin B2 (riboflavin) and B6 (pyridoxine) deficiency. In addition to niacin deficiency, pellagra was also characterized by deficiency in lysine and tryptophan contents. Nixtamalisation was also found to increase the lysine and tryptophan content of maize to some extent, but more importantly, the indigenous Americans had learned long ago to balance their consumption of maize with beans and other protein sources such as amaranth as well as meat and fish, in order to acquire the complete range of amino acids for normal protein synthesis.

Furthermore, most of the niacin in the kernel (63%) occurs in the outermost layer of the endosperm. This layer is often removed with the pericarp during dehulling. In

Mexico, Guatemala, and other countries, maize is treated with an alkaline solution of lime, which releases the niacin and helps prevent pellagra. Furthermore, pellagra seldom occurs among people in Latin America, since they eat tortillas—tortilla preparation greatly increases the bioavailability of the niacin in maize. Therefore, cooking maize in alkaline solutions and complementing maize-based diets with nuts and fish can remove the risk go pellagra.

Maize Quality for Livestock

In most industrialized countries, maize has little significance in human food. Over three quarters of maize produced in these countries is fed to animals particularly cattle, hogs and poultry; the remainder is used mainly as input into a number of extractive industries. Because of its high starch and low fiber content, maize is an extremely concentrated energy source. Readily consumed by the animals, among all major cereals, it gives the highest conversion of dry matter to meat, milk and eggs. In developing countries maize is used primarily as human food and as animal feed. More than 70% of the U.S. corn crop is used in animal feeds either in the United States or at its ultimate export destination; therefore, genetic modifications of corn for animal feed have major profit and development potential. At the same time, the livestock feed industry balances rations to supply nutrients at the least cost. Nutrient sources change, depending on price and availability, so corn that delivers a higher level of nutrients (e.g., protein, energy) will likely substitute for other ingredients in the ration, such as soybean meal and feed fats. Because the demand for feed ingredients is limited by the total demand for meat, milk, and egg products, feed ingredient enhancements may serve only to substitute for other feed ingredients without increasing demand as a whole. However, individual feeders can realize feed cost savings by using corn that delivers higher levels of important nutrients, if that type of corn is readily available at the right price. Modified corn may also prove to be a preferred source of certain nutrients for a particular species and growth stage. Modified corn can replace corn-soybean based feed for swine, poultry and beef cattle. Major traits which need further improvement for livestock feeding are protein content, germ oil content and germ size, glutelin content, protein quality in terms of lysine, methionine and tryptophan content. Higher digestibility of proteins and starch are other major traits for enhancing feed value of corn.

Lysine is the first limiting amino acid in corn-soybean meal diets for swine. In poultry diets, lysine might be limiting if protein sources other than soybean meal are used. Beef cattle rely on the rumen to supply amino acids, so minimum levels of lysine are not set for cattle diets. Methionine is the first limiting amino acid in corn-soybean meal diets for poultry, and is the fourth limiting amino acid for swine. Synthetic methionine is often added to swine starter diets and poultry rations. Methionine also can be converted to cystine; thus, when no cystine is available, methionine can meet the requirement for total sulfur amino acid (TSAA). Increase in TSAA (total methionine and cystine) content is required for proper growth and development. Tryptophan is usually considered the second limiting amino acid in swine diets. Threonine is usually the third limiting amino acid in corn-soybean meal diets for swine. It has been shown to improve performance when added to a corn-based diet and may be added as a crystalline supplement to swine starter diets.

Protein Quality

In addition to productivity gains realized by supplementing poultry rations with MET, optimizing the limiting amino acid (MET) content increases the efficiency of dietary protein utilization. Increasing the N use efficiency of poultry or reducing the total amount of protein fed could potentially decrease N excretion, resulting in diminished N pollution. Therefore improving methionine content in maize grains has a potential for both animal health and the environment. Methionine content can be increased both by conventional breeding and genetic engineering approaches. The maize inbred line with relatively high levels of kernel methionine has been developed. The breeding line has elevated levels of a high-MET 10-kilodalton (kDa) zein protein. The structural gene for the 10-kDa zein has been cloned and found to contain 23% MET. This gene, designated dzs10, is regulated post transcriptionally by a trans-acting regulatory gene, dzr1. In addition, a native landrace, IAPO-13, has been reported to have very high MET levels ranging from two to four times the MET level of normal dent maize. This landrace should be an excellent source of increased MET that may be used as a donor in traditional breeding programs.

Limiting amino acid content can be increased in maize by increasing glutelin fraction. The glutelin protein fraction in corn is a rich source of amino acids such as lysine, methionine, and cystine that may be limiting amino acids in swine and poultry diets. Corn typically contains 25% of its total protein as glutelin proteins, which reside primarily in the germ. Increasing glutelin protein improves overall amino acid balance in corn protein for feed applications. If glutelin content could be doubled with a corresponding increase in protein of 2.2 percentage points, glutelin would then comprise 40% of the protein and provide greater amounts of limiting amino acids (e.g., lysine, methionine) in swine and poultry diets. Zein proteins are found in corn endosperm and are distinctly different in amino acid composition than proteins found in corn germ. They are classified according to amino acid composition and named with letters (A, B, C, etc.). Zein protein represents 52% of total kernel protein in corn. C-zein and D-zein comprise 5.9% of zein protein and are particularly rich sources of methionine, the first limiting amino acid in poultry diets, and of sulfur-containing amino acids.

Transgenic approach is another approach to increase methionine levels. This can be achieved by increasing the expression of the high-MET 10-kDa zein proteins in grain using seed specific promoters. Both whole-kernel and free MET levels can also be influenced by modification of the aspartate pathway, the biosynthetic pathway leading to MET, lysine, isoleucine, and threonine. Aspartate kinase (AK) is the first enzyme of the aspartate pathway leading to the biosynthesis of lysine, MET, threonine, and isoleucine. Methionine biosynthesis is primarily regulated by cystathionine gamma-synthase (CGS). This chloroplast-localised enzyme uses O-phosphohomoserine and cysteine to form cystathionine, the first intermediate metabolite of the Met biosynthesis pathway. In this reaction thiol group of cysteine is transferred to homoserine to form cystathionine. Cystathionine is converted into methionine through homocysteine an intermediate in the methionine biosynthesis (Fig.1).

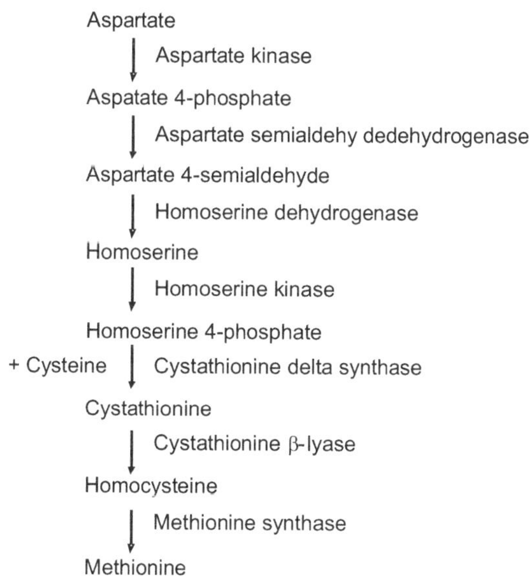

Fig. 1: Methionine biosynthesis pathway.

High Oil Corn

High oil corn (HOC) contains 1.5 to 2 times more oil as well as higher quality proteins than normal yellow dent corn. It is attractive as a livestock feed because it has greater energy value than normal yellow dent corn and can replace more expensive dietary sources of fats and proteins. This is a result of selecting for enlarged size of the germ, where the oil is located, and displacing some of the endosperm, which is largely starch. Since oil has 2.25 times the energy value of starch, optimum high oil corn (OHOC) grain contains 4 to 6% more energy than typical corn (Table 4). Because the composition of the grain is changed but the stover is not altered, high oil corn silage will contain 2 to 4% more energy than typical corn silage. Optimum high oil corn grain has more oil (fat) and protein but less starch than typical corn. Typically, starch content declines by 1.3% for every 1% increase in crude fat. The increased crude protein of optimum high oil corn is composed largely of germ protein, a protein source richer in essential amino acids (lysine, methionine) than protein from the endosperm portion of the corn kernel. Since optimum high oil corn has a higher nutrient density than typical corn, it is most beneficial and valuable in situations where a high nutrient density is needed. This makes optimum high oil corn more useful in diets for lactating cows than for growing heifers or dry cows. Within the lactating herd, optimum high oil corn will produce the greatest benefit for high producing cows, enhancing production and persistency by reducing the drain on tissue energy reserves. Contract production of high oil grain may offer corn growers higher profits through premiums. There are several different seed technologies available for producing high oil corn. Several companies are marketing and developing nutritionally dense corns which contain higher oil levels along with higher protein and/or altered amino or fatty acid profiles.

Table 4: Select grain quality attributes of normal, high oil, high lysine, and waxy corn (Peter R. Thomison *et al.* 2001)

Type	% Oil	% Protein	% Starch	M.E. (Kcal/lb)	% Oleic (18:1)
Normal Dent	4.2-4.8	7.7-8.2	71.3-73.4	1777-1795	23.4-27.7
High Oil	7.2-8.2	8.0-9.0	66.2-67.9	1851-1869	32.5-34.8
High Lysine	4.0-4.5	7.3-8.5	70.5-72.2	1770-1785	25.7-27.5
Waxy	3.2-3.6	8.9-10.1	73.1-73.3	1747-1758	26.7-29.2

Low Phytate Maize

Approximately 20% of the total phosphorus in corn is available to meet the needs of livestock; the remainder is present as undigestible phytate and is not biologically available. Unavailable phosphorus is excreted and contributes to environmental problems. Among the various myo-inositol phosphates, phytic acid is the most abundant form in cells. In plants, phytic acid is distributed widely in different tissues, but accumulates in a large amount in seeds. Plants respond to nutrient phosphorus levels, and seed phytic acid content is correlated positively with nutrient phosphorous levels. Phytic acid serves as a storage form of myo-inositol and phosphorus for utilization during seed germination and seedling growth. Phytic acid also controls inorganic phosphate concentration in developing seeds and seedlings. Phytic acid is biosynthesized from glucose 6-P that is first converted to 1D-myo-inositol 3-phosphate [Ins(3)P] by Ins(3)P synthase (EC 5.5.1.4). Sequential phosphorylation of myo-inositol led to phytic acid production. The defective biosynthetic pathway can led to reduced phytic acid levels. This can be induced by mutagenic agents.

Two low-phytic acid (*lpa*) mutants have been identified in maize and they are defective in phytic acid biosynthesis in developing seed (Raboy *et al.* 2000). The *lpa1* mutant does not accumulate myo-inositol monophosphate or polyphosphate intermediates. It has been proposed that *lpa1* is a mutation in myo-inositol supply, the first part of the phytic acid biosynthesis pathway (Raboy *et al.* 2000). The *lpa2* mutant has reduced phytic acid content in seeds and accumulates myo-inositol phosphate intermediates. Maize *lpa2* gene encodes a myo-inositol phosphate kinase (ZmIPK) that belongs to the Ins(1,3,4)P3 5/6-kinase gene family. The *lpa3* mutant seeds have reduced phytic acid content and accumulate myo-inositol, but not myo-inositol phosphate intermediates. Maize *lpa3* gene was found to encode an MIK (Shi *et al.* 2005). The low phytic mutants are excellent source for reducing phytic acid content in maize seeds.

Maize as Biofuels

The world energy demand is increasing very fast with the increase in human population and progress in economic development. However, fossil fuel, the major source of energy supply, is limited. This necessitates the need to identify renewable energy sources, such as biofuels produced from agricultural biomass. Biofuels are liquid transportation fuels made from plant matter instead of petroleum. Bioethanol and biodiesel—the primary biofuels today—can substitute for gasoline and diesel or be blended with them to reduce toxic air emissions. Currently maize has become very important biofuel crop in the world especially in United States. It represents 31% of the world production of cereals and occupies a little over one fifth of the worldwide cereal-dedicated land. Two key parts of maize plants can be converted into bioethanol: the kernel, which is mainly made of starch, and the stover, which is predominantly made of lignin and cellulosic (cell wall) components. The bioconversion of starch into ethanol is a two-step process. The first step is saccharification, where starch is converted into sugar using an amylolytic microorganism or enzymes such as glucoamylase and α-amylase. The second step is fermentation, where sugar is converted into ethanol using *Saccharomyces cerevisiae*. The use of amylolytic yeasts for the direct fermentation of starch is an alternative to the conventional multistage process which offers poor economic feasibility. Although there are over 150 amylolytic yeast species, their industrial use is limited because of their low ethanol tolerance. Therefore, most research is focused on the development of genetically engineered amylolytic strains of *S. cerevisiae* to make the process economically viable.

Starch is composed of two glucose polymers, amylose and amylopectin. In amylose, glucose units are linked in a linear fashion by α, 1-4 linkages. Amylopectin, by contrast, is more branched and about 5% of its glucose units are linked by α, 1-6 linkages. Normal maize starches contain about 20–30% of amylose and 70–80% amylopectin. The amylose/amylopectin ratio in starch affects its physical and physicochemical properties, such as gelatinization and recrystallization. For example, gelatinization is the first step in bioethanol production from starch. It is conceivable that a modified starch with decreased gelatinization temperature might require less energy for the conversion process. Alteration in starch structure can be achieved by modifying genes encoding the enzymes responsible for starch synthesis, many of which have more than one isoform. Transgenic lines with modified expression of specific starch synthases, starch branching enzymes or starch debranching enzymes are being generated in attempts to produce starch granules with increased or decreased crystallinity, and thus altered susceptibility to enzymatic digestion. Another strategy is to reduce the energy requirements for the starch to ethanol conversion process.

Early research results show that tropical maize, when grown in the Midwest, requires few crop inputs such as nitrogen fertilizer, chiefly because it does not produce any ears. It also is easier for farmers to integrate into their current operations than some other dedicated energy crops because it can be easily rotated with corn or soybeans, and can be planted, cultivated and harvested with the same equipment which the U.S. farmers already have. Finally, tropical maize stalks are believed to require less processing than corn grain, corn stover, switchgrass, *Miscanthus giganteus* and the scores of other plants now being studied for biofuel production. What it does produce, straight from the field with no processing, is 25% or more sugar and mostly sucrose, fructose and glucose. Recent advances in plant genetic engineering could reduce biomass conversion costs by developing crop varieties with less lignin crops that self-produce cellulase enzymes for cellulose degradation and ligninase enzymes for lignin degradation, or plants that have increased cellulose or an overall biomass yield. Recent research showed that expression of a lignocellulosic feedstocks derived from corn stover could be used for conversion to bioethanol, two major limitations to the process are the costs of transport and processing of biomass.

One solution is to produce microbial cellulase enzymes in the plant cells to facilitate the conversion of fermentable sugars in planta during the biomass to bioethanol conversion process. Expression of the catalytic domain of the thermostable 1,4-b-endoglucanase (E1) of *Acidothermus cellulolyticus* in maize proves the concept that maize can be used as a biofactory for cellulose-degrading enzymes. Even though expression of E1 has not achieved desirable levels, targeting the enzymes to specific subcellular compartments or tissues has shown to be effective in allowing the plants to accumulate higher levels of recombinant enzymes. In addition to subcellular targeting of these enzymes, it is also important to express these cell wall degrading enzymes during appropriate developmental stages, rather than over the entire lifetime of the plants. To convert them effectively into fermentable sugars for ethanol production, a range of approaches using genetic engineering have been explored. One strategy is to modify the characteristics and properties of starch or lignocellulose so that they can be converted more readily to the desired products. The other strategy is to introduce biomass conversion enzymes into plants so that they can aid the conversion process more effectively.

Though, bioenergy represents an opportunity to reduce greenhouse gas emissions, there are many negative effects of generation of biofuels. Investments in bioenergy must be managed carefully, at national, regional and local levels to avoid new environmental and social problems.

- Rapid growth in liquid biofuel production will make substantial demands on the world's land and water resources at a time when demand for both food and forest products is also rising rapidly.
- Changes in the carbon content of soils and carbon stocks in forests and peat lands might offset some or all of the benefits of the greenhouse gas reductions.

- Use of large-scale mono-cropping could lead to significant biodiversity loss, soil erosion and nutrient leaching.
- It is noted that soaring palm oil demand has already led to the clearing of tropical forests in Southeast Asia. Such clearings could result in emissions that were even higher than those caused by fossil fuels.
- The diversion of land used to grow food for fuel will increase food prices for basic commodities, putting a strain on the poor. Already, there has been recent steep rises in maize and sugar prices. Biofuels tend to allocate productive resources (e.g. land, labour, capital) away from the production of food crops into the production of feed stocks for biofuels. Biofuels may reduce the availability of food for nutritional use, because 'effective' demand for grains, sugar or oils and other basic food staples as feedstock for fuel production is able to outbid that for food. This new source of demand has been playing an important role in influencing prices.

Summary

Maize (*Zea mays,* or corn as it is known in some countries) is the principal cereal crop in tropical and subtropical regions throughout the world. Virtually every part of maize has economic value. Grains are used as human feeds, fermented to produce a vide range of food and beverages, fed to livestock and industrial uses in the production of starch, oil, sugar, cellulose and ethanol and transformation into plastics and fabrics. Maize ethanol is being used as an additive in gasoline for motor fuels to increase the octane number with lower pollutants, and reduce petroleum use. Translucency of grain is very important trait. Maize kernels with high test weight, maximum hard endosperm, minimum stress cracks, no mold, and large uniform kernel are desired by the processors. Corn-based snack foods, such as corn and tortilla chips, represent one of the fastest growing segments of salty snack food market. Maize is deficient in lysine. To improve lysine content, high lysine mutants called *opaque -2* were generated and used in the development of quality maize protein (QPM). QPM have high nutritional quality, high yields, appearance of normal maize, greater hardness than *opaque-2,* equal or superior pest and disease resistance. The genes for high lysine content in QPM lines have been identified and used as molecular markers in breeding. For the improvement of feed value of maize, efforts are on to develop corm with higher quantity and quality of protein, high oil content and low phytate content. Recent developments in molecular techniques will help in improving nutritional and industrial quality of maize.

References

Babu R, Nair SK, Kumar A, Venkatesh S, Sekhar JC, Singh NN, Srinivasan G, Gupta HS (2005). Two-generation marker-aided backcrossing for rapid conversion of normal maize lines to quality protein maize (QPM). *Theor. Appl. Genet.* 111: 888-897.

Bressani, R. (1991). Protein quality of high-lysine maize for humans. *Cereal Foods World.* 36 (9): 806-811.

Chandrashekar, A and Mazhar, H. (1999). The biochemical basis and implications of grain strength in sorghum and maize. *J. Cereal Sci.* 30: 193-207.

CIMMYT. (2001). Final progress report: The improvement and promotion of quality protein maize in selected developing countries. Mexico City. Mexico.

Civitelli, R., Villareal, D. T., Agnusdei, D., Nardi, P., Avioli, L. V. and Gennari, C. (1992). Dietary L-Lysine and calcium metabolism in humans. *Nutrition.* 8(6): 400-405.

Cordova H (2000). Quality Protein Maize: Improved Nutrition and Livelihoods for the Poor. CIMMYT Maize Research Highlights 1999-2000 *www.cimmyt.org/Research/Maize/results/MzHigh9900/mrhigh99-00.htm*: 27-31.

Graham, G. G., Lembecke, J., Lancho, E. and Morales, E. (1989). Quality protein maize: digestibility and utilization by recovering malnourished infants. *Pediatrics.* 83(3): 416-421.

Hernandez, H.H., Bates, L.S. (1969). A modified method for rapid tryptophan analysis of maize. *CIMMYT Res. Bull.* 13.

Hoseney, R.C. (1994). Principles of Cereal Science and Technology. IInd edition. American Association of Cereal Chemists, Inc. MN, USA.

Huang, S.S., Adams, W.R., Zhou, Q., Malloy, K.P., Voyles, D.A., Anthony, J., Kriz, A.L. and Luethy, M.H. (2004). Improving nutritional quality of maize proteins by expressing sense and antisense zein genes. *J. Agric. Food Chem.* 52: 1958-1964.

Huang, S., Frizzi, A., Florida, C.A., Kruger, D.E. and Luethy, M.H. (2006). High lysine and high tryptophan transgenic maize resulting from the reduction of both 19- and 22-kD a-zeins. *Plant Mol. Bio.* 61: 525–535.

Huang, S., Kruger, D.E., Frizzi., A., D'Ordine, R.L., Florida, C.A., Adams, W.R., Brown, W.E. and Luethy, M.H. (2005). High lysine corn produced by the combination of enhanced lysine biosynthesis and reduced zein accumulation. *Plant Biotechnol.* 3: 555–569.

Moro, G.L., Habben, J.E., Hamaker, B.R., Larkins, B.A. (1996). Characterization of the variability in lysine content for normal and opaque 2 maize endosperm. *Crop Sci.* 36: 1651-1659.

Ortiz-Monasterio, J.I., Palacios-Rojas, N., Meng E., Pixley, K., Trethowan, R. and Pena, R.J. (2007). Enhancing mineral and vitamin content of wheat and maize through plant breeding. *J. Cereal Sci.* 46: 293-307.

Paiva, P., Kriz, A., Peixoto, M. J. V. V. D., Wallace, J. C. and Larkins, B. A. (1991). Quantitation and distribution of γ-zein in the endosperm of maize kernels. *Cereal Chem.* 68(3):276-279.

Prasanna, B.M., Vasal S.K., Kassahun, B. and Singh, N.N. (2001). Quality protein maize. *Current Sci.* 81(10): 1338-1319.

Raboy, V., Gerbasi, P.F., Young, K.A., Stoneberg, S.D., Pickett, S.G., Bauman, A.T., Murthy, P.P.N., Sheridan, W.F. and Ertl, D.S. (2000). Origin and seed phenotype of maize low phytic acid 1–1 and low phytic acid 2–1. *Plant Physiol.* 124: 355–368

Rice, A.L., West, Jr.K.P., and Black, R.E. (2004). Vitamin A deficiency. In: Ezzati, M., Lopez, A.D. Rodgers, A. and Murray, C.J.L. (Eds.), Comparative Quantification of Health Risks: *Global and Regional Durden of Disease Attribution to Selected Major Risks Factors*, Vol. I. World Health Organization, Geneva.

Segal, G., Song, R.T. and Messing, J. (2003). A new opaque variant of maize by a single dominant RNA-interference-inducing transgene. *Genetics* 165: 387-397.

Serna-Saldivar, S. O., Rooney, L. W. and Greene, L. W. (1992b). Effects of lime treatment on the bioavailability of calcium in diets of tortillas and beans: bone and plasma composition in rats. *Cereal Chem.* 69(1): 78-81.

Shewry, P.R. (2007). Improving the protein content and composition of cereal grain. *J. Cereal Science.* 46(3): 239-250.

Shi, J.R., Wang, H.Y., Hazebroek, J., Ertl, D.S. and Harp, T. (2005). The maize low-phytic acid 3 encodes a myo-inositol kinase that plays a role in phytic acid biosynthesis in developing seeds. *Plant J.* 42: 408–419.

Stein, A.J., Meenakshi, J.V., Qaim, M., Nestel, P., sachdev, H.P.S. and Bhutta, Z.A. (2005). Analyzing health benefits of biofortified staple crops by means of the disability-adjusted life years approach – a hand book focusing on iron, zinc and vitamin A. *Harvest Plus Technical Monograph* No.4., HarvestPlus, Washington, D.C.

Thomison, P.R., Geyer, A.B., Dobbels, T. and Siegrist H. Grain Quality Attributes of TopCross® High Oil, High Lysine, Waxy, and Conventional Yellow Dent Corns. In: Ohio State University Extension Factsheet. *D:\Book\maize\specialty corn.htm.*

Thomison, P.R., Geyer, A.B., Dobbels, T., Siegrist, H. (2001). Grain Quality Attributes of TopCross High Oil, High Lysine, Waxy, and Conventional Yellow Dent Corns Ohio State University Extension Factsheet. Horticulture and Crop Science.

Zarkadas, C.G., Hamilton, R.I., Yu, Z.R., Choi, V.K., Khanizadeh, S., Rose, N.G.W. and Pattison, P.L. (2000). Assessment of the protein quality of 15 new northern adapted cultivars of quality protein maize using amino acid analysis. *J. Agric. Food Chem.* 48: 5351-5361.

□□□

INDEX

❐❐❐

MIX

Papier | Fördert gute Waldnutzung

FSC® C083411

Zeitfracht Medien GmbH
Ferdinand-Jühlke-Straße 7
99095 Erfurt, Deutschland
produktsicherheit@kolibri360.de